SYDNEY D'AGVILO

INTERVALIC THEORY

The Intervalic Structures
of Subatomic Particles and
the Last Foundations of Physics

Volume I

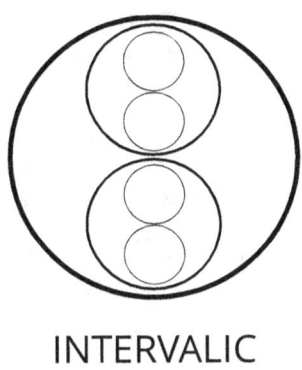

INTERVALIC
PRESS

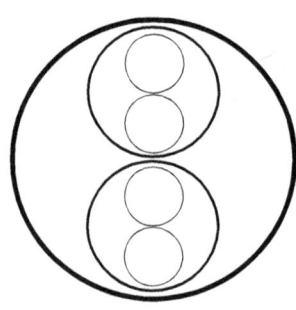

INTERVALIC
PRESS

INTERVALIC THEORY IN PHYSICS
The Intervalic Structures of Subatomic Particles
and the Last Foundations of Physics
(Volume 1)

© 2005 Sydney d'Agvilo

Place: Free Planet Earth
Date: December 2018, 1st. edition

Printed by Amazon
Made in Planet Earth

ISBN 13: 9781791605803

IN MEMORIAM

ALBERT EINSTEIN
(1879-1955)

TABLE OF CONTENTS

Volume I

0.	*ABSTRACT* (BY WAY OF INTRODUCTION)	9
1.	INTERVALIC SYSTEM OF DIMENSIONS	41
2.	INTERVALIC QUANTA, INTERVALIC LIMITS	69
3.	INTERVALIC SYSTEM OF UNITS	95
4.	INTERVALIC GEOMETRY OF THE SPEED OF LIGTH	127
5.	$I = c^{\pm 2} \hbar \, Q^{-2}$: INTERVALIC ENERGY	165
6.	INTERVALINO	185
7.	INTERVALIC DALINO	209
8.	INTERVALIC ELECTRON	227
9.	INTERVALIC GAUDINO	233
10.	INTERVALIC LEPTON-CHARGED MASSIVE BOSON	241
11.	INTERVALIC DYNAMICS OF LEPTON	257
12.	INTERVALIC FRACTIONAL GAUDINO	283
13.	INTERVALIC LISZTINO	299
14.	INTERVALIC ZERO CHARGED MASSIVE BOSON	315
15.	INTERVALIC FRACTIONAL LISZTINO: QUARKS	325
16.	INTERVALIC QUARKS UP AND DOWN	353
17.	INTERVALIC NUCLEON	369
18.	INTERVALIC BARYON	395
19.	INTERVALIC PI MESON	419
20.	INTERVALIC MESON	435

Volume II

21.	INTERVALIC PRIMORDIAL SYNTHESIS (IPS)	465
22.	INTERVALIC SYMMETRY BREAKINGS	495
23.	SYNTHESIS OF INTERVALIC STRUCTURES	523
24.	CHANGELESS —STRONG— INTERVALIC INTERACTION	559
25.	CHANGEFUL —WEAK— INTERVALIC INTERACTION	599
26.	PARTICLES SHARE IN THE INTERVALIC PRIMORDIAL UNIVERSE	617
27.	INTERVALIC GLOBAL CANCELLATION OF ENERGIES	643
28.	INTERVALIC DECAY	655
29.	INTERVALIC DARK MATTER	677
30.	INTERVALIC BINDING ENERGY	695
31.	INTERVALIC NUCLEUS. Intervalic Origin of the Binding Energy in Nucleus and the Intervalic Resonance Frequorce of Nucleus	725
32.	INTERVALIC SPIN ENERGY	751
33.	INTERVALIC NEUTRINO	775
34.	INFORMATIONAL INTERACTION	793
35.	THE INTERVALIC CODE	819
36.	THE END OF THE STANDARD MODEL	827

Abstract (by way of introduction)

INTERVALIC THEORY IN PHYSICS
The Intervalic Structures of Subatomic Particles and the Last Foundations of Physics

The Intervalic Theory is the major paradigm shift that has occurred in the history of mankind, affecting all disciplines. It was originally postulated by Sydney d'Agvilo in 1987 in the field of Music, in 2005 in Physics, and in 2016 in Economics, disciplines that respectively represent the bastions of the three major branches of knowledge: the arts, the hard sciences and social sciences.

From a historical standpoint the Intervalic Theory closes, in the most unexpected way, the long gap existing between relativity and quantum mechanics, two theories of Nature that are mutually incompatible: the first one, of *geometric* nature; the second one, *probabilistic*. Most physicists have been working for decades on trying to discover a theory of everything which includes relativity in the quantum paradigm. On the contrary, the Intervalic Theory has surprisingly formulated a theory of everything absorbing the correct parts of quantum mechanics within the relativistic paradigm, which from now on is also encompassed in a theory of everything more general: the Intervalic Theory, which converts all Physics in geometry. This means the failure of the probabilistic conception of knowledge and of the world, and the triumph of the geometric vision; or in other words, the failure of pseudo scientific *empirical-inductive* method, and the triumph of true scientific method *par excellence*, the *logical-deductive* one.

Intervalic Theory, in any medium, is based on a single axiom: the *intervalic axiom*, which simply states: "an interval exists". In physics, branch that enjoys precedence over all others for obvious reasons, because it is what explains the generation of the Universe, the intervalic axiom means that a one-dimensional space line —an interval—, is the only thing whose existence is postulated before the creation of the Universe, when there was absolutely nothing, that is, neither matter nor energy, time or three-dimensional space, much less the so-called "quantum vacuum". The existence of a linear space interval necessarily implies the existence of mathematics. Well, from the mathematical logic and a space interval of any length there will necessarily generate time, energy, matter and all the Universe we know. That interval is the *intervalic length*, $\hbar$, very similar to the Planck's length value, although it could have taken any other value and the Universe thus created would be identical and indistinguishable to the current one because all physics constants —including the elementary charge— are not fundamental indeed, but they are logically derived from the only two fundamental constants of Nature, which compose the *intervalic dimensional basis* of the intervalic system of physical quantities (and because you can not measure anything from "outside" the Universe).

The interval axiom determines an *intervalic dimensional basis*, which is different and *not equivalent* to the traditional ones (comprising length, mass, time, etc.), being composed exclusively of length, L, and the imaginary number $i = \sqrt{-1}$, both derived from the intervalic axiom. Alternatively, the intervalic dimensional basis can also be deduced logically and necessarily from the definition of *time* as *imaginary space* ($T = iL$), which is the definition of time that was originally postulated in the Intervalic Theory in Music.

The *intervalic dimensional basis* shows the astonishing fact that the equation of dimensions of the two fundamental constants of Nature, namely Planck's constant, and the speed of light, c, respectively coincide with the two elements comprising the intervalic dimensional basis (L, i), being L the dimension of $\hbar$, and i the dimension of c (or i^{-1}, which produces an identical dimensional system). The systematic combination of L and i generates logically the intervalic group of all existing physical quantities, which are exactly 40. As each and every one of the physical quantities thus generated is a certain combination of $\hbar$ —the *quantum* of length— and c —the *limit* of speed and of energy at subatomic scale—, we have got the remarkable result that all physical quantities are simple combinations of the two fundamental constants of Nature, $\hbar$ and c, and therefore all of them are geometrically delimited by a lower mark —*quantum*— or a higher mark —*limit*—. This means that *there can not be infinite values for any physical quantity*, and not only for speed or action, as believed by quantum mechanics. This means, for example, that there are no infinite masses or energies, which implies that singularities can not exist (black holes) as they are described by some science fantasy tales that make an improper use of relativity. In fact, the intervalic cosmology, closely linked to the particle physics theory, shows what is the intervalic structure of very massive stars, which are described using some equations common to subatomic particles, being this study a fascinating branch of the Intervalic Theory which explains a lot of astronomical phenomena in a completely unattainable way for quantum mechanics.

    Traditional systems of physical quantities and units do lack of epistemological range, meaning that two *distinct* physical quantities can have the *same* equation of dimensions, that is to say, the equation of dimensions of a physical quantity is not significant except formally. By contrast, the intervalic system of physical quantities has got an epistemological range, which means that if two physical quantities have the *same* equation of dimensions equation, that means that they are the *same* magnitude, although we have misinterpreted them as different in the past. This happens, for example, with speed and energy, which are actually a single physical quantity, or equally with acceleration and

INTERVALIC UNITS, INTERVALIC QUANTA, INTERVALIC LIMITS

Intervalic dimension	Physical quantity	Definition	Rank *	Value showing the equivalence between intervalic units and SI units
1	Permitivity	$\varepsilon_I = c^{\pm 2} \mu_0^{-1}$	c.f.	$1/4\pi$ (1) = $8.85418781 \cdot 10^{-12}$ (F m^{-1})
	Molar gas constant	R	c.f.	1 (1) = 8.314510 (J mol^{-1} K^{-1})
	Boltzmann constant	$k_B = R/N_A$	c.f.	1 (1) = $1.380658 \cdot 10^{-23}$ (J K^{-1})
	Momentum	$\mathbf{p}_I = \mathbf{m}_I \mathbf{v}_I = c^{-1} c$	s.l.	1 (1) = 1 (kg m s^{-1})
-1	Permeability	$\mu_I = \mu_0$	c.f.	4π (-1) = $4\pi \cdot 10^{-7}$ (H m^{-1})
	Gravitational potential	$\Phi_I = c^{\pm 2}$	s.l.	1 (-1) = $8.98755179 \cdot 10^{16}$ (m^2 s^{-2})
	Antimomentum	$-\mathbf{p}_I = c^{\pm 2}$	c.f.	1(-1) = $8.98755179 \cdot 10^{16}$ (-kg m s^{-1})
i	**Mass**	$\mathbf{m}_I = c^{-1}$	s.l.	1 (i) = $3.335640952 \cdot 10^{-9}$ (kg)
i^{-1}	**Velenergy:** Velocity, Energy	$\mathbf{v}_I = c$	a.l.	1 (i^{-1}) = $2.99792458 \cdot 10^{8}$ (m s^{-1})
		$E_I = c$	s.l.	1 (i^{-1}) = $2.99792458 \cdot 10^{8}$ (J)
	Temperature	$\Theta_I = c\, k_B^{-1}$	a.l.	1 Θ_I (i^{-1}) = $2.17138589 \cdot 10^{31}$ (K)
L	**Length** (real space)	$\mathbf{l}_I = \hbar$	q.	1 (L) = $1.0556363 \cdot 10^{-34}$ (m)
	Action	$S_I = \hbar$	q.	1 (L) = $1.0556363 \cdot 10^{-34}$ (J s)
	Capacitance	$C_I = \hbar$	q.	1 (L) = $1.0556363 \cdot 10^{-34}$ (F)
-L	Antilength	$L_I = c^{\pm 2} \hbar$	c.f.	1 (-L) = $9.487585915 \cdot 10^{-18}$ (m)
iL	**Time** (imagin. space)	$t_I = c^{-1} \hbar$	q.	1 (iL) = $3.521223673 \cdot 10^{-43}$ (s)
i^{-1}L	Antitime	$-t_I = c\hbar$	c.f.	1 (i^{-1}L) = $3.164718011 \cdot 10^{-26}$ (-s)
L^{-1}	Wavevector	$k_I = \hbar^{-1}$	a.l.	1 (L^{-1}) = $9.47295958 \cdot 10^{33}$ (m^{-1})
-L^{-1}	**Gravitational field--Poweration:** Acceleration Power	$g_I = c^{\pm 2} \hbar^{-1}$	a.l.	1 (-L^{-1}) = $8.51387148 \cdot 10^{50}$ (m s^{-2})
		$a_I = c^{\pm 2} \hbar^{-1}$	a.l.	1 (-L^{-1}) = $8.51387148 \cdot 10^{50}$ (m s^{-2})
		$W_I = c^{\pm 2} \hbar^{-1}$	a.l.	1 (-L^{-1}) = $8.51387148 \cdot 10^{50}$ (W)
i L^{-1}	Linear density	$\Delta^1_I = c^{-1} \hbar^{-1}$	a.l.	1 (i L^{-1}) = $3.15983919 \cdot 10^{25}$ (kg m^{-1})
i^{-1}L^{-1}	**Frequorce, φ:** Frequency, Force	$\nu_I = c\,\hbar^{-1}$	a.l.	1 (i^{-1} L^{-1}) = $2.839921837 \cdot 10^{42}$ (s^{-1})
		$F_I = c\,\hbar^{-1}$	a.l.	1 (i^{-1} L^{-1}) = $2.839921837 \cdot 10^{42}$ (N)
	Conductivity	$\sigma_I = c\,\hbar^{-1}$	a.l.	1 (i^{-1} L^{-1}) = $2.839921837 \cdot 10^{42}$ (S m^{-1})
L^2	Area	$S_I = \hbar^2$	q.	1 (L^2) = $1.114367998 \cdot 10^{-68}$ (m^2)
-L^2	Antiarea	$-S_I = c^{\pm 2} \hbar^2$	c.f.	1 (-L^2) = $1.001544009 \cdot 10^{-51}$ (-m^2)
i L^2	Inertia area momentum	$\mathbf{I}^2_I = c^{-1} \hbar^2$	q.	1 (i L^2) = $3.717131529 \cdot 10^{-77}$ (kg m^2)
i^{-1}L^2	?	$c\,\hbar^2$	q.	1 (i^{-1}L^2) = $3.340791212 \cdot 10^{-60}$ (J m^2)
L^{-2}	Viscosity (dynamic)	$\eta_I = \hbar^{-2}$	a.l.	1 (L^{-2}) = $8.973696318 \cdot 10^{67}$ (Pa s)
-L^{-2}	Area power	$W^2_I = c^{\pm 2} \hbar^{-2}$	a.l.	1 (-L^{-2}) = $8.065156039 \cdot 10^{84}$ (-m^{-2})

INTERVALIC THEORY:
The Intervalic Structures of Subatomic Particles and the Last Foundations of Physics

INTERVALIC UNITS, INTERVALIC QUANTA, INTERVALIC LIMITS

Intervalic dimension	Physical quantity	Definition	Rank *	Value showing the equivalence between intervalic units and SI units
$i L^{-2}$	Area density	$\Delta^2_I = c^{-1} \hbar^{-2}$	a.l.	$1 (i L^{-2}) = 2.993302893 \cdot 10^{59}$ (kg m^{-2})
	Inflexion	$i_I = c^{-1} \hbar^{-2}$	a.l.	$1 (i L^{-2}) = 2.993302893 \cdot 10^{59}$ (m s^{-3})
$i^{-1} L^{-2}$	Surface Tension	$\sigma_I = c \hbar^{-2}$	a.l.	$1 (i^{-1} L^{-2}) = 2.690246477 \cdot 10^{76}$ (N m^{-1})
L^3	Volume	$V_I = \hbar^3$	q.	$1 (L^3) = 1.17636731 \cdot 10^{-102}$ (m^3)
$-L^3$	Antivolume	$-V_I = c^{\pm 2} \hbar^3$	c.f.	$1 (-L^3) = 1.057266212 \cdot 10^{-85}$ (-m^3)
$i L^3$	Inertia volume momentum	$I^3_I = c^{-1} \hbar^3$	q.	$1 (i L^3) = 3.923938974 \cdot 10^{-111}$ (kg m^3)
$i^{-1} L^3$	Fermi constant ph. quantity	$c \hbar^3$	q.	$1 (i^{-1} L^3) = 3.526660474 \cdot 10^{-94}$ (kg m^3)
L^{-3}	Fluctuation	$f_I = \hbar^{-3}$	a.l.	$1 (L^{-3}) = 8.50074625 \cdot 10^{101}$ (m s^{-4})
$-L^{-3}$	Volume power	$W^3_I = c^{\pm 2} \hbar^{-3}$	a.l.	$1 (-L^{-3}) = 7.640089715 \cdot 10^{117}$ (-m^{-3})
	Irradiance	$E_{eI} = c^{\pm 2} \hbar^{-3}$	a.l.	$1 (-L^{-3}) = 7.640089715 \cdot 10^{117}$ (W m^{-2})
$i L^{-3}$	Volume density	$\rho_I = c^{-1} \hbar^{-3}$	a.l.	$1 (i L^{-3}) = 2.835543731 \cdot 10^{93}$ (kg m^{-3})
$i^{-1} L^{-3}$	Pressure	$P_I = c \hbar^{-3}$	a.l.	$1 (i^{-1} L^{-3}) = 2.548459613 \cdot 10^{110}$ (Pa)
	Energy-tension density	$u_I = c \hbar^{-3}$	a.l.	$1 (i^{-1} L^{-3}) = 2.548459613 \cdot 10^{110}$ (J m^{-3})
$i^{1/2} L^{1/2}$	**Magnetic charge**	$\theta_I = \sqrt{-(c \hbar)}$	s.l.	$1 (i^{1/2} L^{1/2}) = 1.778965433 \cdot 10^{-13}$ (Wb)
	Magnetic flux	$\Phi_I = \sqrt{-(c \hbar)}$	s.l.	$1 (i^{1/2} L^{1/2}) = 1.778965433 \cdot 10^{-13}$ (Wb)
$i^{-1/2} L^{1/2}$	**Electric charge**	$q_I = \sqrt{-(c^{-1} \hbar)}$	q.	$1 (i^{-1/2} L^{1/2}) = 5.93398995 \cdot 10^{-22}$ (C)
$i^{1/2} L^{-1/2}$	Current	$I_I = \sqrt{-(c \hbar^{-1})}$	a.l.	$1 (i^{1/2} L^{-1/2}) = 1.685206764 \cdot 10^{21}$ (A)
	Electric potential	$V_I = \sqrt{-(c \hbar^{-1})}$	a.l.	$1 (i^{1/2} L^{-1/2}) = 1.685206764 \cdot 10^{21}$ (V)
	Magnetic vector potential	$A_I = \sqrt{-(c \hbar^{-1})}$	a.l.	$1 (i^{1/2} L^{-1/2}) = 1.685206764 \cdot 10^{21}$ (Wb m^{-1})
$i^{-1/2} L^{-1/2}$	Magnetic inverflux	$\Phi^{-1}_I = \sqrt{-(c^{-1} \hbar^{-1})}$	s.l.	$1 (i^{-1/2} L^{-1/2}) = 5.621244694 \cdot 10^{12}$ (Wb^{-1})
$i^{1/2} L^{3/2}$	Bohr magneton ph.quantity	$\mu_{BI} = \sqrt{-(c \hbar^3)}$	q.	$1 (i^{1/2} L^{3/2}) = 1.877940487 \cdot 10^{-47}$ (J T^{-1})
$i^{-1/2} L^{3/2}$	?	$\sqrt{-(c^{-1} \hbar^3)}$	q.	$1 (i^{-1/2} L^{3/2}) = 6.264135195 \cdot 10^{-56}$ (T^{-1})
$i^{1/2} L^{-3/2}$	Electric field strength	$\mathcal{E}_I = \sqrt{-(c \hbar^{-3})}$	a.l.	$1 (i^{1/2} L^{-3/2}) = 1.596389556 \cdot 10^{55}$ (V m^{-1})
	Magnetic field strength	$H_I = \sqrt{-(c \hbar^{-3})}$	a.l.	$1 (i^{1/2} L^{-3/2}) = 1.596389556 \cdot 10^{55}$ (A m^{-1})
	Magnetic flux density	$B_I = \sqrt{-(c \hbar^{-3})}$	a.l.	$1 (i^{1/2} L^{-3/2}) = 1.596389556 \cdot 10^{55}$ (T)
$i^{-1/2} L^{-3/2}$	Electric polarisation	$P_I = \sqrt{-(c^{-1} \hbar^{-3})}$	a.l.	$1 (i^{-1/2} L^{-3/2}) = 5.324982377 \cdot 10^{46}$ (C m^{-2})
$i^{1/2} L^{5/2}$	?	$\sqrt{-(c \hbar^5)}$	q.	$1 (i^{1/2} L^{5/2}) = 1.982422148 \cdot 10^{-81}$ ( )
$i^{-1/2} L^{5/2}$	?	$\sqrt{-(c^{-1} \hbar^5)}$	q.	$1 (i^{-1/2} L^{5/2}) = 6.612648501 \cdot 10^{-90}$ (C^{-1} m^3)
$i^{1/2} L^{-5/2}$	Charge density	$\rho_I = \sqrt{-(c \hbar^{-5})}$	a.l.	$1 (i^{1/2} L^{-5/2}) = 1.512253373 \cdot 10^{89}$ (C m^{-3})
	Current density	$J_I = \sqrt{-(c \hbar^{-5})}$	a.l.	$1 (i^{1/2} L^{-5/2}) = 1.512253373 \cdot 10^{89}$ (A m^{-2})
$i^{-1/2} L^{-5/2}$	?	$\sqrt{-(c^{-1} \hbar^{-5})}$	a.l.	$1 (i^{-1/2} L^{-5/2}) = 5.044334281 \cdot 10^{80}$ ( )

Main differences between the
INTERVALIC SYSTEM of PHYSICAL QUANTITIES
and all other systems of units and dimensions

- It is the unique dimensional system whose **dimensional basis** —(L, i)— is just composed by the single intervalic dimensions of the **last fundamental constants of Nature, ℏ and c**: dim ℏ = (L), dim c (i^{-1}). Of course, the system has two formulations: dim c = (i^{-1}) or dim c = (i), which are absolutely equivalent.
- Existing physical quantities are *generated* by all algebraic combinations between the two dimensional basis —L and i— which makes a *finite* and *ordered* set of **40 physical quantities**. The *number of physical quantities* is given by the formula: $4 + 12n$, being n the number of actual dimensions of space.
- There are no physical quantities whose equation of dimensions have more than the *actual dimensions* of space (3) and time (1), as in other systems, which is absolutely a nonsense or, at least, an inconsistency.
- There are neither different physical quantities with the same equation of dimensions (as in traditional units), nor different dimensions with the same physical quantity (as in the misleading called "geometrized" units, which is the poorest system of units ever made, not even being consistent).
- The intervalic dimensions of all physical quantities can *operate algebraically* with the signs of its corresponding magnitudes in any equation.
- The own definition of all existing physical quantities as an algebraic combination of c (i^{-1}) and ℏ (L) yields unavoidably a *geometric height* for every physical quantity, making the full set of INTERVALIC QUANTA and INTERVALIC LIMITS, which form not only the Intervalic Units, but above all, the foundations of the **underlying fundamental geometry of Nature**, from which are derived the genuine **intervalic symmetries of Nature**, long time searched by Physics.
- Therefore, every physical quantity can not acquire any value, as each of them is *geometrically* closed by its corresponding height: an intervalic quanta or limit, which is another great difference between the intervalic system of units and the remaining systems, which like to play with infinites and singularities.
- Being the intervalic dimensions the truthful units of Nature, the intervalic physical quantities have got an *epistemological rank* by means

of their equations of dimensions have got an *heuristic value*, which is lacked in other systems. The most important example is the merging of two traditional physical quantities into a new one because their intervalic dimensions are identical. That is the case of: velenergy (velocity-energy), frequorce (frequency-force) and poweration (power-acceleration)-gravitational field strength. The merged physical quantities means that they are really the same underlying physical quantity, although in phenomenology there may appear as different in incomplete or false dimensional systems. All this allow to unify intervalic dimensions, physical quantities and units in a unique concept, if desired.

- When applying basic geometry to the intervalic dimensions inside the Argand-like Intervalic Dimensional Space, a full set of invariant **Intervalic Transformations** of physical quantities is *geometrically* derived. The Intervalic Transformations comprise the former Lorentz-Einstein transformations of Special Relativity, which stays as a specific case inside a much wider geometry.
- Contrarily to supposed, the Intervalic System of Dimensions is the unique system which is **not equivalent** to all the remaining dimensional systems of units (which are, from now on, irrelevant in Physics research).
- All results yielded in the Intervalic Theory of Particle Physics are *geometric statements* logical and unavoidably deduced from the *intervalic quanta and limits* of the **Intervalic System of Units** *without using any mathematical formalism*.
- The Intervalic Theory is the unique Physics theory ever postulated which has *no one arbitrary constant*.
- Inasmuch as c and ℏ are universal constants, the *intervalic quanta and limits* are reliable physical quantities of *universal validity*. The Intervalic System of Units is not an arbitrary one but the genuine **system of units of Nature**. It must be noted that the *intervalic symmetries of Nature* can not be deduced by means of any other dimensional system, but only from the intervalic one. Thus its knowledge might be viewed as a clue of the scientific degree of development of a civilization, and so it is also apparent that the Intervalic Units are the unique which could be shared with hypothetical advanced extraterrestrial intelligences.
- Any value expressed in *intervalic units* can be interpreted as a *dimensionless interval or ratio* and Physics really becomes truthful Geometry. Hence the name of the Intervalic Theory.

gravitational field, whose discovery was, in words of Albert Einstein, "the happiest thought in my life", which led him to postulate the general relativity theory.

Algebraic operations between different physical quantities can't be performed on a single axis as in traditional systems, but as a result of the introduction of the imaginary number i in the intervalic dimensional basis, two axes are needed: a real one (representing the real component, L) and an imaginary one (representing the imaginary component, i) —in mathematical terms, an Argand space—. Because of the peculiar properties of the i number, each successive integer powers of the speed of light, c, whose intervalic dimension is i^{-1}, represents a turn of $\pm 90°$ on the dimensional axes of coordinates:

THE GROUP OF GEOMETRIC *TRANSFORMERS* IN THE INTERVALIC DIMENSIONAL SYSTEM		
Value	*Intervalic dimension*	*Rotation*
c^{-1}	i	$+90°$
c^{1}	i^{-1}	$-90°$
$c^{\pm 2}$	$i^{\pm 2} = -1$	$\pm 180°$
c^{-3}	$i^{3} = i^{-1}$	$+270° = -90°$
c^{3}	$i^{-3} = i$	$-270° = +90°$
$c^{\pm 4}$	$i^{\pm 4} = 1$	$\pm 360°$

INTERVALIC THEORY:
The Intervalic Structures of Subatomic Particles and the Last Foundations of Physics

INTERVALIC TRANSFORMATIONS OF VELOCITY IN THE INTERVALIC SYSTEM OF PHYSICAL QUANTITIES

Velocity ratio, $v_c = v/c$ (1)

[Figure: Argand diagram with vertical axis v_c (1), horizontal axes c^{-1} (i) on left labeled "Intervalic mass, $c^{-1} = 1$ (i)" and c (i^{-1}) on right labeled "Speed of light (in intervalic units), $c = 1$ (i^{-1})"; a right triangle with hypotenuse s is shown in the upper-right quadrant.]

From the figure (which is a simple Argand space composed by two axes: a real one and an imaginary one), we have:

$s^2 = c^2 + v_c^2$, that is to say, $s = \sqrt{(c^2 + v_c^2)}$

To make a formulation which looks like the relativistic one, it can be introduced the *intervalic factor xi*, ξ, defined as the inverse of the invariant interval, s:

$\xi \equiv 1/s = 1/\sqrt{(c^2 + v_c^2)}$
$\xi^{-1} \equiv s = \sqrt{(c^2 + v_c^2)}$

The intervalic dimension of ξ is $1/\sqrt{(-1)} = (i^{-1})$. Taking the value of c in intervalic units ($v_c = v/c$ is already an intervalic dimensionless ratio):

$\xi \equiv 1/s = 1/\sqrt{[1(-1) + v_c^2]} = 1/\sqrt{[-1 + v_c^2]}$

And thus the Intervalic Transformations of time, space, mass and momentum *regarding* VELOCITY are:

$L \cdot \xi \, (i^{-1}) = T \, (i \, L)$, time $m \cdot \xi^{-1} \, (i) = p \, (1)$, momentum
$T \cdot \xi^{-1} \, (i) = L \, (L)$, space $p \cdot \xi \, (i^{-1}) = m \, (i)$, mass

For converting $\xi \, (i^{-1})$ into the dimensionless relativistic gamma factor, $\gamma = 1/\sqrt{(1 - v_c^2)}$, it only have to be multiplied dimensionally by i: $\xi \, (i^{-1}) \cdot (i) = \gamma \, (1)$.

Hence multiplying *dimensionally* by (i) or (i^{-1}) the above equations in the intervalic dimensional space are obtained the classic Lorentz-Einstein transformations.

The **Intervalic Transformations** of space, time, energy, mass and momentum *regarding* TEMPERATURE are, already written in the traditional relativistic mode:

$L = L_0 \, \gamma^{-1}(\Theta)$ $T = \gamma(\Theta) \, T_0$ $E = \gamma(\Theta) \, E_0$ $m = \gamma(\Theta) \, m_0$ $p = \gamma(\Theta) \, p_0$

where $\gamma(\Theta)$ is the corresponding new gamma factor regarding temperature: $\gamma(\Theta) = 1/\sqrt{(1 - \Theta_c^2)}$, where it can be seen the dimensionless ratio $\Theta_c = \Theta/\Theta_I$, being Θ_I the intervalic limit of temperature: $\Theta_I = c \, k_B^{-1} = 2.17138589 \cdot 10^{31}$ (K). This geometric limit of temperature plays an important role in the phenomenon of "bounce" of the Universe, between the Big Crunch and the next Big Bang.

The geometric representation of any physical quantity on these dimensional coordinates gives automatically an *invariant* geometric measurement, which is to say that the transformations of Lorentz-Einstein of special relativity are already included in the properties of the intervalic dimensional space, being inherent to the intervalic metric itself, so it affects to all physical quantities, standing out for its simplicity the intervalic transformations of space, time, energy, mass or time *regarding temperature*, as can be seen in the graph.

Some of the most important equations of physics, such as all of special relativity, are derived in an independent way by the Intervalic Theory as mere *geometric statements* of its intervalic equation of dimensions, since in the Intervalic Theory *all equations are relativistic*, so there is no a non-relativistic mechanics as in quantum mechanics. Among these geometric statements it must be stand out what is possibly the most important formula of Physics, the equation of *intervalic energy*, I, which establishes the *equivalence* between *electric charge*, Q, and *energy*:

$$I = c^{\pm 2} \hbar Q^{-2}$$

	Mass	Electric charge
Equivalent energy	$E = c^{\pm 2} m$	$I = c^{\pm 2} \hbar Q^{-2}$
Field energy	$U = G m^2 / r$	$U = (1/4\pi\varepsilon) Q^2 / r$

This equation, painfully unknown by quantum mechanics, plays a role analogous to: $E = c^{\pm 2} m$, which establishes the *equivalence* between *mass* and *energy*, Einstein's famous equation now also deduced in an independent way to relativity, as a simple *geometric statement* by the Intervalic Theory (please note that the geometric constant c is always written ahead of the variables, and their power is ± 2 instead of $+2$, as the -2 power produces an identical rotation (in single intervalic units where the value of c and $\hbar$ is set to the unit).

Similarly is deduced geometrically the equation of the photon momentum, which until now was only an empirical result: $E = c\, p$. It should be noted that these important physical equations are always *invariant* because they are mere *geometric statements* which describe the *dimensional equivalence between two physical quantities*, which does not depend on the frame of reference chosen.

Furthermore, when comparing the equivalent energy of mass and electric charge is further exposed the initial gross mistake on which the standard model is based, whose complex Lagrangian formalism can not hide its serious inconsistency of misapplying the formula of the electromagnetic potential energy to a particle which, according to the

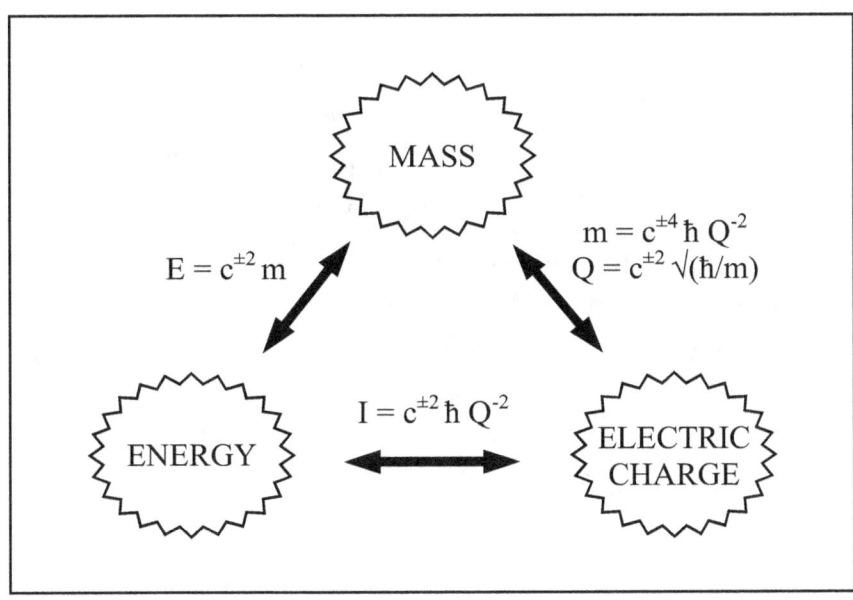

standard model itself, is a mysterious "energy density" lacking of structure. But if this particle is not composed by other sub particles, then, it can not possess electromagnetic energy, since, by definition, this energy only makes sense applied to a particle or to a body *with structure*, as it is well known by any high school student. Clumsy inconsistencies like that, among several others as the unwarranted use of about 20 arbitrary constants set *ad hoc*, were what led Albert Einstein to never accept the validity of the standard model, who always regarded it as merely provisional and baseless. It has been now confirmed with overwhelming logic by the Intervalic Theory.

The systems of physical quantities and units have no physical meaning (beyond the formal one), nor have any logical link with the equations of physics. On the contrary, the entire corpus of physics and all the results derived by the Intervalic Theory —which has not a single arbitrary constant—, are *geometric and logical statements* necessarily deduced from the intervalic *quanta* and *limits* of the *intervalic system of physical quantities and units*, without the mediation of any complex mathematical formalism. This is an outstanding feature of the strong logic economy with what has been deduced the theory. This implies that the intervalic structure of subatomic particles and the fundamental symmetries of Nature will never be able to be discovered by quantum mechanics or any other theory that does not use the *intervalic —or natural— system of physical quantities*, which differs essentially from the rest ones because it includes the imaginary number $i = \sqrt{-1}$ in its dimensional basis, which makes sense when you consider that the i number i appears in most physics equations. Therefore, any dimensional basis that does not contain the i number will not be able to have any logical bond with physical equations (except symbolically), nor to deduct any of them. Hence, quantum mechanics is by his reason in an impasse which impedes it to make further progress, as its last foundations are incomplete and inconsistent.

INTERVALIC THEORY:
The Intervalic Structures of Subatomic Particles and the Last Foundations of Physics

The theoretical definition and the exact geometry value of the elementary charge, e, and of the fine structure constant, α, are straightforwardly deduced starting from the intervalic system of physical quantities. Indeed, the definition and the geometric value of the *intervalic quantum of electric charge*, q_I, which is automatically deduced by simple dimensional analysis, is:

$$q_I = \sqrt{-(c^{-1}\hbar)} = 1 \ (i^{-1/2}L^{1/2}) = 5.93398995 \cdot 10^{-22} \ (C)$$

As it is usually explained in textbooks, the fine structure constant, α, is a measure of the squared value of the elementary charge *in natural units*. This value implies that the value of e is 270 in natural units, which coincides exactly with the value of the elementary charge *in intervalic units*, so we can say that the intervalic units are the genuine natural units, hitherto unknown:

$$e = 270 \ q_I = 270 \ \sqrt{-(c^{-1}\hbar)} = 1.60217733 \cdot 10^{-19} \ (C)$$
$$\alpha = 270^2 \cdot 10^{-7} = 1/137.1742112 \ (1) \ \text{—exact value—}$$

This result, as final as obvious, had not been accepted by quantum mechanics as a result of its absurd prejudice that subatomic particles lack of structure, a quasi religious dogma, completely wrong, on which the standard model is based, that has prevented significant progress in theoretical physics for decades.

From here we can get the exact geometric values of fundamental constants, rather than the traditional empirical values, of which c was set by hand, e was empirically set, and $\hbar$ was derived indirectly from the previous one. Now we know that there is a deviation of about 5/10,000 somewhere in all these traditional empirical values gathered. Since the value of c is taken by definition and e is quite reliable, we conclude that this small deviation is on $\hbar$, which can now be corrected and thus to define the exact geometric values of the two fundamental constants of

Nature (expressed in arbitrary traditional units) from the exact theoretical value of the fine structure constant:

$$c = 270^2 \, \hbar \, e^{-2} = 2.9979246 \cdot 10^8 \text{ (m/s)}$$
$$\hbar = 270^{-2} \, c \, e^2 = 1.0556363 \cdot 10^{-34} \text{ (m)}$$

The fact that the value of the elementary charge is 270 in intervalic or natural units *necessarily implies* that the elementary charge is not the *quantum of electric charge*, as previously thought, but that subatomic particles have got *structure*, i.e., they are composed by other sub particles even more fundamental. This result bankrupts the root paradigm of quantum mechanics, which conceives subatomic particles as "energy densities without structure", while according to the Intervalic Theory they are composite particles, which have an *intervalic structure* —of extraordinary logical economy and mathematical beauty—, of which all their physical properties and also all fundamental interactions of physics are deduced, whose origins were unknown to quantum mechanics.

In this way, subatomic particles are *states of minimum energy* that have become very deep wells of electromagnetic potential energy, which is the reason why they are all *identical*, and not because they obey certain quantum, mysterious and inexplicable numbers, as postulated by the standard model, or because there is an essential difference between the physical nature of the subatomic world and the other one of the macroscopic or cosmological world.

In fact, the results of the Intervalic Theory show that there is not such difference, even between the organic and the inorganic world, as it is shown, for example, by means of the ratio of the constituent energies —intervalic and electromagnetic— of nucleon, which are ruled by the *golden mean* or the Φ number, which also determines the pattern of so-called *harmonious growth* of living beings. Well, it has been found that the Φ number also rules the primordial intervalic synthesis of subatomic particles:

Structural energy ratios of nucleon
$\langle I(N)/U(N) \rangle$ = 1.618829402 ~ Φ
$\langle I(N)/E(N)_{mass} \rangle$ = 0.618143766 ~ Φ^{-1}
$\langle U(N)/E(N)_{mass} \rangle$ = 0.381855365 ~ 1 - Φ^{-1}

Deviation from the golden mean, Φ
$\Delta[\langle I(N)/U(N) \rangle]$ = +0.0491593%
$\Delta[\langle I(N)/E(N)_{mass} \rangle]$ = +0.0177623%
$\Delta[\langle U(N)/E(N)_{mass} \rangle]$ = -0.0289759%

Among the thousand first natural numbers, 270 is what most classes or symmetries possesses, each one of them indicates the existence of a well of electromagnetic potential energy under the *principle of minimum energy* and also under the *intervalic principle of minimum information*, being the former one a simple corollary. From the point of view of chaos theory, the number 270 would be the principal attractor at subatomic scale, and its divisors the secondary attractors. These 16 classes are natural dividers of 270, which determine the values of the electrical charges of subatomic particles allowed by the intervalic symmetries of the elementary charge, and are as follows (expressed in natural units: e = 270):

1, 2, 3, 5, 6, 9, 10, 15, 18, 27, 30, 45, 54, 90, 135, 270.

Of these, half are unstable below its threshold temperature of synthesis, being stable: 2, 3, 5, 6, 18, 30, 45 and 270. Although these intervalic symmetries has been uncovered by the logical-deductive way, which is the scientific method *par excellence*, the truth is that it could have also been discovered via empirical-inductive, as far as it is an empirical evidence that massive subatomic particles detected till today have got masses whose values are proportional to the inverse square of 270, 45, 30, 18, 6 and 5, as anyone can easily check. If this incontrovertible empirical data has not led to the discovery of the Intervalic Theory previously, this is due to the absurd prejudice of quantum mechanics to believe that subatomic particles are punctual entities *without structure*, childish dogma, comparable to medieval believing about the Earth as the centre of the Universe, which have prevented the advance of Physics for decades. Certainly, in the annals of Physics, after the missed belief in a flat Earth, the standard model of particles without structure will be history as the most ridiculous mistake in Physics.

The synthesis of subatomic particles is always made obeying one of the most fundamental principles of Nature: the *spin-statistics theorem*, of mathematical origin, which states that any identical particles have a degree of freedom, what means that the *constituent* particles of a subatomic particle can only be in a *symmetric or antisymmetric state under interchange*, being also generated both states in accordance with the general principle of *intervalicity*, which allows only logical and necessary relationships between the elements of the theory (or under the so-called *universality assumption* of quantum mechanics, which would be a corollary of the above).

The fundamental law that rules the intervalic structure and the physical properties of all subatomic particles is the *intervalic principle of energy balance*, which states that the total energy or *structural energy*, A, of all subatomic particles is composed by three factors: intervalic energy, I, electromagnetic energy, $U \approx I^{-1}$ —whose magnitude is in inverse proportion to the above—, and the spin energy, E_J —being the latter the only one not manifested as mass but as kinetic energy—, according to the following basic equation, also written in developed form below:

$$I - I^{-1} - E_J = 0$$
$$c^{\pm 2} h Q^{-2} - [\tfrac{1}{2}(1/4\pi\varepsilon_0)Q^2 / r] - m r^2 \omega_J^2 = 0$$

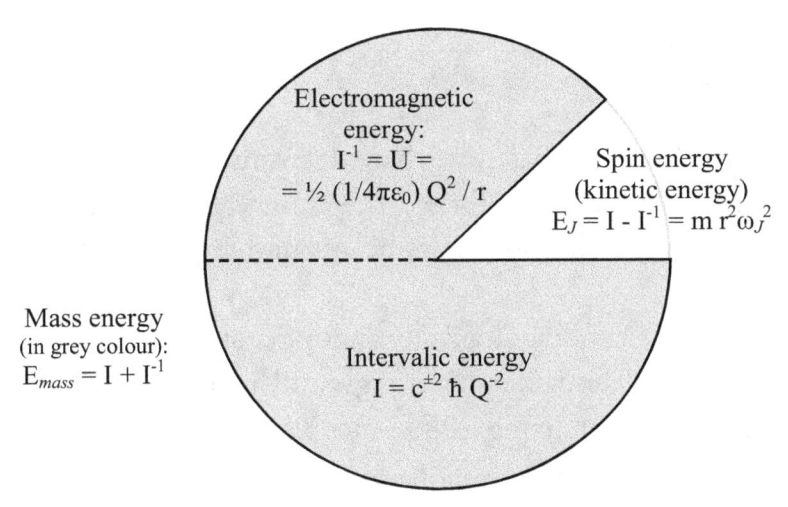

The intervalic energy, I, represents half of the structural energy, A, of the subatomic particle, which coincides precisely with the Schwarzschild's mass for the particle:

$$A = 2\,I$$
$$A = I + I^{-1} + E_J = c^{\pm 2}\, m_{Schw}$$

If all the structural energy was manifested as mass, all particles would have got exactly the minimum mass to be black holes. This never happens because the spin energy does not contribute to the mass but it is manifested as kinetic energy, being able to define the spin energy according to these values:

$$E_J = c^{\pm 2}\, (m_{Schw} - m)$$

This further gives us a new definition of the Schwarzschild's mass for subatomic particles, since its magnitude is simply twice the intervalic energy:

$$I = \tfrac{1}{2}\, c^{\pm 2}\, m_{Schw}$$

This is the energy balance that determines the fundamental underlying architecture that forms the intervalic structure of all subatomic particles with extraordinary beauty and logical simplicity.

Well, from here, it follows logically and inevitably the synthesis of all subatomic particles that make the energy and matter in the universe, both *visible* and *dark*, totalizing 108 *visible* particles (without including baryons and mesons, which are monteverdinos) and 39 *dark* particles, whose intervalic structures and physical properties have been described with full geometric precision by the Intervalic Theory. The way that Nature has made this, starting exclusively from the existence of an interval, is a way plenty of logical elegance and disarming economy. Although the description of the *primordial intervalic synthesis of subatomic particles* can not be condensed into a short essay, we will try to offer some tables that show graphically the intervalic structure of the most important subatomic particles.

The primordial intervalic synthesis of intervalic structures, which originated from nothing all subatomic particles and fundamental interactions of physics, has seven consecutive phases. In each phase it is synthesized an intervalic structure in both symmetric and antisymmetric state under interchange, being one of them *visible* matter or energy — particles which in turn are the constituent sub particles of the next phase, which is still synthesizing a new type of intervalic structure in symmetric and antisymmetric state under interchange—, and other one *dark* matter or energy —which does not continue with the process of synthesis, causing a symmetry breaking, so that at each stage only one of the two states continues the process of synthesis that leads to the next phase—.

INTERVALIC THEORY:
The Intervalic Structures of Subatomic Particles and the Last Foundations of Physics

Along with singularities (black holes), antimatter universes and parallel universes, another one of the misleading myths that the Intervalic Theory invalidates is that one of the Big Bang considered as a haphazard and random explanation of the origin of the Universe. On the contrary, the Intervalic Theory shows what the origin of the Big Bang, which was a huge release of energy that came in the 4th phase of the primordial intervalic synthesis: the dalino-synthesis —where dalinos were synthetized, which are the first particles with electromagnetic energy—, a phase that began between $1.521843955 \cdot 10^{-10}$ and $1.979589614 \cdot 10^{-9}$ seconds *after the beginning of time* —which came into existence just at the beginning of the 2nd phase or photon-synthesis. The dalino-synthesis released the fabulous amount of energy (in form of photons and neutrinos) of:

$$E_B(D_{270}) = m(270\, I) - m(D_{270}) = 5{,}621{,}244.136 \ (MeV/c^2)$$
$$E_B(D_{45}) = m(45\, I) - m(D_{45}) = 936{,}855.694 \ (MeV/c^2)$$

respectively for each electron (D_{270}) and each nucleonic dalino (D_{45}) synthesized. The dalino-synthesis began as early as the temperature of the primordial Universe fell below the threshold temperatures of synthesis for each dalino:

$$\Theta_B(D_{270}) < E_B(D_{270}) / k_B = 6.523179514 \cdot 10^{16} \ (K)$$
$$\Theta_B(D_{45}) < E_B(D_{45}) / k_B = 1.087175336 \cdot 10^{16} \ (K),$$

since above these threshold temperatures the 16 dalinos allowed by the intervalic symmetries are in a state of recombination (symmetries that will permanently last below those temperatures, but not as *real* states but as *virtual* states):

$D_{270} \ \ 2\,D_{135} \ \ 3\,D_{90} \ \ 5\,D_{54} \ \ 6\,D_{45} \ \ 9\,D_{30} \ \ 10\,D_{27} \ \ 15\,D_{18} \ \ 18\,D_{15} \ \ 27\,D_{10}$
$30\,D_9 \ \ 45\,D_6 \ \ 54\,D_5 \ \ 90\,D_3 \ \ 180\,D_2 \ \ 270\,D_1$

This extraordinary release of energy in the primordial Universe is what is known as the Big Bang.

INTERVALIC PRIMORDIAL ASSEMBLY OF INTERVALIC STRUCTURES
WHICH ORIGINATED ALL DEGREES OF FREEDOM AND INTERACTIONS

Tree level	1	2	3	4	5	6	7	8
Assemblies are made in symmetric and antisymmetric states under intervalic change, with identical shares. The appearance of each new intervalic structure assembled makes automatically the introduction of a new degree of freedom with its corresponding fundamental interaction in the Universe. When some intervalic structures do not make further assemblies, there is a *symmetry breaking* in the tree and that branch of the Intervalic Primordial Assembly ends.	**INTERVALIC STRING** —quantum of space— (Share = 1) Intervalic string state $S = \{\uparrow, \downarrow\}$ Intervalic string radius $r_S = \frac{1}{2} \hbar$ Intervalic string spin $J_S = \frac{1}{2} \hbar$ Intervalic string length $l_S = \pi \hbar = \pi \hbar$ Number of intervalic strings $n(S) = 4 n(\gamma) = 4 \cdot 4 \cdot c^3 = c^3$	**PHOTON** —quantum of light— (Share = 1/2) = Symmetric assembly of Intervalic Strings: $\gamma = \{\uparrow\downarrow\}$ $2^{-\frac{1}{2}} (\ket{\uparrow\downarrow}+\ket{\downarrow\uparrow})$ Photon radius $r_\gamma = 1.0556363 \cdot 10^{-34}$ (m) Frequency of primordial photon $\varphi_\gamma = \varphi_I = \Theta_\gamma c k_B^{-1}$ Temperature of primordial photon $\Theta_\gamma = \Theta_I = c k_B^{-1}$ $= 2.17138589 \cdot 10^{31}$ (K) Intervalic-relativistic transformations of time regarding temperature $t = t_0/\sqrt{1-|\Theta/\Theta_I|^2}$ Total energy of primordial photons $\Sigma E(\gamma) = \frac{1}{4} c^2 \hbar^2 \varphi_\gamma^2 = \frac{1}{4} c^9$ $= 4.890196776 \cdot 10^{75}$ (J) Since the energy of primordial photons was exactly the intervalic velenergy, $c = 2.99792458 \cdot 10^8$ (J), we can surprisingly know with astonishing simplicity the *number* of primordial photons assembled from intervalic strings at the beginning of the Universe: $n(\gamma) = \Sigma E(\gamma)/c = \frac{1}{4} c^8$ $= 1.631194063 \cdot 10^{67}$ **Neutrino** Neutrino is defined as an intervalic string that came to light. The majority of neutrinos have been made at the Intervalic Primordial Assembly of gaudinos and dalinos, which began between 1.521843955 $\cdot 10^{-20}$ (s) and 1.979589614 $\cdot 10^{-9}$ (s) after the intervalic structure level No. 2 (synthesis of photon and chi) which marked the *beginning of time*.	**INTERVALINO** —quantum of matter— (Share = 1/4) = Antisymmetric assembly of Photons Intervalino state $\mathbf{I} = 2^{-\frac{1}{2}} (\ket{\uparrow\downarrow}-\ket{\downarrow\uparrow})$ Intervalino radius $r(\mathbf{I}) = 2^{-\frac{1}{2}}/\omega(\mathbf{I}) = 2h$ $= 2.1112726 \cdot 10^{-34}$ (m) Intervalino spin $J(\mathbf{I}) = 0$ Intervalino charge $\mathbf{q}_I = \sqrt{\omega(c^3\hbar)} = 1 (c^{-2}\hbar^2)^{1/2}$ $= 5.93398995 \cdot 10^{-22}$ (C) Intervalino intervalic energy $I(\mathbf{I}) = c^2 \hbar \mathbf{q}_I = c^{-1}$ $= 20,819.42423$ (MeV/c^2) Intervalino electromagnetic potential energy: $U(\mathbf{I}) = 0$ Intervalino mass: $m(\mathbf{I}) = I(\mathbf{I})$ Intervalino spin energy $E_s(\mathbf{I}) = I(\mathbf{I}) - U(\mathbf{I}) = c^{-1}$ $= 20,819.42423$ (MeV/c^2) Intervalino linear velocity on surface: $v(\mathbf{I}) = 0$ Total energy of primordial intervalinos and gravitons $\Sigma E(\mathbf{I}) = \Sigma E(g) = \frac{1}{2} \Sigma E(\gamma)$ $= 2.445098338 \cdot 10^{75}$ (J) Number of primordial intervalinos assembled at the IPA $n(\mathbf{I}) = \Sigma E(\mathbf{I})/I(\mathbf{I}) =$ $= (1/8) c^{10} = 7.330220558 \cdot 10^{83}$ Threshold temperature of annihilation-materialization of intervalinos, $\gamma\gamma \leftrightarrow \mathbf{I}$ $\Theta_m(\mathbf{I}) \geq c^{-2} m(\mathbf{I}) / k_B = 4\pi \Theta_m =$ $= 2.415992632 \cdot 10^{14}$ (K) Threshold frequence of photons at the annihilation-materialization of intervalinos, $\gamma\gamma \leftrightarrow \mathbf{I}$: $\varphi_m = 4\pi \varphi_m = 1(c^{-1}h) = 3.159831911 \cdot 10^{25}$ (s^{-1}) (Big Bang origin)	**DALINOS** —quanta of electric charge— (Share = 1/8) = Symmetric assembly of Intervalinos $D^{(\pm 1, 2, 3, 5, 6, 9, 10, 15, 18, 27, 30, 45, 54, 90, 135, 270)}$ —16 electric charged dalinos geometrically allowed by the intervalic symmetries of Nature— **Electron** $e^- = G_1 = D_{270} = 270 \mathbf{I} = 540 \gamma = 1080 S$ Electron charge $e = 270 \mathbf{q}_I = 1 (c^{-2}\hbar^2)^{1/2}$ Structural energy balance $1 \cdot \mathbf{I}^{-1} \cdot E_s(e^-) =$ $- [c^{-2}\hbar (270 \varphi_m)^2] -$ $- [\frac{1}{2} (1/4\pi\epsilon_0) e^2 / r_e] -$ $m_e \omega_e(e)^2 r_e^2 = 0$	**GAUDINOS** (Share = 1/16) = Symmetric assembly of Dalinos $G^{(\pm 1, \pm 2)}$: Nucleonic gaudinos $G^{(0)}$: Leptons-Charged Massive Bosons —16 elementary charged gaudinos geometrically allowed by the intervalic symmetries of Nature— **Muon, Tau,** $Z^\pm, W^\pm, Y^\pm, X^\pm$	**LISZTINOS**($^{\pm\frac{1}{3}, \pm\frac{2}{3}}$) (Share = 1/32) = Symmetric assembly of gaudinos **Quarks** (49 quarks = 7 lisztinian families $\times$ 7 dalinian symmetries) Inasmuch the vast majority of ZCMB decayed into quarks, and as the vast majority of primordial quarks decayed finally into the isoquark 1_3q_u)$\mathbf{182q_u}^{(u)}$—former quarks up and down— we can deduce the maximum number of nucleonic isoquarks generated at the primordial Universe (taking roughly $m(q) \sim 313$ MeV/c^2): $n(q) \leq \Sigma m(_1q_u, _1q_d)/\Sigma m(q) =$ $1.218937472 \cdot 10^{83}$ **LISZTINOS**(^{0}) = Antisymmetric assembly of Gaudinos **Bileptons-Z.C.M. Bosons:** Z^0, W^0, Y^0, X^0	**MONTEVERDINOS** = Assembly of 3 lisztinos($^{\pm\frac{1}{3}, \pm\frac{2}{3}}$) **Baryons** **MONTEVERDINOS** = Assembly of 2 Lisztinos($^{\pm\frac{1}{3}, \pm\frac{2}{3}}$) **Mesons**	**PALESTRINOS** Nuclei: $1 < A \leq 3$ **PSEUDO-PALESTRINOS** Nuclei: $A > 3$ 7th. symmetry breaking
		CHI (Share = 1/2) = Antisym. assembly of S $\not\gamma = 2^{-\frac{1}{2}}(\ket{\uparrow\uparrow}-\ket{\downarrow\downarrow})$ Dark energy	**GRAVITON** (Share = 1/4) = Symmetric assembly of $\not\gamma$ $g = \ket{\not\gamma\not\gamma}_S = [\ket{\uparrow\uparrow\gamma\gamma}, \ket{\uparrow\uparrow\downarrow\downarrow} + \ket{\downarrow\downarrow\uparrow\uparrow}, \ket{\downarrow\downarrow\gamma\gamma}]$ Dark energy	**ZERO CHARGED DALINOS** (Share = 1/8) = Antisymmetric assembly of $\mathbf{I}$ —8 zero charged dalinos— $D_{270}^0, D_{90}^0, D_{54}^0, D_{30}^0, D_{18}^0, D_{10}^0, D_{6}^0, D_{2}^0$ Dark matter	**ZERO CHARGED GAUDINOS** (Share = 1/16) = Antisymmetric assembly of D —25 zero charged gaudinos— Dark matter			
					6th. symmetry breaking			
				ELEMENTARY CHARGE ATTRACTOR 5th. symmetry breaking				
				Degree of freedom: **ELECTRIC CHARGE** → **ELECTROMAGNETIC INTERACTION.** 3rd. symmetry breaking				
			INTERVALIC CHANGEFUL INTERACTION (formerly weak interaction). 4th. symmetry breaking					
			Degree of freedom: **MASS** → GRAVITATIONAL INTERACTION. DARK MATTER UNIVERSE. 2nd. symmetry breaking					
		Degree of freedom: **SPIN** → INTERVALIC CHANGELESS INTERACTION (formerly strong interaction). LIGHT UNIVERSE. 1st. symmetry breaking						
INTERVALIC LENGTH								
	Degree of freedom: **INFORMATION** → INFORMATIONAL INTERACTION. INFORMATION UNIVERSE (TIMELESS). Zero symmetry breaking							

INTERVALIC STRUCTURES
OF SUBATOMIC PARTICLES
ALLOWED BY THE INTERVALIC SYMMETRIES (mass in MeV)

DALINAR SYMMETRY	LEPTONS-MASSIVE BOSONS AND NEUTRINOS	QUARKS (FRACTIONAL LISZTINOS)
{D270}	$G_1D_{270}^{(\pm)}$ (0.5) = $e^\pm$ electron $v_{D270} = v_e$ neutrino $(1.1833119 \cdot 10^{-14})$	–
{D135}	–	–
{D90}	–	–
{D54}	–	–
{D45}	$G_6D_{45}^{(\pm)}$ (106) = $\mu^\pm$ muon $v_{D45} = v_\mu$ neutrino $(2.0005108 \cdot 10^{-7})$	$L_{1/3}\tfrac{1}{3}G_6 2D_{45}^{(1/3)}$ (35) *last radiant decay quark* $L_{2/3}\tfrac{2}{3}G_6 4D_{45}^{(1/3, 2/3)}$ (69) *constituent quark of π meson* $L_1 1G_6 6D_{45}^{(1/3, 2/3)}$ (104) $L_2 2G_6 12D_{45}^{(1/3, 2/3)}$ (207) **$L_3 3G_6 18D_{45}^{(1/3, 2/3)}$ (311)** *former quarks up, down* $L_4 4G_6 24D_{45}^{(1/3, 2/3)}$ (414) $L_5 5G_6 30D_{45}^{(1/3, 2/3)}$ (518) *former quark strange*
{D30}	$G_9D_{30}^{(\pm)}$ (373) – v_{D30} –	$L_{1/3}\tfrac{1}{3}G_9 3D_{30}^{(1/3)}$ (117) $L_{2/3}\tfrac{2}{3}G_9 6D_{30}^{(2/3)}$ (233) $L_1 1G_9 9D_{30}^{(1/3)}$ (350) $L_2 2G_9 18D_{30}^{(1/3)}$ (699) $L_3 3G_9 27D_{30}^{(1/3)}$ (1,049) **$L_4 4G_9 36D_{30}^{(2/3)}$ (1,399)** *former quark charm* $L_5 5G_9 45D_{30}^{(1/3)}$ (1,748)
{D27}	–	–
{D18}	$G_{15}D_{18}^{(\pm)}$ (1,777) = $\tau^\pm$ tau $v_{D18} = v_\tau$ neutrino $(2.6777745 \cdot 10^{-4})$	$L_{1/3}\tfrac{1}{3}G_{15} 5D_{18}^{(1/3)}$ (540) $L_{2/3}\tfrac{2}{3}G_{15} 10D_{18}^{(2/3)}$ (1,079) $L_1 1G_{15} 15D_{18}^{(1/3)}$ (1,619) $L_2 2G_{15} 30D_{18}^{(1/3)}$ (3,238) **$L_3 3G_{15} 45D_{18}^{(1/3)}$ (4,857)** *former quark bottom* $L_4 4G_{15} 60D_{18}^{(1/3)}$ (6,476) $L_5 5G_{15} 75D_{18}^{(1/3)}$ (8,095)
{D15}	–	
{D10}		
{D9}		
{D6}	$G_{45}45D_6^{(\pm)}$ (46,565) $Z^\pm$ *massive boson* $L_2 2G_{45} 90D_6^{(0)}$ (91,188) Z^0 **massive boson** v_{D6} neutrino	$L_{1/3}\tfrac{1}{3}G_{45} 15D_6^{(1/3)}$ (14,571) $L_{2/3}\tfrac{2}{3}G_{45} 30D_6^{(2/3)}$ (29,141) $L_1 1G_{45} 45D_6^{(1/3)}$ (43,712) $L_2 2G_{45} 90D_6^{(1/3)}$ (87,426) $L_3 3G_{45} 135D_6^{(1/3)}$ (131,135) **$L_4 4G_{45} 180D_6^{(2/3)}$ (174,846)** *former quark top* $L_5 5G_{45} 225D_6^{(1/3)}$ (218,558)
{D5}	$G_{54}54D_5^{(\pm)}$ (80,423) $W^\pm$ **massive boson** $L_2 2G_{54} 108D_5^{(0)}$ (160,928) W^0 massive boson v_{D5} neutrino	$L_{1/3}\tfrac{1}{3}G_{54} 18D_5^{(1/3)}$ (25,178) $L_{2/3}\tfrac{2}{3}G_{54} 36D_5^{(1/3, 2/3)}$ (50,356) $L_1 1G_{54} 54D_5^{(1/3)}$ (75,534) $L_2 2G_{54} 108D_5^{(1/3, 2/3)}$ (151,068) **$L_3 3G_{54} 162D_5^{(1/3, 2/3)}$ (226,601)** $L_4 4G_{54} 216D_5^{(1/3, 2/3)}$ (302,134) $L_5 5G_{54} 270D_5^{(1/3, 2/3)}$ (377,668)
{D3}	$G_{90}90D_3^{(\pm)}$ (372,518) $Y^\pm$ *massive boson* $L_2 2G_{90} 180D_3^{(0)}$ (745,037) Y^0 **massive boson** v_{D3} neutrino	$L_{1/3}\tfrac{1}{3}G_{90} 30D_3^{(1/3)}$ (116,564) $L_{2/3}\tfrac{2}{3}G_{90} 60D_3^{(2/3)}$ (233,128) $L_1 1G_{90} 90D_3^{(1/3, 2/3)}$ (349,693) $L_2 2G_{90} 180D_3^{(1/3, 2/3)}$ (699,384) $L_3 3G_{90} 270D_3^{(1/3, 2/3)}$ (1,049,078) **$L_4 4G_{90} 360D_3^{(1/3, 2/3)}$ (1,398,771)** $L_5 5G_{90} 450D_3^{(1/3, 2/3)}$ (1,748,463)
{D2}	$G_{135}135D_2^{(\pm)}$ (1,257,249) $X^\pm$ **massive boson** $L_2 2G_{135} 270D_2^{(0)}$ (2,514,499) X^0 massive boson v_{D2} neutrino	$L_{1/3}\tfrac{1}{3}G_{135} 45D_2^{(1/3)}$ (393,404) $L_{2/3}\tfrac{2}{3}G_{135} 90D_2^{(1/3, 2/3)}$ (786,808) $L_1 1G_{135} 135D_2^{(1/3, 2/3)}$ (1,180,213) $L_2 2G_{135} 270D_2^{(1/3, 2/3)}$ (2,360,424) $L_3 3G_{135} 405D_2^{(1/3, 2/3)}$ (3,540,638) $L_4 4G_{135} 540D_2^{(1/3, 2/3)}$ (4,720,850) $L_5 5G_{135} 675D_2^{(1/3, 2/3)}$ (5,901,063)
{D1}	–	–

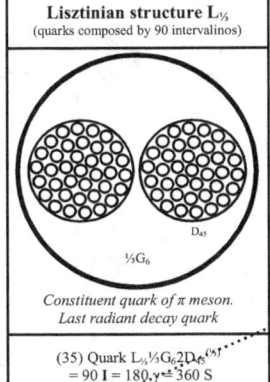

Lisztinian structure $L_{1/3}$
(quarks composed by 90 intervalinos)

Constituent quark of π meson. Last radiant decay quark

(35) Quark $L_{1/3}\tfrac{1}{3}G_6 2D_{45}^{(1/3)}$
= 90 I = 180 γ = 360 S

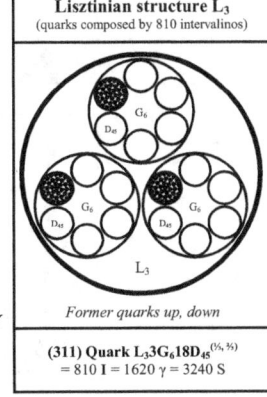

Lisztinian structure L_3
(quarks composed by 810 intervalinos)

Former quarks up, down

(311) Quark $L_3 3G_6 18D_{45}^{(1/3, 2/3)}$
= 810 I = 1620 γ = 3240 S

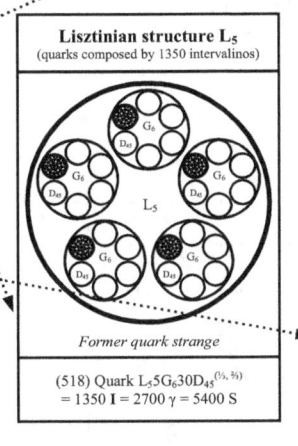

Lisztinian structure L_5
(quarks composed by 1350 intervalinos)

Former quark strange

(518) Quark $L_5 5G_6 30D_{45}^{(1/3, 2/3)}$
= 1350 I = 2700 γ = 5400 S

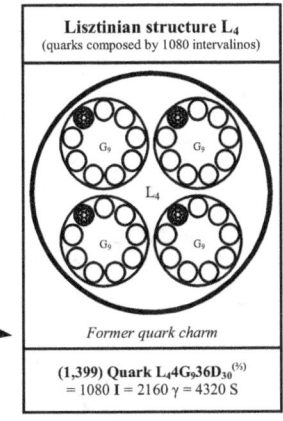

Lisztinian structure L_4
(quarks composed by 1080 intervalinos)

Former quark charm

(1,399) Quark $L_4 4G_9 36D_{30}^{(2/3)}$
= 1080 I = 2160 γ = 4320 S

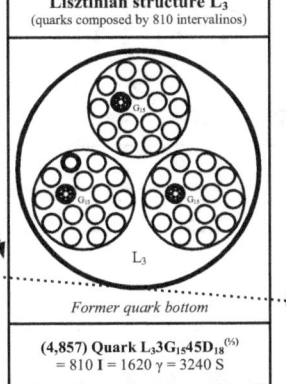

Lisztinian structure L_3
(quarks composed by 810 intervalinos)

Former quark bottom

(4,857) Quark $L_3 3G_{15} 45D_{18}^{(1/3)}$
= 810 I = 1620 γ = 3240 S

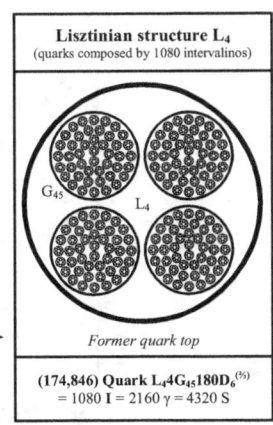

Lisztinian structure L_4
(quarks composed by 1080 intervalinos)

Former quark top

(174,846) Quark $L_4 4G_{45} 180D_6^{(2/3)}$
= 1080 I = 2160 γ = 4320 S

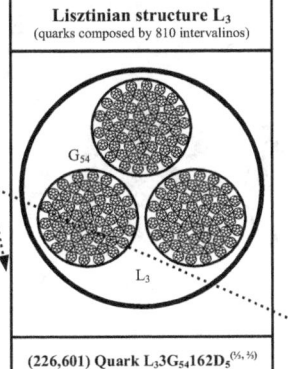

Lisztinian structure L_3
(quarks composed by 810 intervalinos)

(226,601) Quark $L_3 3G_{54} 162D_5^{(1/3, 2/3)}$
= 810 I = 1620 γ = 3240 S

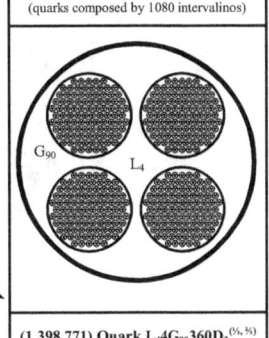

Lisztinian structure L_4
(quarks composed by 1080 intervalinos)

(1,398,771) Quark $L_4 4G_{90} 360D_3^{(1/3, 2/3)}$
= 1080 I = 2160 γ = 4320 S

Dimensional Basis of the Intervalic System of Units: (L, i) $\dim(\hbar) = L$ $\dim(c) = i^{-1}$	INTERVALIC LEPTONS-CHARGED MASSIVE BOSONS		
	ELECTRON (0.51099906 MeV/c^2)	**MUON (105.658389 MeV/c^2)**	**TAU (1,777 MeV/c^2)**
	$e = D_{270} = 270\ I = 540\ \gamma = 1080\ S$	$\mu = G_6 = 6\ D_{45} = 270\ I = 540\ \gamma = 1080\ S$	$\tau = G_{15} = 15\ D_{18} = 270\ I = 540\ \gamma = 1080\ S$
Intervalic structure	D_{270}	G_6 / D_{45}	G_{15} / D_{18}
Intervalic energy, I	$I(e) = c^{*2} \hbar\, e^{-2} = 4.575639166 \cdot 10^{-14}$ (J) = $= 0.285588809$ (MeV/c^2)	$I(\mu) = \sum(c^{*2}\hbar\, Q^{-2}) = c^{*2}\hbar\, e^{-2} +$ $+ 6\,(c^{*2}\hbar\,(45\,q_I)^{-2}) = 61.9727717$ (MeV/c^2)	$I(\tau) = \sum(c^{*2}\hbar\, Q^{-2}) = c^{*2}\hbar\, e^{-2} +$ $+ 15\,(c^{*2}\hbar\,(18\,q_I)^{-2}) = 964.1478215$ (MeV/c^2)
Electromagnetic energy, U	$U(e) = c^{*2}m(e) - I(e) = 3.611471925 \cdot 10^{-14}$ (J) = $= 0.22541025$ (MeV/c^2)	$U(\mu) = c^{*2}m(\mu) - I(\mu) = 6.999210568 \cdot 10^{-12}$ (J) = $= 43.6856173$ (MeV/c^2)	$U(\tau) = c^{*2}m(\tau) - I(\tau) = 1.302333333 \cdot 10^{-10}$ (J) = $= 812.8521785$ (MeV/c^2)
Spin energy, E_J	$E_J(e) = I(e) = 9.64167241 \cdot 10^{-15}$ (J)	$E_J(\mu) = I(\mu) - U(\mu) = 2.929926422 \cdot 10^{-12}$ (J)	$E_J(\tau) = I(\tau) - U(\tau) = 2.42402449 \cdot 10^{-11}$ (J)
Radius, r	$r_e = \tfrac{1}{2}(1/4\pi\varepsilon_0)\,e^2 / U(e) = 3.194098699 \cdot 10^{-15}$ (m)	$r_\mu = \tfrac{1}{2}(1/4\pi\varepsilon_0)(270\,q_I)^2 / U(\mu) = 1.648099834 \cdot 10^{-17}$ (m)	$r_\tau = \tfrac{1}{2}(1/4\pi\varepsilon_0)(270\,q_I)^2 / U(\tau) = 8.857484858 \cdot 10^{-19}$ (m)
Angular velocity due to spin, ω_J	$\omega_J(e) = (E_J(e)/m_e\,r_e^2)^{1/2} = 3.220944289 \cdot 10^{22}$ (s^{-1})	$\omega_J(\mu) = (E_J(\mu)/m_\mu\,r_\mu^2)^{1/2} = 7.567601581 \cdot 10^{24}$ (s^{-1})	$\omega_J(\tau) = (E_J(\tau)/m_\tau\,r_\tau^2)^{1/2} = 9.87597042 \cdot 10^{25}$ (s^{-1})
Linear velocity due to spin on surface, v_J	$v_J(e) = \omega_J(e)\,r_e = 1.028801396 \cdot 10^{8}$ (m s^{-1}) = $= 0.343171206\ c$	$v_J(\mu) = \omega_J(\mu)\,r_\mu = 1.247216291 \cdot 10^{8}$ (m s^{-1}) = $= 0.416026573\ c$	$v_J(\tau) = \omega_J(\tau)\,r_\tau = 8.747625846 \cdot 10^{7}$ (m s^{-1}) = $= 0.29178939\ c$
Intervalic quantum of charge $q_I = \sqrt{(c^{-1}\hbar)} = 1\,(i^{-1/2}L^{1/2}) =$ $5.93398995 \cdot 10^{-22}$ (C)	**W$^\pm$ MASSIVE BOSON (80,423 MeV/c^2)**	**Y$^\pm$ MASSIVE BOSON (372,518 MeV/c^2)**	**X$^\pm$ MASSIVE BOSON (1,257,249 MeV/c^2)**
	$W^\pm = G_{54} = 54\ D_5 = 270\ I = 540\ \gamma = 1080\ S$	$Y^\pm = G_{90} = 90\ D_3 = 270\ I = 540\ \gamma = 1080\ S$	$X^\pm = G_{135} = 135\ D_2 = 270\ I = 540\ \gamma = 1080\ S$
Intervalic structure			
Intervalic energy, I	$I(W^\pm) = \sum(c^{*2}\hbar\,Q^{-2}) = c^{*2}\hbar\,e^{-2} +$ $+ 54\,(c^{*2}\hbar\,(5\,q_I)^{-2}) = 44{,}970.24192$ (MeV/c^2)	$I(Y^\pm) = \sum(c^{*2}\hbar\,Q^{-2}) = c^{*2}\hbar\,e^{-2} + 90\,(c^{*2}\hbar\,(3\,q_I)^{-2}) =$ $3.335645528 \cdot 10^{-8}$ (J) = 208,194.5279 (MeV/c^2)	$I(X^\pm) = \sum(c^{*2}\hbar Q^{-2}) = c^{*2}\hbar\,e^{-2} + 135\,(c^{*2}\hbar\,(2q_I)^{-2}) =$ $= 1.125779279 \cdot 10^{-7} = 702{,}655.8532$ (MeV/c^2)
Electromagnetic energy, U	$U(W^\pm) = c^{*2}m(W^\pm) - I(W^\pm) = 5.6801605 \cdot 10^{-9}$ (J) = $= 35{,}452.76$ (MeV/c^2)	$U(Y^\pm) = c^{*2}m(Y^\pm) - I(Y^\pm) = 2.632758703 \cdot 10^{-8}$ (J) = $= 164{,}323.802$ (MeV/c^2)	$U(X^\pm) = m(X^\pm) - I(X^\pm) = 8.88557208 \cdot 10^{-8}$ (J) = $= 554{,}593.547$ (MeV/c^2)
Spin energy, E_J	$E_J(W^\pm) = I(W^\pm) - U(W^\pm) = 1.524869713 \cdot 10^{-9}$ (J) = $= 9{,}517.484017$ (MeV/c^2)	$E_J(Y^\pm) = I(Y^\pm) - U(Y^\pm) = 7.02886825 \cdot 10^{-9}$ (J) = $= 43{,}870.726$ (MeV/c^2)	$E_J(X^\pm) = I(X^\pm) - U(X^\pm) = 2.37222071 \cdot 10^{-8}$ (J) = $= 148{,}062.307$ (MeV/c^2)
Radius, r	$r_W = \tfrac{1}{2}(1/4\pi\varepsilon_0)(270\,q_I)^2 / U(W^\pm) = 2.0308225 \cdot 10^{-20}$ (m)	$r_Y = \tfrac{1}{2}(1/4\pi\varepsilon_0)(270\,q_I)^2 / U(Y^\pm) = 4.381482156 \cdot 10^{-21}$ (m)	$r_X = \tfrac{1}{2}(1/4\pi\varepsilon_0)(270\,q_I)^2 / U(X^\pm) = 1.298215261 \cdot 10^{-21}$ (m)
Angular velocity due to spin, ω_J	$\omega_J(W^\pm) = (E_J(W^\pm)/m_W\,r_W^2)^{1/2} = 5.0783155 \cdot 10^{27}$ (s^{-1})	$\omega_J(Y^\pm) = (E_J(Y^\pm)/m_Y\,r_Y^2)^{1/2} = 2.34808191 \cdot 10^{28}$ (s^{-1})	$\omega_J(X^\pm) = (E_J(X^\pm)/m_X\,r_X^2)^{1/2} = 7.92474925 \cdot 10^{28}$ (s^{-1})
Linear velocity due to spin on surface, v_J	$v_J(W^\pm) = \omega_J(W^\pm)\,r_W = 1.03131574 \cdot 10^{8}$ (m s^{-1}) = $= 0.34400990\ c$	$v_J(Y^\pm) = \omega_J(Y^\pm)\,r_Y = 1.02880790 \cdot 10^{8}$ (m s^{-1}) = $= 0.343173375\ c$	$v_J(X^\pm) = \omega_J(X^\pm)\,r_X = 1.02880304 \cdot 10^{8}$ (m s^{-1}) = $= 0.343171755\ c$
Structural energy balance for subatomic particles $I - I^{-1} - E_J = 0$ $c^{*2}\hbar\,Q^{-2} - [\tfrac{1}{2}(1/4\pi\varepsilon_0)Q^2/r] - m\,r^2\omega_J^2 = 0$	INTERVALIC BILEPTONS-ZERO CHARGED MASSIVE BOSONS		
	Z^0 MASSIVE BOSON (91,188 MeV/c^2)	**Y^0 MASSIVE BOSON (745,037 MeV/c^2)**	**X^0 MASSIVE BOSON (2,514,499 MeV/c^2)**
	$Z^0 = L_2 = 2\,G_{45} = 90\,D_6 = 540\,I = 1080\,\gamma = 2160\,S$	$Y^0 = L_2 = 2\,G_{90} = 180\,D_3 = 540\,I = 1080\,\gamma = 2160\,S$	$X^0 = L_2 = 2\,G_{135} = 270\,D_2 = 540\,I = 1080\,\gamma = 2160\,S$
Intervalic structure	$G_{45}^{(A)}$ / $L_2^{(0)}$	$G_{90}^{(A)}$ / $L_2^{(0)}$	$G_{135}^{(A)}$ / $L_2^{(0)}$

INTERVALIC THEORY:
The Intervalic Structures of Subatomic Particles and the Last Foundations of Physics

NUCLEON AND ITS CONSTITUENT ELECTROMAGNETIC PARTICLES

	PROTON	NEUTRON	QUARK UP
	$p = M_3 = 3 L_3 = 9 G_6 = 54 D_{45} =$ $= 2430\, I = 4860\, \gamma = 9720\, S$	$n = M_3 = 3 L_3 = 9 G_6 = 54 D_{45} =$ $= 2430\, I = 4860\, \gamma = 9720\, S$	$u^{+\frac{2}{3}} = L_3^{+\frac{2}{3}} = G_6^{+\frac{2}{3}} + G_6^{-\frac{1}{3}} + G_6^{+\frac{1}{3}} =$ $= 810\, I = 1620\, \gamma = 3240\, S$
Intervalic Structure (figured representation)	(figure: M_3^+)	(figure: M_3^0)	(figure: $L_3^{+\frac{2}{3}}$)
Intervalic energy I	$I(p)_{MLGD} = I(p)_M + I(u+u+d)_L + I[2(G_6^{\pm\frac{2}{3}}) + 7 (G_6^{\pm\frac{1}{3}})]_G + 54 I(D_{45})_D = 578.6029324\ (MeV/c^2)$	$I(n)_{MLGD} = I(n)_M + I(u+d+d)_L + I[(G_6^{\pm\frac{2}{3}}) + 8(G_6^{\pm\frac{1}{3}})]_G + 54 I(D_{45})_D = 582.1727926\ (MeV/c^2)$	$I(u)_{LGD} = I(L_3^{+\frac{2}{3}}) + I(G_6^{+\frac{2}{3}}) + 2I(G_6^{\pm\frac{1}{3}}) + 18 I (D_{45}) = 191.4872982\ (MeV/c^2)$
Electromagnetic energy U	$U(p)_{MLGD} = U(p)_M + U(u+u+d)_L + U[2(G_6^{\pm\frac{2}{3}}) + 7 (G_6^{\pm\frac{1}{3}})]_G + 54 U(D_{45})_D = 359.6693703\ (MeV/c^2)$	$U(n)_{MLGD} = U(n)_M + U(u+d+d)_L + U[(G_6^{\pm\frac{2}{3}}) + 8 (G_6^{\pm\frac{1}{3}})]_G + 54 U(D_{45})_D = 357.3928374\ (MeV/c^2)$	$U(u)_{LGD} = U(L_3^{+\frac{2}{3}}) + U(G_6^{+\frac{2}{3}}) + 2UI(G_6^{\pm\frac{1}{3}}) + 18 U(D_{45}) = 120.6486320\ (MeV/c^2)$
Spin energy E_J	$E_J(p)_M = I(p)_M - 0 = 4.5756390 \cdot 10^{-14}\ (J) = 0.285588809\ (MeV/c^2)$	$E_J(n)_M = 0 - E_q(n)_M = 5.58841376 \cdot 10^{-14}\ (J) = 0.34880120\ (MeV/c^2)$	$E_J(u)_L = I(u)_L - E_q(u)_L = 2.573797104 \cdot 10^{-13}\ (J) = 1.60643710\ (MeV/c^2)$
Mass energy m	$m(p) = 938.2723027\ (MeV/c^2)$	$m(n) = 939.5656300\ (MeV/c^2)$	$m(u) = 312.1359302\ (MeV/c^2)$
Radius r	$r_N \approx \frac{1}{2}(r_{int} + r_{ext}) = 1.237448636 \cdot 10^{-15}\ (m)$	$r_N \approx \frac{1}{2}(r_{int} + r_{ext}) = 1.237448636 \cdot 10^{-15}\ (m)$	$r_u = 6.88054386 \cdot 10^{-16}\ (m)$
Angular velocity due to spin, ω_J	$\omega_J(p) = (E_J(p)_M / m_p r_p^2)^{\frac{1}{2}} = 4.252280621 \cdot 10^{21}\ (s^{-1})$	$\omega_J(n) = (E_J(n)_M / m_n r_n^2)^{\frac{1}{2}} = 4.696141808 \cdot 10^{21}\ (s^{-1})$	$\omega_J(u) = (E_J(u)_L / m_u r_u^2)^{\frac{1}{2}} = 3.125776462 \cdot 10^{22}\ (s^{-1})$
Linear velocity due to spin on surface, v_J	$v_J(p) = \omega_J(p)\, r_N = 5.230305164 \cdot 10^6\ (m\,s^{-1}) = 0.01744642\, c$	$v_J(n) = \omega_J(n)\, r_N = 5.776254424 \cdot 10^6\ (m\,s^{-1}) = 0.01926751\, c$	$v_J(u) = \omega_J(u)\, r_u = 2.150704204 \cdot 10^7\ (m\,s^{-1}) = 0.07173977\, c$

	QUARK DOWN	NUCLEONIC GAUDINOS	NUCLEONIC DALINO
	$d^{-\frac{1}{3}} = L_3^{-\frac{1}{3}} = G_6^{+\frac{1}{3}} + G_6^{-\frac{1}{3}} + G_6^{-\frac{1}{3}} =$ $= 810\, I = 1620\, \gamma = 3240\, S$	$G_6^{+\frac{2}{3}} = 5 D_{+45} + 1 D_{-45} = 270\, I = 540\, \gamma$ $G_6^{-\frac{2}{3}} = 2 D_{+45} + 4 D_{-45} = 270\, I = 540\, \gamma$ $G_6^{+\frac{1}{3}} = 4 D_{+45} + 2 D_{-45} = 270\, I = 540\, \gamma$	$D_{\pm 45} = 45\, I = 90\, \gamma = 180\, S$
Intervalic Structure (figured representation)	(figure: $L_3^{\pm\frac{1}{3}}$)	(figure: G_6)	(figure: I)
Intervalic energy I	$I(d)_{LGD} = I(L_3^{-\frac{1}{3}}) + I(G_6^{-\frac{1}{3}}) + 2 I(G_6^{\pm\frac{1}{3}}) + 18 I (D_{45}) = 195.3427472\ (MeV/c^2)$	$I(G_6^{\pm\frac{1}{3}})_{GD} = 64.25748329\ (MeV/c^2)$ $I(G_6^{\pm\frac{2}{3}})_{GD} = 62.32975882\ (MeV/c^2)$	$I(D_{45})_D = c^{\pm 2}\hbar\,(45\, q_i)^{-2} = c^{\pm 2}\hbar\,[45\sqrt{-(c^{-1}\hbar)}]^{-2} = 1.6472301 \cdot 10^{-12}\ (J) = 10.281197\ (MeV/c^2)$
Electromagnetic energy U	$U(d)_{LGD} = U(L_3^{-\frac{1}{3}}) + U(G_6^{-\frac{1}{3}}) + 2U(G_6^{\pm\frac{1}{3}}) + 18 U(D_{45}) = 118.3721063\ (MeV/c^2)$	$U(G_6^{\pm\frac{1}{3}})_{GD} = 41.34633762\ (MeV/c^2)$ $U(G_6^{\pm\frac{2}{3}})_{GD} = 39.41861307\ (MeV/c^2)$	$U_{G6\pm\frac{1}{3}}(D_{45})_D = 6.409125138\ (MeV/c^2)$ $U_{G6\pm\frac{2}{3}}(D_{45})_D = 6.24848144\ (MeV/c^2)$
Spin energy E_J	$E_J(d)_L = I(d)_L - E_q(d)_L = 5.662353564 \cdot 10^{-13}\ (J) = 3.53416158\ (MeV/c^2)$	$E_J(G_6^{\pm\frac{1}{3}})_G = I(G_6^{\pm\frac{1}{3}})_G - E_q(G_6^{\pm\frac{1}{3}})_G = 5.133952133\ (MeV/c^2)$	$E_J(G_6\pm\frac{1}{3}D_{45})_D = I(D_{45})_D - U_{G6\pm\frac{1}{3}}(D_{45})_D = 3.873072012\ (MeV/c^2)$
Mass energy m	$m(d) = 313.7148535\ (MeV/c^2)$	$m(G_6^{\pm\frac{1}{3}}) = 105.6038198\ (MeV/c^2)$	$m_{G6\pm\frac{1}{3}}(D_{45}) = 16.69032229\ (MeV/c^2)$
Radius r	$r_d = 6.88054386 \cdot 10^{-16}\ (m)$	$r_{G6} = 8.299740193 \cdot 10^{-17}\ (m)$	$r_{G6\pm\frac{1}{3}}(D_{45}) = 3.12047512 \cdot 10^{-18}\ (m)$
Angular velocity due to spin, ω_J	$\omega_J(d) = (E_J(d)_L / m_d r_d^2)^{\frac{1}{2}} = 4.624593827 \cdot 10^{22}\ (s^{-1})$	$\omega_J(G_6^{\pm\frac{1}{3}}) = (E_J(G_6^{\pm\frac{1}{3}})_G / m_{G6}\, r_{G6}^2)^{\frac{1}{2}} = 7.964878478 \cdot 10^{23}\ (s^{-1})$	$\omega_{J\text{-}G6\pm\frac{1}{3}}(D_{45}) = (E_J(D_{45})_D / m_{D45}\, r_{D45}^2)^{\frac{1}{2}} = 4.628021876 \cdot 10^{25}\ (s^{-1})$
Linear velocity due to spin on surface, v_J	$v_J(d) = \omega_J(d)\, r_d = 3.181972066 \cdot 10^7\ (m\,s^{-1}) = 0.10613916\, c$	$v_J(G_6^{\pm\frac{1}{3}}) = \omega_J(G_6^{\pm\frac{1}{3}})\, r_{G6} = 6.610642204 \cdot 10^7\ (m\,s^{-1}) = 0.22050729\, c$	$v_{J\text{-}G6\pm\frac{1}{3}}(D_{45}) = \omega_J(D_{45})\, r_{D45} = 1.444162712 \cdot 10^8\ (m\,s^{-1}) = 0.48172083\, c$

INTERVALIC STRUCTURE OF NUCLEON

$p = M_3^+ = uud = 3\,L_3\,(L_3^{+\frac{2}{3}},\,L_3^{-\frac{1}{3}},\,L_3^{+\frac{2}{3}}) = 9\,G_6 = \{(2G_6^{-\frac{1}{6}},\,4G_6^{-\frac{1}{6}},\,3G_6^{-\frac{1}{6}}),\,(3G_6^{-\frac{1}{6}},\,3G_6^{+\frac{1}{3}},\,2G_6^{-\frac{1}{6}}),\,(3G_6^{-\frac{1}{6}},\,3G_6^{-\frac{1}{6}},\,3G_6^{+\frac{1}{3}})\} = 54\,D_{45}\,(30D_{-45},\,24D_{45}) = 2430\,I = 4860\,\gamma = 9720\,S$

$n = M_3^0 = dud = 3\,L_3\,(L_3^{-\frac{1}{3}},\,L_3^{+\frac{2}{3}},\,L_3^{-\frac{1}{3}}) = 9\,G_6 = \{(1G_6^{-\frac{1}{6}},\,5G_6^{-\frac{1}{6}},\,3G_6^{-\frac{1}{6}}),\,(3G_6^{-\frac{1}{6}},\,3G_6^{+\frac{1}{3}},\,3G_6^{-\frac{1}{6}}),\,(3G_6^{-\frac{1}{6}},\,3G_6^{-\frac{1}{6}},\,3G_6^{+\frac{1}{3}})\} = 54\,D_{45}\,(27D_{-45},\,27D_{45}) = 2430\,I = 4860\,\gamma = 9720\,S$

Intervalic structure levels: 0: Point, 1: Intervalic String (S), 2: Photon (γ), 3: Intervalino (I), 4: Dalino (D), 5: Lisztino (L), 6: Gaudino (G), 7: Monteverdino (M), 8: Palestrino (P)

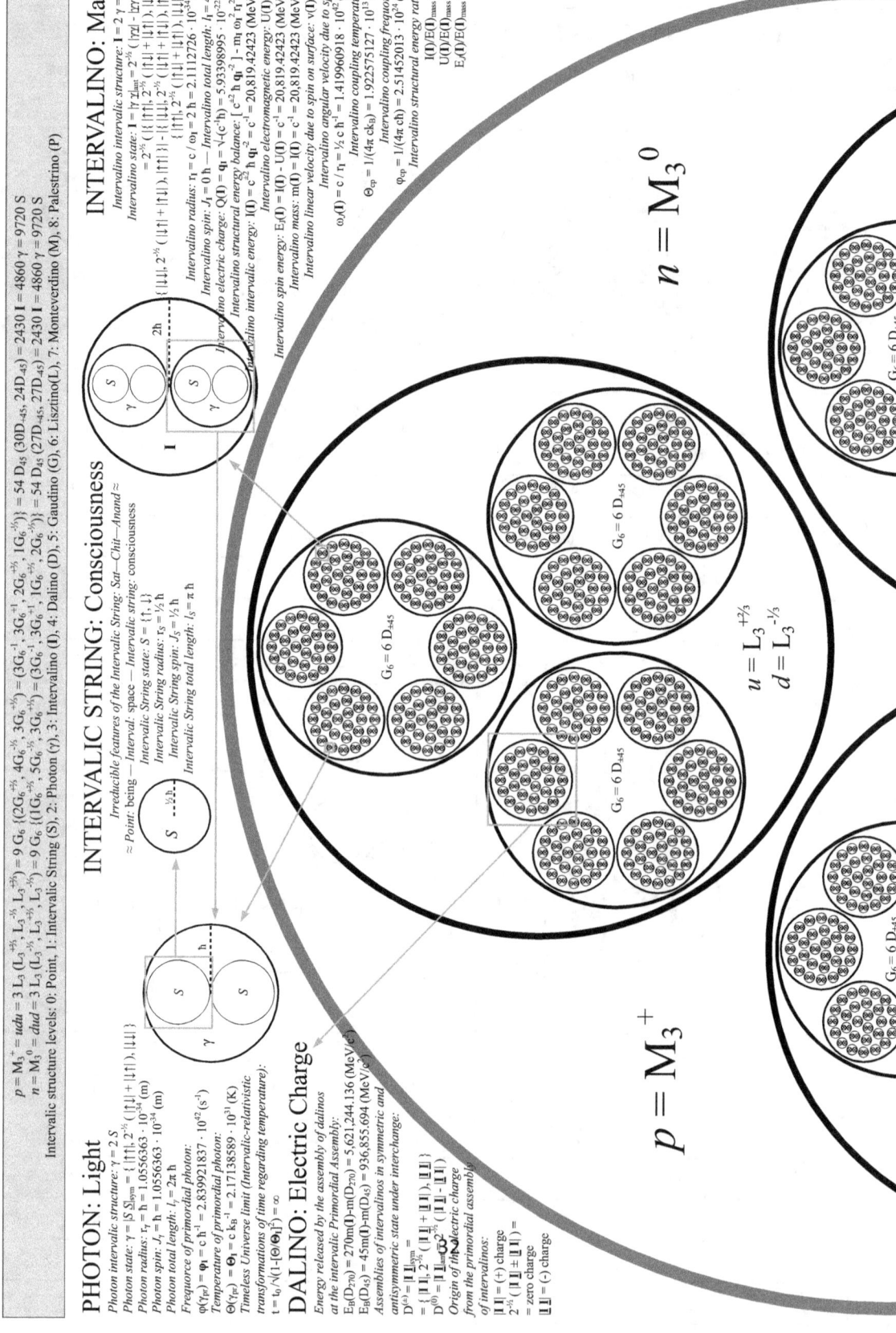

PHOTON: Light

Photon intervalic structure: $\gamma = 2\,S$
Photon state: $\gamma = |S\,\underline{S}|_{sym} = \{|\uparrow\uparrow|,\,2^{-\frac{1}{2}}(|\uparrow\downarrow| + |\downarrow\uparrow|),\,|\downarrow\downarrow|\}$
Photon radius: $r_\gamma = \hbar = 1.0556363 \cdot 10^{-34}$ (m)
Photon spin: $J_\gamma = \hbar = 1.0556363 \cdot 10^{-34}$ (m)
Photon total length: $l_\gamma = 2\pi\,\hbar$
Frequence of primordial photon:
$\phi(\gamma_{pr}) = \phi_1 = c\,\hbar^{-1} = 2.839921837 \cdot 10^{42}$ (s^{-1})
Temperature of primordial photon:
$\Theta(\gamma_{pr}) = \Theta_1 = c\,k_B^{-1} = 2.17138589 \cdot 10^{31}$ (K)
Timeless Universe limit (Intervalic-relativistic transformations of time regarding temperature):
$t = t_0 / \sqrt{(1 - [\Theta/\Theta_1]^2)} = \infty$

DALINO: Electric Charge

Energy released by the assembly of dalinos at the intervalic Primordial Assembly:
$E_B(D_{-70}) = 270\,m(I) - m(D_{-70}) = 5{,}621{,}244.136$ (MeV/c^2)
$E_B(D_{45}) = 45\,m(I) - m(D_{45}) = 936{,}855.694$ (MeV/c^2)
Assemblies of intervalinos in symmetric and antisymmetric state under interchange:
$D^{(+)} = |\underline{\mathbf{I}}\,\underline{\mathbf{I}}|_{sym} =$
$= \{|\underline{\mathbf{I}}\,\underline{\mathbf{I}}|,\,2^{-\frac{1}{2}}(|\underline{\mathbf{I}}\,\underline{\underline{\mathbf{I}}}| + |\underline{\underline{\mathbf{I}}}\,\underline{\mathbf{I}}|),\,|\underline{\underline{\mathbf{I}}}\,\underline{\underline{\mathbf{I}}}|\}$
$D^{(0)} = |\underline{\mathbf{I}}\,\underline{\mathbf{I}}|_{antisym} = 2^{-\frac{1}{2}}(|\underline{\mathbf{I}}\,\underline{\underline{\mathbf{I}}}| - |\underline{\underline{\mathbf{I}}}\,\underline{\mathbf{I}}|)$
Origin of the electric charge from the primordial assembly of intervalinos:
$2^{-\frac{1}{2}}(|\underline{\mathbf{I}}\,\underline{\underline{\mathbf{I}}}| \pm |\underline{\underline{\mathbf{I}}}\,\underline{\mathbf{I}}|) =$
$|\underline{\mathbf{I}}\,\underline{\mathbf{I}}| = (+)$ charge
$= $ zero charge
$|\underline{\underline{\mathbf{I}}}\,\underline{\underline{\mathbf{I}}}| = (-)$ charge

INTERVALIC STRING: Consciousness

Irreducible features of the Intervalic String: Sat–Chit–Anand ≈
≈ Point: being — Interval: space — Intervalic string: consciousness
Intervalic String state: $S = \{\uparrow,\,\downarrow\}$
Intervalic String radius: $r_S = \frac{1}{2}\hbar$
Intervalic String spin: $J_S = \frac{1}{2}\hbar$
Intervalic String total length: $l_S = \pi\,\hbar$

INTERVALINO: Mass

Intervalino intervalic structure: $\mathbf{I} = 2\,\gamma = 4\,S$
Intervalino state: $\mathbf{I} = |\gamma\,\underline{\gamma}|_{antisym} = 2^{-\frac{1}{2}}(|\gamma\underline{\gamma}| - |\underline{\gamma}\gamma|) =$
$= 2^{-\frac{1}{2}}(\{|\uparrow\uparrow|,\,2^{-\frac{1}{2}}(|\uparrow\downarrow| + |\downarrow\uparrow|),\,|\downarrow\downarrow|\}$
$- \{|\uparrow\uparrow|,\,2^{-\frac{1}{2}}(|\downarrow\uparrow| + |\uparrow\downarrow|),\,|\downarrow\downarrow|\})$
$\{|\uparrow\uparrow|,\,2^{-\frac{1}{2}}(|\uparrow\downarrow| + |\downarrow\uparrow|),\,|\downarrow\downarrow|\}$
Intervalino radius: $r_I = c / \omega_I = 2\,\hbar = 2.1112726 \cdot 10^{-34}$ (m)
Intervalino spin: $J_I = 0\,\hbar$ — Intervalino total length: $l_I = 4\pi\,\hbar$
Intervalino electric charge: $Q(I) = \sqrt{(c^2\,\hbar\,\mathbf{q}_I^{-2})} = 5.9339895 \cdot 10^{-22}$ (C)
Intervalino structural energy balance: $[c^2\,\hbar\,\mathbf{q}_I^{-2}] - m_I\,\omega_I^2\,r_I^2 = 0$
Intervalino intervalic energy: $I(I) = c^{-2}\,\hbar\,\mathbf{q}_I^{-2} = c^{-1} = 20{,}819.42423$ (MeV/c^2)
Intervalino electromagnetic energy: $U(I) = 0$
Intervalino spin energy: $E_J(I) = I(I) - U(I) = c^{-1} = 20{,}819.42423$ (MeV/c^2)
Intervalino mass: $m(I) = I(I) = c^{-1} = 20{,}819.42423$ (MeV/c^2)
Intervalino linear velocity due to spin on surface: $v(I) = c$
Intervalino angular velocity due to spin: $\omega_I(I) = c / r_I = \frac{1}{2}\,c\,\hbar^{-1} = 1.41996018 \cdot 10^{42}$ (s^{-1})
Intervalino coupling temperature:
$\Theta_{cp} = I / (4\pi\,c\,k_B) = 1.922575127 \cdot 10^{13}$ (K)
Intervalino coupling frequence:
$\phi_{cp} = I / (4\pi\,c\,\hbar) = 2.51452013 \cdot 10^{24}$ (s^{-1})
Intervalino structural energy ratios:
$I(I)/E(I)_{mass} = 1$
$U(I)/E(I)_{mass} = 0$
$E_J(I)/E(I)_{mass} = 1$

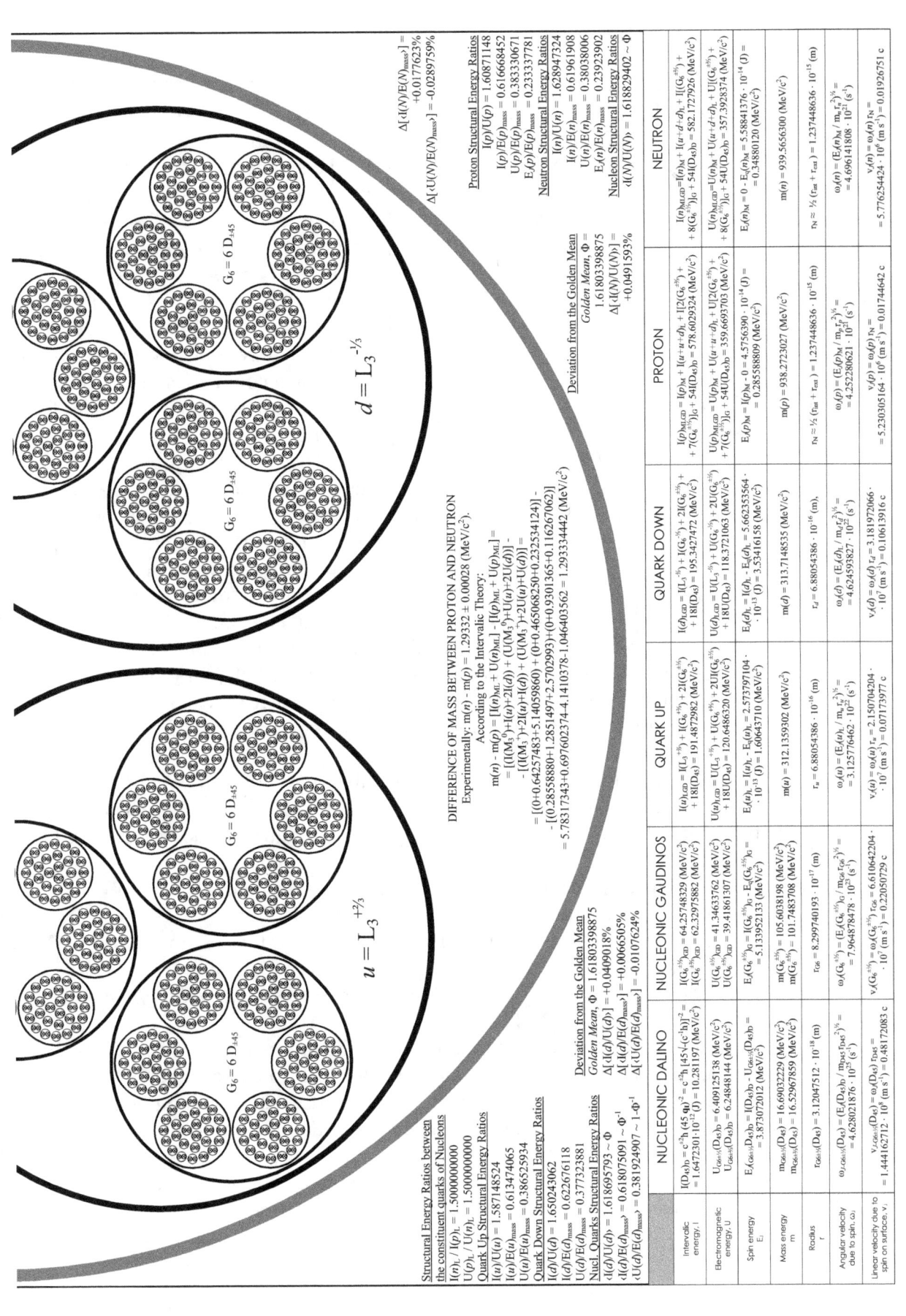

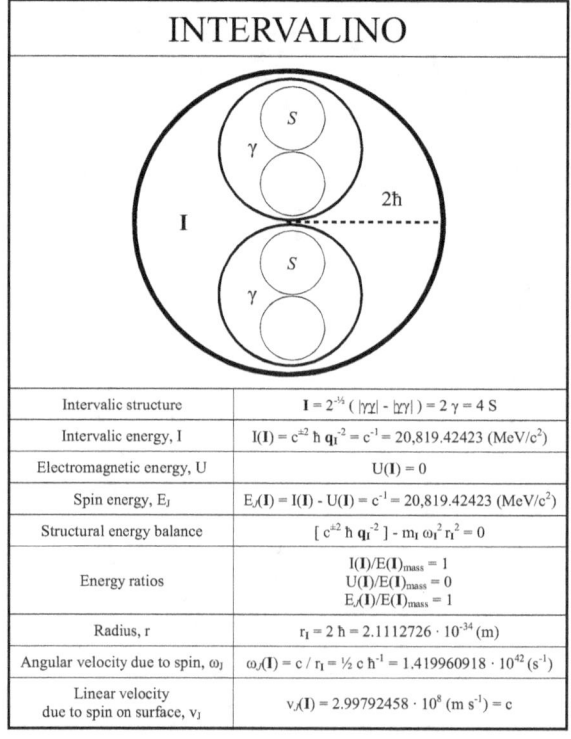

INTERVALINO

Intervalic structure	$I = 2^{-\frac{1}{2}}(|\gamma\gamma| - |\gamma\gamma|) = 2\gamma = 4S$
Intervalic energy, I	$I(I) = c^{\pm 2}\hbar q_I^{-2} = c^{-1} = 20{,}819.42423$ (MeV/c^2)
Electromagnetic energy, U	$U(I) = 0$
Spin energy, E_J	$E_J(I) = I(I) - U(I) = c^{-1} = 20{,}819.42423$ (MeV/c^2)
Structural energy balance	$[c^{\pm 2}\hbar q_I^{-2}] - m_I \omega_J^2 r_I^2 = 0$
Energy ratios	$I(I)/E(I)_{mass} = 1$ $U(I)/E(I)_{mass} = 0$ $E_J(I)/E(I)_{mass} = 1$
Radius, r	$r_I = 2\hbar = 2.1112726 \cdot 10^{-34}$ (m)
Angular velocity due to spin, ω_J	$\omega_J(I) = c/r_I = \frac{1}{2}c\hbar^{-1} = 1.419960918 \cdot 10^{42}$ (s^{-1})
Linear velocity due to spin on surface, v_J	$v_J(I) = 2.99792458 \cdot 10^{8}$ (m s^{-1}) = c

The intervalino is the first particle with *mass*, generated by the synthesis of two photons in antisymmetric state under interchange, whose threshold frequency —*coupling frequency of matter*— must be greater than: $\varphi_{cp} = 1/(4\pi c \hbar) = 2.51452013 \cdot 10^{24}$ (s^{-1}) for the synthesis may occur, so that intervalinos could eventually be synthesized artificially in the laboratory.

The physical properties of intervalino are extraordinary, because it is the only particle with an electric charge that has no electromagnetic energy —which only particles *with structure* can have, being this a basic definition of the physical laws that has forgotten the standard model—, so it is logically necessary the existence of a fundamental particle, for the electromagnetic interaction, with these extraordinary physical features: an indivisible electric charge and, by this reason, without electromagnetic potential energy.

The electric charge of the intervalino is the *intervalic quantum of electrical charge*: $\mathbf{q_I} = \sqrt{-(c^{-1}\hbar)} = 1$ ($i^{-1/2}L^{1/2}$) = $5.93398995 \cdot 10^{-22}$ (C), which is the fundamental charge of Nature, whose geometric value is exactly 1/270 of the elementary charge, e.

Its mass energy, which comes exclusively from the equivalent energy of the previous intervalic quantum of electric charge, $I(\mathbf{I}) = c^{\pm 2}\hbar \mathbf{q_I}^{-2} = c^{-1} = 20{,}819.42423$ (MeV/c^2), is also the *intervalic quantum of mass*: $\mathbf{m_I} = 1$ (i) = $c^{-1} = 20{,}819.42423$ (MeV/c^2), while its spin energy is equally c^{-1}.

Its radius is twice the photon radius, i.e., twice the *intervalic quantum of length*: $\mathbf{r_I} = 2\,\mathbf{l_I} = 2\hbar = 2.1112726 \cdot 10^{-34}$ (m).

The antisymmetric state of the two constituents photons of the intervalino can be visualized as two photons traveling in opposite directions which are coupled tangentially, so that opposite ends to the coupling point of each photon —which are situated in the centre of the intervalino— continue moving at the speed of light, c, as it can not be otherwise, since all non massive particle always moves at the speed of light in intervalic space-time, hence the linear velocity on the "surface" of the intervalino is precisely c.

INTERVALIC THEORY:
The Intervalic Structures of Subatomic Particles and the Last Foundations of Physics

	INTERVALIC PRIMORDIAL ASSEMBLY				
Structure Level	Intervalic structure	Degree of freedom	Interaction introduced	Particles assembled	Symmetry breaking
0	Point	∅	∅	Intervalic String	goes to next structure level
				Intervalic Length	-
1	Intervalic String	Space	Informational	Photon	goes to next structure level
				Chi	dark energy
2	Photon	Spin	Intervalic Changeless —strong—	Intervalino	goes to next structure level
				Graviton	dark energy
3	Intervalino	Mass	Gravitational	16 Dalinos: *electron*	goes to next structure level
				Bintervalino	dark matter
4	Dalino	Electric charge	Electromagnetic	40 Gaudinos: *nucl. Gs, leptons-CMBs*	goes to next structure level
				Bidalino	dark matter
5	Gaudino	Electric charge structure	Intervalic Changeful —weak—	Lisztinos: *49 quarks*	goes to next structure level
				Lisztinos: *bileptons-ZCMBs*	decay
6	Lisztino	Elementary charge	Elementary charge attractor	Monteverdinos: *baryons*	goes to next structure level
				Monteverdinos: *mesons*	decay
7	Monteverdino	∅	∅	Palestrinos: *nuclei: $1 < A \leq 3$*	-
				Pseudopalestrinos: *nuclei: $A > 3$*	-

THE TOTAL ENERGY OF THE INTERVALIC UNIVERSE:
SHARE OF PARTICLES AT THE INTERVALIC PRIMORDIAL ASSEMBLY

At each level of the Intervalic Primordial Assembly the two branches of particles assembled vanish between themselves (one branch is in symmetric state under interchange and the other is antisymmetric). This remarkable logical economy of the Intervalic Universe reminds the perfect Yin-Yang balance of Nature according to *Tao te King*, which traditionally was believed to ruling only on the organic world.

Particles share at the Intervalic Primordial Assembly:

- Photons: > 0
- Chis: $½ = 50\%$
- Intervalinos: ~ 0
- Gravitons: $¼ = 25\%$
- Dalinos: ~ 0
- Zero charged dalinos: $< 1/8 = 12.5\%$
- Gaudinos: ~ 0
- Zero charged gaudinos: $< 1/16 = 6.25\%$
- Lisztinos (quarks): $< 1/16 = 6.25\%$

Grouping them in terms of the darkness of matter:

- Dark energy: $½ = 50\%$
- Dark matter: $< 3/16 = 18.75\%$
- Visible matter: $< 1/16 = 6.25\%$
- Gravitons: $¼ = 25\%$
- Photons: > 0

Please note the fine match of the relations:
- Dark matter + Visible matter = Gravitons
- Dark matter + Visible matter + Gravitons = Chis

***

The Intervalic Theory explains in a different way from the standard model all known experimental results to date. This is done with crushing logic, with geometric precision, and without using a single arbitrary constant. In this way are deduced all the physical properties of the 147 subatomic particles of Nature, both visible as dark. Besides this, it can be highlighted the following physical explanations, among others:

- The deduction of the annihilation ratio and the partial decay widths of all leptons-charged massive bosons (that reach up to the value 1/70) and bileptons-zero charged massive bosons (that reach up to the value 1/30), both detected to date as like as undetected yet.
- The shape of the curve of the binding energy per nucleon and its value ~ 8 (MeV/c^2), since according to the Intervalic Theory *all nuclei with mass number A > 3 are not composed by nucleons but by quarks*, being those ~ 8 (MeV/c^2) the structural energy required to synthesize or "arm" the structure of the nucleon at the last level —the monteverdic one— from three quarks to leaving the nucleus synthetized in a nucleon. This "arming" energy for synthetizing that monteverdic structure is constant regardless of the mass number. *Only nuclei with mass number A ≤ 3 are composed by nucleons.* This explains simply and clearly why deuterons release a huge amount of energy when merging to make a helium nucleus: because they lose the last level of the intervalic structure, whose energy per nucleon is ~ 8 (MeV/c^2), as the Helium is a monteverdino composed by quark, while the deuteron is a palestrino composed by nucleons. That amount of energy is what is released in fusion reactions.
- The *neutron to proton ratio* of the Universe, which is derived exclusively from the intervalic structure of nucleon in two independent ways, both yielding virtually the same result: ~ 22%.

- The *intervalic structure of stars*, whose life in the twilight go through the same phases of the primordial intervalic synthesis, only in reverse, thus having quarks stars, gaudinos stars and dalinos stars. Intervalinos star is only reached at the Big Crunch.

- The *intervalic cosmology*, which establishes a model of indefinitely oscillating Universe, that "bounces" between a Big Crunch and a Big Bang.

- The nature of the former weak and strong nuclear interactions, which are now, respectively, the *changeful* and *changeless intervalic interactions*, as well as the explanation of the intervalic structures involved in the beta decay, the exchange intervalic structure that explains the inner dynamic state of nucleon and π meson, etc.

- All the apparent paradoxes of quantum mechanics, which are but pseudo problems born of the ignorance of symmetries and of ultimate foundations of physics, including from the uncertainty principle assumption to the experiment of double slit, etc., which now have finally been revealed by the Intervalic theory.

- The nature of *vacuum* and of three-dimensional space, which is not a *continuum* at all and not even a discrete space, but simply the folding of a one-dimensional physical space —the only real physical substance— in a three-dimensional mathematical space, just as the nature of time is also virtual, mathematical: *imaginary space*. Since all massive particles is composed by intervalinos, and intervalino is composed by two photons, and photon is composed by two intervalic strings —whose volume and surface are zero (as intervalic strings are but one-dimensional space intervals in a finite number)—, we reach to the surprising conclusion that both actual total volume and actual total area of the Universe are just zero.

- The nature and intervalic symmetries of antigravity, and the theoretical way to generate an antigravity field.

- The nature of *information* and the discovery of the geometric equivalence between energy and information, which forms a further branch of the Intervalic Theory.

- Being constituted all subatomic particles, ultimately, by the synthesis of intervalic strings, its intervalic structure can alternatively be expressed as a quantum-informational state that is named the *intervalic code* of subatomic particles. For example, the intervalic code of the particles synthesized before the dalino-synthesis (which released the energy of the Big Bang) is:

THE INTERVALIC CODE
of first subatomic particles

Intervalic string:
$S = \{\uparrow, \downarrow\}$

Photon = synthesis of intervalic strings in symmetric state under interchange:
$\gamma = \{\,|\uparrow\uparrow|,\ 2^{-\frac{1}{2}}(|\uparrow\downarrow| + |\downarrow\uparrow|),\ |\downarrow\downarrow|\,\}$

Chi = synthesis of intervalic strings in antisymmetric state under interchange:
$\cent = 2^{-\frac{1}{2}}(|\uparrow\downarrow| - |\downarrow\uparrow|)$

Intervalino = synthesis of photons in antisymmetric state under interchange:
$I = 2^{-\frac{1}{2}}(|\gamma\gamma| - |\gamma\gamma|) = 2^{-\frac{1}{2}}(\,|\{\,|\uparrow\uparrow|,\ 2^{-\frac{1}{2}}(|\uparrow\downarrow| + |\downarrow\uparrow|),\ |\downarrow\downarrow|\,\}\{\,|\downarrow\downarrow|,\ 2^{-\frac{1}{2}}(|\downarrow\uparrow| + |\uparrow\downarrow|),\ |\uparrow\uparrow|\,\}| - |\{\,|\downarrow\downarrow|,\ 2^{-\frac{1}{2}}(|\downarrow\uparrow| + |\uparrow\downarrow|),\ |\uparrow\uparrow|\,\}\{\,|\uparrow\uparrow|,\ 2^{-\frac{1}{2}}(|\uparrow\downarrow| + |\downarrow\uparrow|),\ |\downarrow\downarrow|\,\}|\,)$

Graviton = synthesis of photons in symmetric state under interchange:
$g = |\gamma\gamma|_s = [\,|\gamma\gamma|,\ 2^{-\frac{1}{2}}(|\gamma\gamma| + |\gamma\gamma|),\ |\gamma\gamma|\,] = [\,|\{\,|\uparrow\uparrow|,\ 2^{-\frac{1}{2}}(|\uparrow\downarrow| + |\downarrow\uparrow|),\ |\downarrow\downarrow|\,\}\{\,|\uparrow\uparrow|,\ 2^{-\frac{1}{2}}(|\uparrow\downarrow| + |\downarrow\uparrow|),\ |\downarrow\downarrow|\,\}|,\ 2^{-\frac{1}{2}}(|\{\,|\uparrow\uparrow|,\ 2^{-\frac{1}{2}}(|\uparrow\downarrow| + |\downarrow\uparrow|),\ |\downarrow\downarrow|\,\}\{\,|\downarrow\downarrow|,\ 2^{-\frac{1}{2}}(|\downarrow\uparrow| + |\uparrow\downarrow|),\ |\uparrow\uparrow|\,\}| + |\{\,|\downarrow\downarrow|,\ 2^{-\frac{1}{2}}(|\downarrow\uparrow| + |\uparrow\downarrow|),\ |\uparrow\uparrow|\,\}\{\,|\uparrow\uparrow|,\ 2^{-\frac{1}{2}}(|\uparrow\downarrow| + |\downarrow\uparrow|),\ |\downarrow\downarrow|\,\}|),\ |\{\,|\downarrow\downarrow|,\ 2^{-\frac{1}{2}}(|\downarrow\uparrow| + |\uparrow\downarrow|),\ |\uparrow\uparrow|\,\}\{\,|\downarrow\downarrow|,\ 2^{-\frac{1}{2}}(|\downarrow\uparrow| + |\uparrow\downarrow|),\ |\uparrow\uparrow|\,\}|\,]$

INTERVALIC THEORY:
The Intervalic Structures of Subatomic Particles and the Last Foundations of Physics

The Intervalic Theory performs a complete reformulation of physics without introducing a single arbitrary constant, deducting all, in a logical and necessary way, from the intervalic system of physical quantities, which is the *natural* system of units, from which emerge the fundamental symmetries of physics and the underlying fundamental geometry of Nature, without whose knowledge no further progress is possible. Therefore, the heuristic key lies in the discovery of the intervalic dimensional basis of the intervalic or natural system of physical quantities (and not their units as they themselves are secondary). From an epistemological point of view it is interesting to note that the intervalic or natural dimensional basis could have been uncovered by different logical ways:

- From the definition of *time* as *imaginary space*, $T = iL$, which is the route used by the Intervalic Theory in Music, that historically led to the postulation of the Intervalic Theory.

- From the physical interpretation of the Planck's quantum of action as a pattern of length, which was warned by Lancelot Law Whyte in his *Critique of Physics:* "Planck's constant (in appropriate combinations) determines the linear scale of the structure of matter and of radiation, in terms of the selected unit of length".

- From the classic formulation of special relativity by Minkowski in 1908, where he introduces the imaginary number, i, in the temporal component of the equations of relativity, which in his own words, leads to define the speed of light as: $c = i$. In fact, the formulation of Einstein-Minkowski does work because it uses —although partial and unconsciously— the dimensions of the intervalic system of physical quantities.

- From the definition of the fine structure constant, since we know that is value is the square of the elementary charge *in natural units*, which necessarily implies that, in natural units, $e = 270$. From here the logical deduction of the intervalic or natural dimensional basis is immediate.

- From the empirical data showing conclusively that the masses of massive subatomic particles detected to date are proportional to the inverse square of 270, 45, 30, 18, 6 and 5.

- From the finding that the *frames of reference* used by the methodology of the classical-quantum metric (usually an *observer* external to the system), are *arbitrary* and *privileged*, which is a serious epistemological inconsistency. This methodological paradigm —subliminal and unconscious, and therefore much more difficult to detect and to correct— results in partial or erroneous measurements and interpretations, being therefore replaced by the intervalic metric, where there are no arbitrary and privileged observers, but each particle has its own reference system (in fact, the physical features of subatomic particles can only be expressed with and from this intervalic metric). Henceforth it is understood that the Intervalic Theory could be seen as the development and logical continuation of the philosophy that emanates from relativity, so it can be said that there has been a historical progression towards intervalicity, which has gone through three progressive stages: classical mechanics, relativistic mechanics and intervalic mechanics. A somewhat surprising way, this is the route used by the Intervalic Theory in Economics, which is, likewise, the *natural* theory of economics.

Any of these six ways implies, when being developed with the blind ruthlessness of logic, the deduction of the intervalic dimensional basis and, consequently, of the Intervalic Theory, which is deduced in full from the intervalic or natural system of physical quantities. The fact that this rigorous logical deduction has not been made previously by some of these ways, indicates how difficult it is for the human mind to get out of the classical-quantum current paradigm, whose false dogmas prevent even start the developing of something as implacable as it is logic.

Chapter 1

INTERVALIC SYSTEM OF PHYSICAL QUANTITIES
The Dimensional Foundations of Physics

Physics thinking is expressed within a determined *dimensional system*, which is like the base signs of a language. It was supposed that all physical languages were equivalent, but the mere existence of the Intervalic System of Physical Quantities *demonstrates* that all dimensional systems are *not* equivalent since they yield *different* Physics. From now, the physical knowledge and the quest for a faultless theory should begin by choosing the most reliable —and presumably *universal*— dimensional system of physical quantities. There must be only *one* dimensional system which is based on the *last fundamental units* of the Universe, still unknown. The Intervalic System of Physical Quantities claims to be this *natural* dimensional system.

INTRODUCTION TO DIMENSIONAL SYSTEMS OF PHYSICAL QUANTITIES

The dimensional basis (L, M, T, I, Θ, J, 1) of the traditional system of units is really a mere *assumption* whose logical arguments are in no way preferable to other hypothetical assumptions, such as, i.e., dimensional basis with three, five or eight proper dimensions. It can be pointed two precisions regarding the usual dimensional system (L, M, T, I, Θ, J, 1): first, we have added the dimensionless (1) to its basis since it really appears in some physical quantities (like cycles, radians or the Avogadro constant), and its existence can't be freely supposed; and second, the intensity (J) should not be put along with the rest of dimensions because it is more a psychophysical quantity than a physical one.

It is quite disconcerting the little attention bestowed to the epistemological research on dimensional systems of physical quantities since they conform the foundations of the Physics building. Nowadays, when String Theory is trying to derive all physical quantities from vibrations, that is to say, from space-time, it should be questioned what number of physical quantities would be preferable for a dimensional basis. If a system with one proper dimension (like the one used in String Theory) is sufficient for representing all physical quantities, it should implies, by Occam's razor, that the rest of supposed proper dimensions are superfluous. Unfortunately the *degenerate dimensions* —wrongly called "geometrized" since they are geometryless— used in String Theory only can work if the calculus of its final results are translated into traditional physical quantities and units. This means that degenerate dimensions are not in practice an *independent* system of dimensions. The question now is: could be proposed a reliable system of physical quantities one-proper-dimension-based that really works?

Other *tacit assumptions* that underlies on all known dimensional systems of physical quantities up to date are: first, the proper dimensions must be composed by powers of real numbers; and second, if any, there is one dimensionless physical quantity, (1). Once more, there is no logical reason to prefer these assumptions to any others. Moreover, due to the growing and outstanding role played by imaginary or complex numbers in Physics, it could be expected that these numbers will do appear in the dimensional basis of a system of physical quantities. Once more, the problem is: how could they appear —if possible— in a consistent, independent and useful system of physical quantities?

INTERVALIC THEORY:
The Intervalic Structures of Subatomic Particles and the Last Foundations of Physics

Although mathematics involved in this essay are very simple, the right comprehensiveness of the transcendence of the intervalic dimensional system is far to be immediate, and it is only achieved when the intervalic dimensions are known by heart. Then, and only then, the traditional becomes as absurd as irrelevant. Before describing it, or in order to describe it, lets go to comment briefly some logical foundations of its mother: the Intervalic Theory (abbreviated as IT).

LOGICAL FOUNDATIONS OF THE INTERVALIC THEORY

> In fact we have not yet made enough direct experiments to know even whether the dimensional system which is used for electrons is correct. Since no electron velocity has ever been directly measured we cannot be sure that the dimensions of the new constant 'h' —called Planck's constant— are really what we suppose, energy multiplied by time.
>
> LANCELOT L. WHYTE
> *Archimedes or The Future of Physics*

All postulates and every step in the Intervalic Theory is intended to be the most logically simple. Thus the theory postulates principles of *simplicity* and *economy* as some of the highest rules for its own logical development. In this way, all Physics should be based on the last and simplest combinations of the ultimate units, and in a successful theory the theoretical laws of Physics should have the epistemological rank of *geometrical statements*.

According to that, we can comment some epistemological and aesthetic *desiderata* that we would hope to find in a theory of everything: the theory shouldn't have arbitrary constants; all phenomena *allowed* by an elegant theory should exist in the Universe, since we think the demiurge does not waste time; if a physical quantity or dimension has a physical meaning, it is expected that its *inverse* value also should have a proper physical meaning; the conservation laws of energy, charge, etc. not only are verified, but the total sum of any of these quantities in the Universe must be zero (this is called the *intervalic zero assumption*); the preferable qualitative number of elements for the last foundations of physics —such as the qualitative number of proper dimensions of the dimensional basis— it is expected to be related with the elemental logical values: zero (0), one (1) or infinite (∞).

These logical constrains involve that, first of all, we need to determinate which are the last and simplest reliable units of physics, and we also need to find a dimensional system of physical quantities that verifies the above *desiderata*. Since the last blocks or units of physics are to be expressed in some dimensional system of physical quantities, the faithful determination of such a system may be the most important task for any Physics theory that aspires to be final —in some degree—. Although the String Theory appears to satisfy these *desiderata*, the truth is that it satisfy neither of them, as we will see along this paper.

The analysis of dimensions is perhaps one of the most forgotten field in Physics. It is often put sideways in text books as a not important theme. This judgment is based on the assumption that all dimensional systems of physical quantities are *equivalent* for describing the nature of physic world. If all of them are equivalent, then it is irrelevant the question about its trueness or falseness.

Well, since according to this, we can choose any dimensional system of physical quantities without affecting the meaning of physics, now we are going to define and to choose one system: the *intervalic* one. As stated below, it is supposed that this detail is irrelevant for the physics knowledge and can not affect physics foundations and principles anyway.

The intervalic system of physical quantities is based simply on a statement which belongs to (and now is shared with) the Intervalic Theory in Music. It says that *time is imaginary space*, that is to say, in mathematical form:

$$T = i\,L$$

where T is time, L is longitude and i is $\sqrt{-1}$. If we define time as: $T = i^{-1}L$, instead of: $T = i\,L$, we obtain an equivalent —really an identical— dimensional system of physical quantities.

Hence, if now we are going to reformulate the whole Physics within the *intervalic system of physical quantities*, we will obtain, surprisingly... an entirely new Physics! This unexpected fact demonstrates that our previous assumption about the equivalence of all systems of physical quantities was completely wrong. Indeed, the mere existence of the intervalic system proves that not all dimensional systems of physical quantities are equivalent. This is a little earthquake in the foundations of Physics, since in this new situation the question about the trueness or falseness of the dimensional systems of physical quantities becomes absolutely relevant and transcendent. Really, according to scientific methodology, we would have to take a decision about which is the most reliable dimensional system of physical quantities for the description of the physical world, since the intervalic one yields a Physics substantively different from the other dimensional systems of physical quantities known to date. It is very important to point out that all the Intervalic Theory of Particle Physics is based solely on a purely change of the dimensional system of physical quantities. From here, the whole theory is *logically* and *necessarily* derived, without introducing a single arbitrary constant nor any assumption or adjustment made by hand.

If the Intervalic Theory in Physics would be false and finish here, its logical consequence shall be similar for Physics as the Gödel's theorem about the formally undecided propositions was for Mathematics. But the Intervalic Theory don't finish here; it starts from this point. Moreover, the intervalic system of physical quantities claims to be not only the more reliable among all known dimensional systems, but the only truthful system of physical quantities: in other words: the unique and genuine *natural* system of physical quantities.

By this reason, and although the task could looks somewhat painful at first sight, the correct scientific aim and the serious application of the scientific method request us ineluctably to *reformulating* the whole body of Physics according to the new dimensions of the intervalic system of physical quantities. We have noticed —or simply remember, since it was already known from the XXth. century— that the huge building of Physics have feet of clay, and now we cannot look to another place, but to solve the problem. We can be sure that it is worth it.

INTERVALIC DIMENSIONS

When reformulating all physical quantities within intervalic dimensions ($L = L$, $T = iL$, $M = T/L = i$), where $i = \sqrt{-1}$, we obtain the following finite set composed by 40 quantized dimensions, named the *intervalic group* (some of these physical quantities will be briefly commented at due course):

1	Permitivity, momentum-inertia, entropy, Boltzmann constant
-1	Permeability, gravitational potential
i	Invervelocity-mass, impedance
i^{-1}	c, velocity-energy, temperature, specific heat capacity, conductance
L	ℏ, length, action, capacitance
-L	Antilength, inductance
i L	Time
i^{-1}L	Antitime
L^{-1}	Wavevector
$-L^{-1}$	Acceleration, power, gravitational field
$i L^{-1}$	Linear perdensity, antifrequorce

INTERVALIC THEORY:
The Intervalic Structures of Subatomic Particles and the Last Foundations of Physics

$i^{-1}L^{-1}$	Frequorce (frequency-force), linear tension, conductivity
L^2	Area
$-L^2$	Antiarea
$i L^2$	Inertia area momentum
$i^{-1}L^2$	?
L^{-2}	Viscosity
$-L^{-2}$	Area power
$i L^{-2}$	Inflexion, area perdensity
$i^{-1}L^{-2}$	Surface tension
L^3	Volume
$-L^3$	Antivolume
$i L^3$	Inertia volume momentum
$i^{-1}L^3$	Fermi weak interaction constant
L^{-3}	Fluctuation
$-L^{-3}$	Volume power, irradiance, Stefan-Boltzmann constant
$i L^{-3}$	Density volume perdensity
$i^{-1}L^{-3}$	Pressure, energy-tension density
$i^{1/2}L^{1/2}$	Magnetic charge, magnetic flux
$i^{-1/2}L^{1/2}$	Electric charge
$i^{1/2}L^{-1/2}$	Current, electric potential, magnetic vector potential
$i^{-1/2}L^{-1/2}$	Magnetic inverflux
$i^{1/2}L^{3/2}$	Bohr magneton
$i^{-1/2}L^{3/2}$	?
$i^{1/2}L^{-3/2}$	Electric field, magnetic field, magnetic flux density
$i^{-1/2}L^{-3/2}$	Electric polarization
$i^{1/2}L^{5/2}$	?
$i^{-1/2}L^{5/2}$	?
$i^{1/2}L^{-5/2}$	Charge density, current density
$i^{-1/2}L^{-5/2}$	?

The *intervalic dimensional basis* that corresponds —and generates— this system of physical quantities is, as will be explained later:

DIMENSIONAL COMPOSITION AND SYMMETRIES OF THE INTERVALIC GROUP

The 40 physical quantities and dimensions of the Universe in bidimensional representation: semigroup (−1) at left, semigroup (+1) at right. (Please note that physical quantities can not exist in the shadowed squares of the axes)

(L, i)

As we can see, the set of intervalic dimensions have some differences with the other dimensional systems:

1. The number of physical quantities is a *finite* and *ordered* set.
2. There does not exist physical quantities whose equation of dimensions have more than the real or unfolded dimensions of real space (3) and time (1). It can't be more than *one* time dimension in the intervalic system because of the proper definition of time as iL —it is determined by the mathematical properties of the i number—.
3. There are no different physical quantities with the same equation of dimensions (as in the traditional system), nor different dimensions with the same physical quantity (as in the wrong called "geometrized" units system since they do not define any kind of geometry).

All differences between the intervalic system of physical quantities and units and the remaining dimensional systems will be described at the end of chapter No. 3.

Although systems showing the opposite to these properties was tolerated or put slanted for the Classic Physics, in the Intervalic Theory this situations is not allowed, and the theory asserts that those systems are logically inconsistent and physically wrong. Thus, the intervalic system of physical quantities incorporates an *epistemological rank* previously unknown.

Moreover, with this new vistas there has no sense to maintain the traditional separation between physical quantities and its corresponding equation of dimensions, and between those and the system of physical units. By means of its new epistemological rank, the intervalic system of physical quantities reaches a remarkable unification of all these three concepts in a unique one.

THE DIMENSIONAL BASIS OF THE INTERVALIC SYSTEM OF UNITS: c (t^{-1}) AND ℏ (L)

In the intervalic system of physical quantities all physical quantities are consistently defined as a systematic and nontrivial *combination of the two reliable fundamental constants of Nature: c and ℏ*. Every physical quantity is defined as a combination of the intervalic dimensions of c (t^{-1}) and ℏ (L). The intervalic dimensions of the two fundamental constants are *just* those that conform the *dimensional basis* of the intervalic system of dimensions: (t^{-1}) —or (t) which would yield an entirely equivalent system— and (L), respectively.

The extraordinary fact that the dimensional basis of the intervalic system of dimensions is just composed only by the two fundamental constants, c and ℏ, allows that any physical quantity can be expressed as a *universal* dimensionless ratio in intervalic units.

It can be said that the criterion to drawn the dimensional basis of the intervalic units is just opposite to the criterion followed by the wrongly called "geometrized units", which eliminates the two fundamental constants of Nature making them dimensionless, yielding a misleading and useless system.

MERGING OF PHYSICAL QUANTITIES

In the intervalic system of physical quantities we can affirm that if two physical quantities have the same intervalic dimensions, the are indeed the same underlying physical quantity —although probably viewed in different *scales of observation*, which make the illusion of distinct phenomenology—. By this way, some remarkable merges of physical quantities have been discovered, such as velocity-energy, acceleration-power, frequency-force, etc.

The interpretation of those unifications of physical quantities is a beautiful task for a Physics class. Let us think i.e. the inner nature of velocity-energy and frequency-force regarding the huge differences between its coupled magnitudes at quantum scale and at macroscopic scale. It can be seen that magnitudes of velocity-energy at quantum scale are relatively small, while the acceleration-power and the frequency-force at the same quantum scale are enormous magnitudes. We have to conclude that the macroscopic physical quantities are anthropological concepts — rather deformed— of the genuine quantum physical quantities. Another intuitive example which makes a lot of sense by itself: why does not have been described and correctly named a magnitude as important as the power of a gravitational field? The answer is because they are the same underlying physical quantity, $(-L^{-1})$, so the *acceleration* of a gravitational field is identical to its *power*.

DIMENSIONLESSNESS OF FUNDAMENTAL CONSTANTS IN THE INTERVALIC DIMENSIONAL SPACE

A crucial challenge to any dimensional system of physical quantities is the dimensions that adopt the fundamental constants of physics when written in such a system. Of course, the vast majority of fundamental constants should be no other thing that mere *proportionality constants* in a Physics theory geometrically consistent, and therefore they should be *dimensionless*.

Really, this simple proof can be definitive for the epistemological rank bestowed to a dimensional system of physical quantities, because probably it may be necessary and sufficient to be a reliable system of physical dimensions.

It must be noted the remarkable exception of the Planck's constant which is not a *proportionality* constant by no means, since it establishes a fundamental *quantum* instead of a relation of proportionality between physical quantities. In this way, it can be expected that the Planck's constant should ideally have the single dimension of the dimensional basis of the system, as it is in the intervalic system of physical quantities.

The proportionality constants in the traditional system of physical quantities are not dimensionless. This fact means that we are compelled to choose logically between two options: the traditional systems of physical quantities are wrong, or the intervalic system is wrong. It is not necessary to say that we have chosen the first option. Therefore, the traditional dimensions shall be derivations or combinations of a pristine and basic set of dimensions: those than compose the dimensional basis of the intervalic system of physical quantities.

Of course, this crucial proof can't be accepted if it is passed in a trivial manner, such as, i.e., the degenerate system used in Quantum Field Theory or in String Theory. Moreover, the degenerate — "geometrized"— dimensional system, although naive and inconsistent, doesn't even pass this proof in a trivial way.

The equation of dimensions of the principal proportionality constants when written in traditional and in intervalic dimensions are shown in the table. As can be easily seen, all of them are dimensionless with the exception of the weak interaction and the gravitation constants. This notable fact may mean that we have not found yet the reliable dimensionless constant for weak interaction and gravitation, or that those interactions, having not a dimensionless proportionality constant, are physically quantized someway. Meanwhile it can be a useful guidance for the discovering of genuine theories of gravitational and weak interactions, since now we know that the truthful dimension of Newton constant is, surprisingly, *time* (iL), instead of the insignificant traditional dimension ($L^3 M^{-1} T^{-2}$); and the present dimension of Fermi constant is a still unknown physical quantity: ($i^{-1}L^3$).

The discussion about the intervalic dimensions of the coupling constants of the four traditional forces of Nature (grouped at the end of the table), which corresponding theories are renormalizable only if their intervalic dimension is dimensionless, will be suitably treated in other site.

DIMENSIONS OF PROPORTIONALITY CONSTANTS

Name	Symbol	Traditional dimension	Intervalic dimension
Speed of light	c	$L\,T^{-1}$	i^{-1}
Permitivity of vacuum	ε_0	$L^{-3}\,M^{-1}\,T^{4}\,I^{2}$	1
Permeability of vacuum	μ_0	$L\,M\,T^{-2}\,I^{-2}$	-1
Impedance of vacuum	Z_0	$L^{2}\,M\,T^{-3}\,I^{-2}$	i
Electric constant	k	$L^{3}\,M\,T^{-4}\,I^{-2}$	1
Magnetic constant	k_m	$L\,M\,T^{-2}\,I^{-2}$	-1
Boltzmann constant	k_B	$L^{2}\,M\,T^{-2}\,\Theta^{-1}$	1
Electromagnetic interaction coupling constant	α	1	1
Strong interaction coupling constant	α_s	1	1
Weak interaction coupling constant	$G_w/(c\hbar)^3$	$L^{-4}\,M^{-2}\,T^{4}$	-1
Weak interaction coupling constant	G_w	$L^{5}\,M\,T^{-2}$	$i^{-1}L^{3}$
Gravitational interaction coupling constant	G_N	$L^{3}\,M^{-1}\,T^{-2}$	iL

INTERVALIC DIMENSION OF MASS

The intervalic dimension of mass, (T/L = i), can be deduced from a constellation of facts which have in common the symmetry and economy laws of the intervalic geometry. Perhaps one of these postulates considered alone should not be sufficient for a rigorous deduction of the intervalic dimension of mass, but considered as a whole they conform a puzzle where all pieces fit in their just place. Since the description of these kinds of symmetry of the Intervalic Space cannot be resumed in a few lines, we are going to expose only a very simple deduction "by exclusion" of the intervalic dimension of mass. Thus, if we construct a dimensional system with dimensions (L), (T = iL), (M), and write down a table containing all physical quantities —like our table with the 40 physical quantities of the Intervalic Group— we obtain a strange picture in which there is easily seen that the intervalic dimension of mass only can be (i) —i.e., there is a blank square for the basic dimension (i), indicating that it should correspond necessarily to a fundamental physical quantity, and it can only be the mass, since we don't think that we have no evidence of all fundamental physical quantities at present—. All these reasons allow us to deduce the intervalic dimensions of mass joining several symmetric laws in the Intervalic Space, instead of postulating its dimension by definition in an axiomatic way: M = T/L = iL/L = i. In any case, the physical analysis of the final result would be the same.

Please note that another way in which the intervalic dimensions of mass could have been deduced is directly to taking an assumption almost noticed by Lancelot L. Whyte, when he commented extensively in his excellent book **Critique of Physics** that the genuine dimension of Planck constant was not action but length. Really, if we take by definition: dim. $\hbar$ = L, we have:

dim. $\hbar$ = L = $L^2 M T^{-1}$
M = $L^{-1}T$

Of course, the omitted important detail that allows this relation makes sense is the luminous idea named the first intervalic principle about the nature of time, (T = iL), whose almost insulting simplicity does not make us to forget that the last fundamental and underlying principles of Physics must be of an extreme simplicity, as have been pointed out by the most outstanding physicist in the history of Physics.

As can be easily seen, the intervalic dimension of mass is just the inverse of the intervalic dimension of velocity-energy, that is to say, invervelocity. By this reason we can unify both physical quantities and call mass as invervelocity-mass. Expressing this result in differential language, we can say that if velocity-energy is the change of space by unit of time, invervelocity-mass is the change of time by unit of space. When String Theory is trying to derive subatomic matter from vibrations in space-time, we reach a more general and simple result, based on a new dimensional analysis, and valid in microcosmic scale as well as in macrocosmic scale. Both velocity-energy and invervelocity-mass are, to say this way, imaginary features in space-time: the first, (L/T) = (i^{-1}), and the second, (T/L) = (i).

An important thing we have reached with the intervalic dimensions is to solve one of the most awkward problems of Physics: to find a logical explanation for the dimension of matter that *makes sense*. The ancient Natural Philosophy considered the question of matter as unsolvable *par excellence*: "Only two things can't be defined anyway: God and matter" (J. Scoto Eriugena, ***Periphyseon de divisione naturæ***). The same James Clerk Maxwell said in his article *Atom* for the ***Encyclopedia Britannica*** (1875) that "the first desideratum of a whole theory of matter is to explain: first, mass, and second, gravitation. But to explain mass seems an absurd achievement".

Moreover, invervelocity-mass is a physical quantity that can be geometrically derived as a ±180° rotation —by means of the antizator $c^{\pm 2}$(-1)— in the Intervalic Dimensional Space starting from the generator velocity-energy, being its equivalence:

$$E = c^{\pm 2}\, m$$

This equation is no more than a *geometrical statement,* a geometrical equivalence between physical quantities that we call the *intervalic principle of equivalence between mass and energy* which obviously is similar — but not equal due to the physical interpretation of its intervalic dimensions— to the famous Einstein equation, $E = mc^2$.

Finally, the most rigorous way to deduce the intervalic dimension of mass is simply through the dimensional analysis of the above equation, which give us immediately the intervalic dimension of mass. Showing the dimension of each physical quantity between brackets, we have:

$$E\,(i^{-1}) = m\,(x)\,c^2(-1)$$
$$(i^{-1}) = (x)\,(-1)$$
$$(x) = (i^{-1}) / (-1) = (-i^{-1}) = (i)$$

INTERVALIC DIMENSION OF TEMPERATURE

Temperature is a physical quantity that does not have strong dimensional bonds with the rest of dimensions. Richard Feynman (**Quantum Electrodynamics**) has pointed out as one of the important step in the development of Physics the discovering about the nature of heat phenomena. It was understood that they could be explained in terms of movement. In this view, temperature would be similar to radiant energy: both phenomena could be explained like a sort of movement —or vibration, that is to say, movement— at subatomic scale of observation. Therefore, if we hope that a system of physical quantities be meaningful, it is reasonable to expect that temperature would have just the same dimension as velocity-energy.

When writing physical equations that incorporate that assumption in intervalic dimensions, we find that all makes sense. The traditional and the intervalic dimensions of basic thermodynamics physical magnitudes can be seen in the table. We find that Stefan-Boltzmann constant and molar gas constant are dimensionless, as expected. Due to its relations within *information theory*, the entropy is a physical quantity that must be necessarily dimensionless, as it is in intervalic dimensions. The specific heat capacity has intervalic dimension of energy, which is a meaningful fact since this physical magnitude is a measurement of the "thermal internal energy" inside the matter of chemical substances. The dimension of the Stefan-Boltzmann constant shows us that it is a sort of "voluminic power" that probably merits further discussion. The intervalic dimension of temperature is the same as velocity-energy one, as commented below. And finally, we see that thermal conductivity has the same intervalic dimension of surface tension, a result that introduces new light on thermodynamics phenomena and also merits further commentaries in other sites.

A very important think about the intervalic dimensions of the intervalic system of physical quantities is that they are, for first time, epistemologically consistent. This means that they have an estimable heuristic value for physics research because now they are meaningful, instead of the traditional dimensions, whose internal and logical relations among them was absolutely insignificant.

INTERVALIC GROUP

The whole set of intervalic physical quantities shows clearly a well ordered structure and conforms the *intervalic group*, a finite set integrated by 40 quantized o limited dimensions. Of them, 28 have got gravitational nature —those with *integers* powers— and 12 have got electromagnetic nature —those with *half* powers—. Since two of the first are dimensionless —(1) and (-1)—, there are 26 proper gravitational physical quantities (a meaningless coincidence with the number of 26 topological

TRADITIONAL AND INTERVALIC DIMENSIONS OF THERMODYNAMICS PHYSICAL QUANTITIES

Name	Symbol	Traditional dimension	Intervalic dimension
Boltzmann constant	k_B	$L^2 M T^{-2} \Theta^{-1}$	1
Entropy	S	$L^2 M T^{-2} \Theta^{-1}$	1
Heat capacity	C	$L^2 M T^{-2} \Theta^{-1}$	1
Molar gas constant	R	$L^2 M T^{-2} \Theta^{-1}$	1
Specific heat capacity	c	$L^2 T^{-2} \Theta^{-1}$	i^{-1}
Stefan-Boltzmann constant	σ	$M T^{-3} \Theta^{-4}$	$-L^{-3}$
Temperature	T	Θ	i^{-1}
Thermal conductivity	λ	$L M T^{-3} \Theta^{-1}$	$i^{-1} L^{-2}$

dimensions in String Theory).

On the other hand, a total of 26 physical quantities have *imaginary* numbers in their equation of dimensions (another completely meaningless coincidence with the number of 26 topological dimensions in String Theory), and the remaining 14 quantities have *real* numbers. We can name them respectively time-like and space-like physical quantities.

In the same way as the classification of the elements by Mendeleiev on a periodic table, the intervalic group makes a classification of the physical quantities, similar to another periodic table (although more complex because the organization of the physical quantities have the aspect of a symmetric Lie algebra), and shows us that we only know 35 of the 40 allowed or existing physical quantities, and 5 of them have not been described yet.

There may have some representations for the symmetries of the intervalic group. We have think up a bidimensional representation into two planes (as for equations with imaginary numbers) and a tridimensional representation. The first picture can be seen below in the chapter titled 'Intervalic Dimensions' and the later can be seen nearby these lines.

It can be seen now that the number of physical quantities is finite and it is univocally *determined* by the unfolded dimensions of real space. Thus, if our Universe would have only 2 real spatial dimensions, the intervalic group will have 28 physical quantities; if the Universe would have 4 real spatial dimensions, the intervalic group will have 52 physical quantities; etc., being obviously the recurrent formula:

$$Q = 4 + 12D$$

where Q is the number of physical quantities and D the number of real spatial dimensions.

Apart from this, we have made a table with an individual description of the 40 intervalic physical quantities, which are related with the tridimensional representation of the intervalic group.

(i)
(i^1/2)
(-1)
(1)
(i^-1/2)
(i^-1)

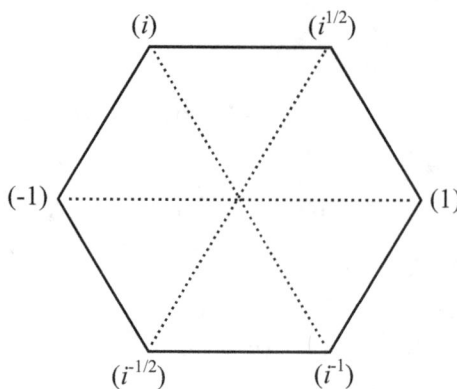

Although it is irrelevant for the body of the theory, and as a simple curiosity, we have that according to the *intervalic zero assumption* —a *desideratum* of logical economy— the sum of all physical quantities and dimensions of the Universe should be *zero*. Only the Intervalic Theory could verify this *desideratum*. However the Intervalic Theory is not based on that assumption, which is unnecessary for deducing the theory. In any of the three groups of hexagons below, the opposite intervalic dimensions vanishes between themselves, being the total result zero; or the magnitudes vanish, being the total sum neutral.

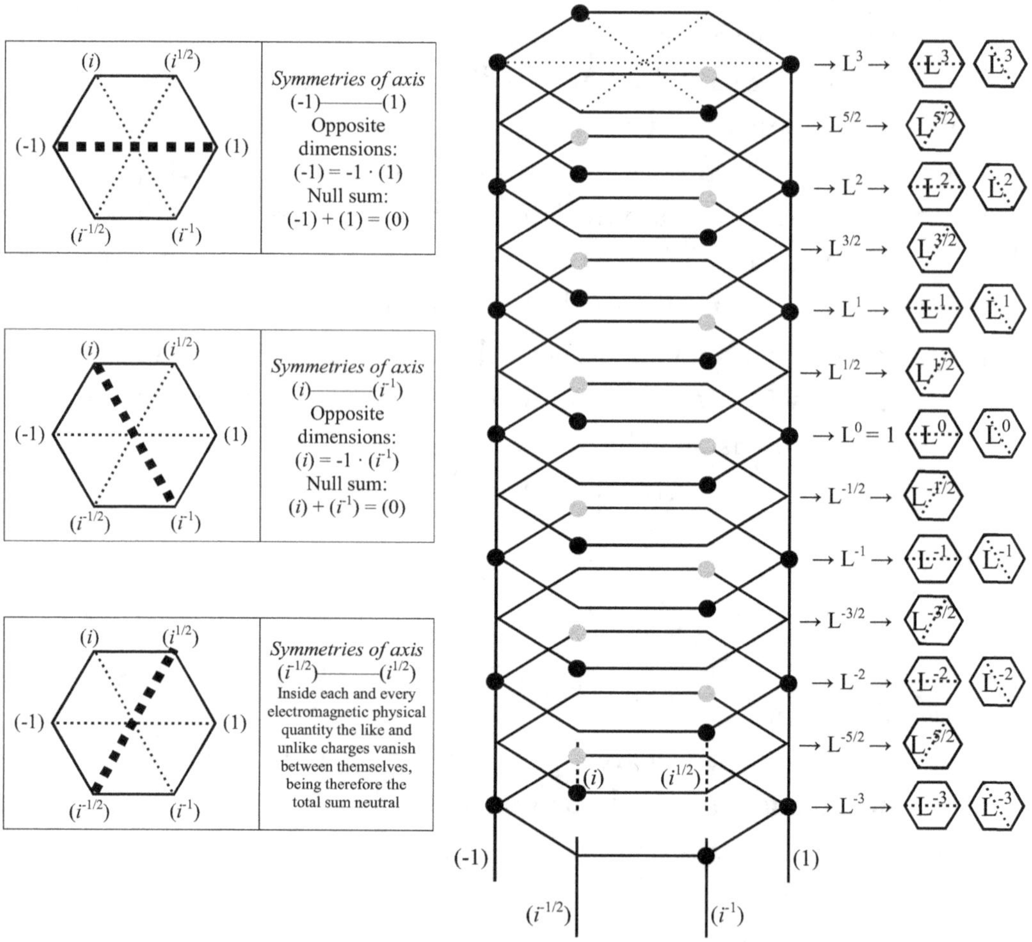

Symmetries of axis
(-1)———(1)
Opposite dimensions:
(-1) = -1 · (1)
Null sum:
(-1) + (1) = (0)
Symmetries of axis
(i)———(i^-1)
Opposite dimensions:
(i) = -1 · (i^-1)
Null sum:
(i) + (i^-1) = (0)
Symmetries of axis
(i^-1/2)———(i^1/2)
Inside each and every electromagnetic physical quantity the like and unlike charges vanish between themselves, being therefore the total sum neutral
→ L^3 →
→ L^5/2 →
→ L^2 →
→ L^3/2 →
→ L^1 →
→ L^1/2 →
→ L^0 = 1
→ L^-1/2 →
→ L^-1 →
→ L^-3/2 →
→ L^-2 →
→ L^-5/2 →
→ L^-3 →
(-1)
(1)
(i^-1/2)
(i^-1)

Symmetries of the 40 physical quantities of the Intervalic Group in tridimensional representation

CORRESPONDENCE BETWEEN THE NUMBER OF REAL SPATIAL DIMENSIONS IN THE UNIVERSE AND THE INTERVALIC GROUP	
Number of dimensions of real space	*Corresponding number of physical quantities in the intervalic group*
0	0
1	16
2	28
3	40
4	52
5	64
6	76
7	88
8	100
9	112
10	124

FRACTAL DIMENSIONS OF PHYSICAL QUANTITIES

(This paragraph is a mere curiosity and may be skipped if desired).

The fractal geometry developed by Benoît Mandelbrot from the concept of Hausdorff-Besicovitch dimension can be applied to the physical quantities in the Intervalic Dimensional Space since they now are geometrically related. According to Mandelbrot, the fractal dimension —of a mathematical function— can be viewed as a measurement of the "recovering degree" of that function on a n-plane with respect to its topological dimension, n. Applying this analogy to the matter-energy *ratio* of the physical quantities we can obtain the Hausdorff-Besicovitch dimension of all intervalic physical quantities. As defined by Mandelbrot, the

CLASSIFICATION OF PHYSICAL QUANTITIES IN THE INTERVALIC GROUP		
Criterion of classification	*Number of physical quantities*	*Feature of the set of physical quantities*
Integer vs. half integer powers	26	Gravitational
	14	Electromagnetic
Real vs. imaginary numbers	14	Space-like
	26	Time-like
Integer vs. fractal dimensions	10	Integer
	30	Fractal
Dimensionful vs. dimensionless	38	Dimensionful
	2	Dimensionless

dimension of a function is *fractal* if, and only if, its dimension is *lesser* than its topological dimension. This ratio can be obtained from the intervalic principle of equivalence between mass, energy and electric charge, as have been explained in other site. In this line, the equivalent energy ratio between the electric charge and the mass of electron is:

$$I(e) / E(m_e) = c^{\pm 2}\hbar\, e^{-2} / c^{\pm 2} m_e = 0.558320$$

Therefore this ratio can be applied also to its respective dimensions, and since we can operate algebraically with fully consistence with the intervalic dimensions, we have:

$\dim (i^{-1/2} L^{1/2} / i) = 0.558320$
$\dim (i^{1/2}) = 1.116640$
$\dim (i^{-1/2}) = 0.895544$

And the fractal dimension of (i) would be:

dim (i) = 0.801999
dim (i^{-1}) = 1.246885

The topological and fractal dimensions of all physical quantities can be seen in the last table. The only integers —no fractal— physical quantities are those with intervalic dimension $L^{\pm n}$ (with n = 1, 2, 3) and the dimensionless ones (1) and (-1), that is to say, only 10 of the whole set of physical quantities of the intervalic group have an integer —non fractal— dimension (another one curious and meaningless coincidence with the number of 10 topological dimensions in the traditional String Theory).

It is intriguing to see that *time* have not and integer Hausdorff-Besicovitch dimension:

dim (T = iL) = 0.801999

The interpretation of this result is that physical time does not cover completely the whole topological dimension of time. It only could be possible if the speed of light was infinite. In this hypothetical and unreal case the electromagnetic energy would have a greater share in the structural energy of subatomic particles and the preceding electron energy ratio would reach just the following limit:

I(e) / E(m_e) = $c^{\pm 2} h \, e^{-2} / c^{\pm 2} m_e$ = ½

PHYSICAL QUANTITIES AND PARTITION OF SPACE IN REGULAR POINT-LATTICES

(This paragraph is a mere curiosity and may be skipped if desired).

We have seen some apparently meaningless coincidences between the number of physical quantities in the Intervalic Theory and the number of topological dimensions in String Theory. There is another curious coincidence between the number of physical quantities and the regular partitions of space. We can think that it is meaningless, as it appears to be, but perhaps someone could think that it is an still unknown property of the Nature's forms.

There are only one regular isotropic partition of space: the *cuboctahedral* point-lattice, and there is only one regular non-isotropic partition of space: the *cubical* point-lattice. Both of them are well known in crystallography. The first is the natural close-packing of *spheres* and the second the natural close-packing of *cubes*. To obtain a perfect close-packing of a sphere with spheres, 12 spheres are needed (the centres of these 13 tangent spheres coincide with the vertices of a cuboctahedron); in a similar way, to recover perfectly a cube with cubes takes 26 cubes.

Thus, the number of electromagnetic physical quantities (12) coincides with the number of spheres needed for the only possible close-packing in a regular isotropic partition of space: 12; and the number of gravitational physical quantities (26) coincides with the number of cubes needed for the only possible close-packing in a regular non-isotropic partition of space: 26. Moreover, the central sphere and the central cube of these regular point-lattices matches with the two remaining dimensionless physical quantities, being its total sum —(12+1) + (26+1) = 40— just the number of physical quantities of the intervalic group: 40.

Someone may think that there are too many coincidences to be all of them meaningless. Although by now we can't make further speculations, it appears to exist some unknown deep relation between the *electromagnetic* phenomena and the *cuboctahedral* point-lattice, and between the *gravitational* phenomena and the *cubical* point-lattice, and surely it would be very interesting to discover such possible relation. In other case, this would be another curious and irrelevant coincidence without any significance.

INTERVALIC THEORY:
The Intervalic Structures of Subatomic Particles and the Last Foundations of Physics

THE 40 PHYSICAL QUANTITIES OF THE INTERVALIC GROUP

Representation	Intervalic dimension	Physical quantity	Traditional dimension	Symbol	Fractal dimension
L^0 (hexagon)	1	Permitivity	$L^{-3} M^{-1} T^4 I^2$	ε	
		Momentum (momentum-inertia)	$L M T^{-1}$	p	0
		Boltzmann constant	$L^2 M T^{-2} \Theta^{-1}$	k_B	
		Entropy	$L^2 M T^{-2} \Theta^{-1}$	S	
	-1	Permeability	$L M T^{-2} I^{-2}$	μ	
		Gravitational potential	$L^2 T^{-2}$	Φ	0
		Antimomentum	-	p_a	
L^0 (hexagon)	i	Mass (invervelocity-mass)	M	m	
		Invervelocity (invervelocity-mass)	$L^{-1} T$	v^{-1}	0.801999
		Impedance	$L^2 M T^{-3} I^{-2}$	Z	
	i^{-1}	Velocity (velocity-energy)	$L T^{-1}$	v	
		Energy (velocity-energy)	$L^2 M T^{-2}$	E	1.246885
		Temperature	Θ	T	
		Specific heat capacity	$L^2 T^{-2} \Theta^{-1}$	c	
		Conductance	$L^{-2} M^{-1} T^3 I^2$	G	

Representation	Intervalic dimension	Physical quantity	Traditional dimension	Symbol	Fractal dimension
⬡ L^1	L	Length (real space)	L	L	1
		Planck constant, action	$L^2 M T^{-1}$	$\hbar$	
		Capacitance	$L^{-2} M^{-1} T^4 I^2$	C	
⬡ L^1	$-L$	Antilength	-	L_a	1
		Inductance	$L^2 M T^{-2} I^{-2}$	L	
⬡ L^{-1}	iL	Time (imaginary space)	T	T	0.801999
		Gravitational constant	$L^3 M^{-1} T^{-2}$	G	
	$i^{-1}L$	Antitime	-	T_a	1.246885
	L^{-1}	Wavevector	L^{-1}	k	1
⬡ L^{-1}	$-L^{-1}$	Acceleration (acceleration-power)	$L T^{-2}$	a	1
		Gravitational field	$L T^{-2}$	g	
		Power (acceleration-power)	$L^2 M T^{-3}$	W	
	iL^{-1}	Antifrequorce	-	φ_a	0.801999
		Linear Perdensity	-	Δ^l	
		Frequorce (frequency-force)	-	φ	
	$i^{-1}L^{-1}$	Frequency	T^{-1}	ν	1.246885
		Force	$L M T^{-2}$	F	
		Linear Tension	T^{-1}	σ	
		Conductivity	$L^{-3} M^{-1} T^3 I^2$	σ	
⬡ L^2	L^2	Area	L^2	S	2
	$-L^2$	Antiarea	-	S_a	2

INTERVALIC THEORY:
The Intervalic Structures of Subatomic Particles and the Last Foundations of Physics

Representation	Intervalic dimension	Physical quantity	Traditional dimension	Symbol	Fractal dimension
L²	iL^2	Inertia area momentum	$L^2 M$	I^2	1.603998
L²	$i^{-1}L^2$	?	-	-	2.493769
-L²	L^{-2}	Viscosity (dynamic)	$L^{-1} M T^{-1}$	η	1/2
-L²	$-L^{-2}$	Area Power	-	W^2	1/2
-L²	iL^{-2}	Area Perdensity	-	Δ^2	0.400999
-L²	$i^{-1}L^{-2}$	Inflexion	$L T^{-3}$	i	0.623442
L³	L^3	Surface Tension	$M T^{-2}$	σ^2	3
L³	$-L^3$	Volume	L^3	V	3
L³	iL^3	Antivolume	-	V_a	2.405997
L³	$i^{-1}L^3$	Inertia volume momentum	$L^5 M T^{-2}$	I^3	3.740654
-L³	L^{-3}	Fermi weak interact. constant	-	G_w	1/3
-L³	$-L^{-3}$	Fluctuation	$L T^{-4}$	f	1/3
-L³		Voluminic Power	-	W^3	1/3
-L³		Irradiance	$M T^{-3}$	E_e	
-L³		Stefan-Boltzmann constant	$M T^{-3} \Theta^{-4}$	σ	
-L³	iL^{-3}	Density	$L^{-3} M$	ρ	0.267333
-L³		Voluminic Perdensity	$L^{-3} M$	Δ^3	
-L³	$i^{-1}L^{-3}$	Pressure	$L^{-1} M T^{-2}$	P	0.415628
-L³		Energy-tension density	$L^{-1} M T^{-2}$	u	

Representation	Intervalic dimension	Physical quantity	Traditional dimension	Symbol	Fractal dimension
$L^{1/2}$	$i^{1/2}L^{1/2}$	Magnetic charge	$L^2 M T^{-2} I^{-1}$	θ	0.558320
	$i^{-1/2}L^{1/2}$	Magnetic flux	$L^2 M T^{-2} I^{-1}$	Φ	0.447772
$L^{-1/2}$	$i^{1/2}L^{-1/2}$	Electric charge	$T I$	q	2.233280
		Current	I	I	
$L^{3/2}$	$i^{1/2}L^{-1/2}$	Electric potential	$L^2 M T^{-3} I^{-1}$	V	2.233280
		Magnetic vector potential	$L M T^{-2} I^{-1}$	A	
	$i^{-1/2}L^{-1/2}$	Magnetic inverflux	-	Φ^{-1}	1.791088
	$i^{1/2}L^{3/2}$	Bohr magneton	$L^2 I$	μ_B	1.674960
	$i^{-1/2}L^{3/2}$	?	-	-	1.343316
$L^{-3/2}$	$i^{1/2}L^{-3/2}$	Electric field strength	$L M T^{-3} I^{-1}$	$\mathcal{E}$	0.744427
		Magnetic field strength	$L^{-1} I$	H	
	$i^{-1/2}L^{-3/2}$	Magnetic flux density	$M T^{-2} I^{-1}$	B	0.597029
$L^{5/2}$		Electric polarization	$L^{-2} T I$	P	
	$i^{1/2}L^{5/2}$	?	-	-	2.791600
	$i^{-1/2}L^{5/2}$	?	-	-	2.238859
$L^{-5/2}$	$i^{1/2}L^{-5/2}$	Charge density	$L^{-3} T I$	ρ	0.446656
		Current density	$L^{-2} I$	J	
	$i^{-1/2}L^{-5/2}$	?	-	-	0.358218

Chapter 2

INTERVALIC QUANTA, INTERVALIC LIMITS
The Geometrical Heights of Nature

The intervalic system of physical quantities determines necessarily the *quantization* or *limitation* of all physical quantities, which conform a finite set with remarkable symmetries named the intervalic group. Intervalic quanta and intervalic limits have not been arranged by hand, like the traditional Planck's physical quantities. On the contrary, both their magnitudes and their dimensions are exclusive combinations derived univocally from $\hbar$ and c —the last fundamental constants of Nature— in a systematic theoretical way. These new heights *geometrize* the mathematical conception of physical quantities and are a part of the underlying and comprehensive *Intervalic Geometry* which makes the foundations of a new Physics.

In this paper are described the fundamental quanta and limits of Nature, being commented some of them individually, such as the intervalic mass, the intervalic velenergy, the intervalic frequorce, the intervalic length, the intervalic electric charge, the intervalic magnetic charge or the intervalic temperature. Also are explained the geometric equivalences among intervalic quanta and limits, and the introduction of the intervalic quanta as a reliable scale for a universal system of units.

INTERVALIC QUANTA AND INTERVALIC LIMITS

> There has always been an attempt to find a unifying theoretical basis for all these [various branches of Physics]... from which all the concepts and relationships among the individual disciplines might be derived by a logical process. This is what we mean by the search of a foundation of the whole of Physics. The confident belief that this ultimate goal may be reached is the wellspring of the passionate devotion that has always motivated the researcher.
>
> ALBERT EINSTEIN
> *The Fundamentals of Theoretical Physics,*
> *"Science" 91, 24.05.1940*

When using intervalic dimensions ($L = L$, $T = iL$, $M = T/L = i$), with $i = \sqrt{-1}$, a new set of fundamental quanta —the term 'quanta' is here used in geometric sense— is necessarily yielded: the *intervalic quanta*. This logical fact can be considered as unexpected and almost magic, since no quanta nor any other physic phenomena should be obtained from a mere change in the dimensional system of physical quantities used.

Unlike the so named Planck quanta which involves complex combinations of three constants, the intervalic ones are reliably *fundamental* since there are composed only by the simplest combinations of the two genuine constants of Nature: the speed of light, c, and the intervalic length, ħ —which replace the action quantum, h—. On the contrary, Planck's quanta are dimensional combinations *ad hoc* of c, h, and the gravitational constant, G, carefully prepared by hand to obtain the desired quantum of longitude, time, mass, etc. Those combinations may have had some value as an *empirical* approximation to the value of the primordial scale of the physical forces, since till the postulation of IT we didn't have other way to determine it. But by means of the intervalic dimensions we can determine faithfully the exact *theoretical* value of that scale. Moreover, the intervalic quanta can do that because they have the truly epistemological rank of *quantum*, whereas Planck's quanta do not have really that logical rank, although they have been freely treated as if they would have it.

Besides, the fact that Planck's quanta are scarcely different from the intervalic quanta means that some of the involved constants (c, ℏ or G) have slightly wrong magnitudes. Since the magnitude of c is taken by definition, we will see that the deviated magnitude is that one of G, being despicable —from an theoretical point of view— the deviation in the magnitude of h. On the contrary, the deviation in the magnitude of G, although small, might be physically meaningful as will be explained in other site.

The comparison of magnitudes between different physical quantities has no sense in the traditional system of dimensions nor in the degenerate dimensions. This is correct because these systems do not have — and they can't know whether they have— a genuine *geometrical relation* among its physical quantities. On the contrary, the so named Intervalic Dimensional Space is like an extended Argand space where *all physical quantities are generated* by successive *rotations* of $\pm 90°$ by means of the geometrical *transformer*: the fundamental constant of proportionality between real and imaginary axes, the speed of light, c, whose intervalic dimension is $L/T = i^{-1}$, and its inverse transformer, c^{-1}, with intervalic dimension $T/L = i$. So we can compare any *real* physical quantity with its corresponding *imaginary* one, that is to say, we can compare any physical quantity with intervalic dimension X, with its corresponding $i^n X$ (where $n = \pm 1, \pm 2, \pm 3, \pm 4$); and vice versa.

Best physicists suspect that we don't know yet the genuine geometry of physics, the underlying fundamental geometry of the Universe. It is hoped that this sort of geometry should be based on principles of astonishing simplicity and, of course, leaded continuously by the signs of logical inevitability and mathematical elegance. In short, it should transform all or a great number of physical equations and formulae into pure *geometric statements* (as partially did General Relativity with gravitation). As we will understand immediately, the inner nature of the *intervalic quanta and limits* is purely *geometrical*, since the intervalic dimensions define a new type of *geometrical* relations among physical quantities.

This situation is very different from the traditional dimensional system or from the degenerate —wrongly called 'geometrized'— system (dim. c = 1, dim. ħ = 1), where, if the dimensional logic would have to be rigorously applied, we would obtain absolutely absurd results such as, i.e., the inverse value of Planck mass should be the quantum of length, and vice versa, since in that trivial system $L = M^{-1}$. In any way it can be qualified as fully consistent or elegant.

Regarding the dimensions of c and h, it is evident that its traditional dimensions (LT^{-1} and L^2MT^{-1}) are not single or elemental dimensions, being this desideratum an important and almost definitive heuristic rule for investigation. In degenerate units c and h are both dimensionless, (1), which is an arbitrary supposition that finally would end in grave inconsistence. Really, it is a little surprising that such dimensional system have been seriously used. In this system the dimensions of all physical quantities are *trivial* and lack of a truthful physical meaning. Anyway, none of them do yield reliable fundamental dimensions for c and ħ.

Besides, as a result of the geometrical conformation of the Intervalic Dimensional Space and the exclusive logical rank of the intervalic dimensions, we can say that *all physical quantities are intervalically quantized or limited*. This have to be interpreted in the sense that the macroscopic physical laws are only a comfortable approximation of the pristine subatomic physical laws, since the quantum values have no relevance in the measurement of the macroscopic world. But we ever must remember that all these last measurements are *multiple* of the quantum values.

This can be easily understood when dealing with the *minimum*

INTERVALIC THEORY:
The Intervalic Structures of Subatomic Particles and the Last Foundations of Physics

INTERVALIC QUANTA AND LIMITS
(shadowed traditional quanta are shown for comparison)

Name	Intervalic dimension	Geometric rank	Intervalic value	Value SI	Definition
Intervalic mass	i	subatomic limit	1	$3.3356409 \cdot 10^{-9}$ (kg)	$m_I = c^{-1}$
Planck mass	i	subatomic limit	6.525616	$2.176711 \cdot 10^{-8}$ (kg)	$m_P = \sqrt{(hc/G)}$
Intervalic energy	i^{-1}	subatomic limit	1	$2.9979246 \cdot 10^{8}$ (J) $1.8711569 \cdot 10^{21}$ (MeV)	$E_I = c$
Planck energy	i^{-1}	subatomic limit	6.525616	$1.956330 \cdot 10^{9}$ (J) $1.2210449 \cdot 10^{22}$ (MeV)	$E_P = \sqrt{(hc^5/G)}$
Intervalic length	L	quantum	1	$1.0556363 \cdot 10^{-34}$ (m)	$l_I = \hbar$
Planck length	L	quantum	0.153088	$1.616051 \cdot 10^{-35}$ (m)	$l_P = \sqrt{(hG/c^3)}$
Intervalic time	$i\,$L	quantum	1	$3.5212226 \cdot 10^{-43}$ (s)	$t_I = c^{-1}\hbar$
Planck time	$i\,$L	quantum	0.153088	$5.390563 \cdot 10^{-44}$ (s)	$t_P = \sqrt{(hG/c^5)}$
Intervalic electric charge	$i^{-1/2}\,L^{1/2}$	quantum	1	$5.9339900 \cdot 10^{-22}$ (C)	$q_I = \sqrt{-(c^{-1}\hbar)}$
Intervalic thermal mass	i	quantum	k_B	$4.6053793 \cdot 10^{-32}$ (kg)	$m_\Theta = c^{-1}k_B$
Intervalic velocity	i^{-1}	absolute limit	1	$2.9979246 \cdot 10^{8}$ (m/s)	$v_I = c$
Intervalic frequency	$i^{-1}\,L^{-1}$	absolute limit	1	$2.8399227 \cdot 10^{42}$ (s^{-1})	$\varphi_I = c\hbar^{-1}$
Intervalic temperature	i^{-1}	absolute limit	k_B^{-1}	$2.1713738 \cdot 10^{31}$ (K)	$\Theta_I = c\,k_B^{-1}$
Elementary charge	$i^{-1/2}\,L^{1/2}$	subatomic quantum	270	$1.602177 \cdot 10^{-19}$ (C)	$e = 270\,q_I$
Intervalic magnetic charge	$i^{1/2}\,L^{1/2}$	subatomic quantum	1	$1.7789652 \cdot 10^{-13}$ (Wb)	$\Phi_I = \sqrt{-(c\hbar)}$
Magnetic flux quantum	$i^{1/2}\,L^{1/2}$	subatomic quantum	$\pi/270$	$2.069963 \cdot 10^{-15}$ (Wb)	$\Phi_0 = \pi\hbar/e$

quanta like the intervalic length, the intervalic time or the intervalic electric charge. Each macroscopic measurement of length, time or electric charge must be necessarily a multiple value of its corresponding intervalic quanta. But we also have logically defined other *maximum* limits, like the intervalic velocity, the intervalic mass, the intervalic energy, the intervalic frequency or the intervalic temperature. What is the meaning of these limits, since they are not the smallest bricks of Nature? Of course, they are the greatest bricks allowed in Nature. Thus, the intervalic mass and the intervalic energy are the greatest values allowed for a single —not composite— subatomic particle, and likewise the intervalic temperature, the intervalic frequency and the intervalic velocity are the maximum values allowed for a single subatomic particle —and *a fortiori*, for that physical quantity—.

We have done a meaningful table with the principal intervalic quanta and limits. Now lets go to comment briefly some of these new quanta. Their three possible geometric ranks —subatomic limit, quantum, absolute limit— are shown in the table.

INTERVALIC MASS
$m_I = c^{-1} = 3.3356409 \cdot 10^{-9}$ (kg) $= 20,819.42423$ (MeV/c²)
Intervalic dimension: $(T/L) = (iL/L) = (i)$

If the quantum of mass would be a reliable value of Nature it is expected it must be a single value, but not a combination of other constants. In this way the comparison between the definitions of the intervalic mass and the Planck mass speaks by themselves. The last one is not a single constant but a combination of other three constants (c, h, G, where only c and h are truly fundamental constants). On the contrary, the intervalic mass is the inverse value —and also has the inverse dimension, $1/i^{-1} = i$— of one of the two reliable constants of nature, c, the speed of light (being the other ℏ). It is difficult to imagine more simplicity and logical economy:

$m_I = c^{-1}$
$m_P = \sqrt{(hc/G)}$

Besides, the value of the intervalic mass is much more suited to expected for quantum gravity than the value of Planck mass. As have been pointed out by Roger Penrose, 'one graviton' effects will appear when mass magnitude is of order 10^{-7} g. It is hoped that some formula leads to a smaller mass than Planck's one. (Roger Penrose, *Newton, Quantum Theory and Reality,* in **300 Years of Gravitation**, edited by S. W. Hawking & W. Israel). But this value is nearer the intervalic mass, which is around seven times smaller than the Planck mass. And, above all, it is a truthful fundamental magnitude, instead of Planck mass, which is not by no means. As mentioned, its magnitude and its dimension can't be more simple indeed: the inverse value of the speed of light, $c^{-1}(i)$ = 20,819.42423 (MeV/c²).

INTERVALIC VELENERGY
$v_I = c = 2.9979246 \cdot 10^8$ (m s⁻¹)
$E_I = c = 2.9979246 \cdot 10^8$ (J)
Intervalic dimension: (i^{-1})

Briefly, since the intervalic dimension of the speed of light is velenergy or velocity-energy, the fundamental geometrical constant of the Intervalic Dimensional Space, c, can be also interpreted as the geometric height of energy, E_I, which is a subatomic limit, being the magnitude of this energy equivalent to that of the intervalic mass:

$E_I = c^2 m_I = c^2 c^{-1} = c$

INTERVALIC FREQUORCE
$\varphi_I = c h^{-1} = 2.8399227 \cdot 10^{42} (s^{-1})$
Intervalic dimension: $(i^{-1} L^{-1})$

Starting from the value of the intervalic energy limit, the value of the intervalic frequorce —frequency-force— limit is straightforwardly yielded:

$$E_I \; c = \hbar \, \varphi_I$$
$$\varphi_I = c \, \hbar^{-1}$$

INTERVALIC LENGTH
$l_I = \hbar = 1.0556363 \cdot 10^{-34}$ (m)
$S_I = \hbar = 1.0556363 \cdot 10^{-34}$ (J s)
Intervalic dimension: (L)

"Since no electron velocity has ever been directly measured [inside the atom] we cannot be sure that the dimensions of the new constant 'h' —called Planck's constant— are really that we suppose, energy multiplied by time". (Lancelot Law Whyte, ***Archimedes or The Future of Physics***). Continuing with Whyte, in the excellent book ***Critique of Physics***, he spends some pages to remark the truthful function and meaning of the Planck's quantum, *h*. Lets hear him:

"The appearance of *h as a constant having the dimensions of action* is an arbitrary historical fact determined by the general form of the physical theories current at the time it was introduced, whereas its *empirical* function, i.e. in connection with empirical data, is in all cases except that of statistical averages to supply in combination with the other constants a universal standard of *length*. [...]

"The primary function of Planck's constant, *h*, in combination with the atomic constants, is to provide physical theory with standards of length in terms of which all directly measurable lengths of theoretical importance can be described... Wherever h appears in a fundamental equation (not referring to statistical assemblies), the empirical content of the equation is the correlation of a *measured length* with other quantities. Equations that contain h and do not at first sight appear to determine a measurable length have to be transformed into assertions about measurable lengths (wave lengths, crystal lattice constants, etc.) before they have a direct relation to experiment. This fact is concealed by the original introduction of the constant as a unit of action, but any revision of physical concepts must render explicit the primary function of h:

"PLANCK'S CONSTANT (IN APPROPRIATE COMBINATIONS) DETERMINES THE LINEAR SCALE OF THE STRUCTURE OF MATTER AND OF RADIATION, in terms of the selected unit of length".

The fact that the intervalic dimension of h is length, (L), is of capital importance since the intervalic quantum of length is not a combination of other quantities or constants (as in the formula of the Planck's length) but it is an absolutely original and fundamental length, not a derived one. Of course, since we only admit two constants to be reliable fundamental, c and h, it was necessary that the quanta of length was just h, but till now no dimensional system have been capable to get this result in a faithfully simple and radical way. Indeed, when using intervalic dimensions h is directly formulated in units of length: $\dim h = (L^2 M T^{-1}) = (L^2 i\ i^{-1} L^{-1}) = (L)$

Perhaps some readers may have seen that any equation in which appears $c^{\pm n}$ can be interpreted in terms of rotations in the Intervalic Dimensional Space. But probably the reader don't believe that all physic equations, and not only those involving the factor $c^{\pm n}$, can be interpreted in a geometrical way within the Intervalic Dimensional Space. The famous Einstein's equation for the energy of photon, $E = h\nu$, is a good example to show the astonishing simplicity in which the theory transforms all physics formulae. If we postulate the existence of a absolute minimal element of space, the quantum of length, l_I, then, any other longitude of the Universe, s, will have to be multiple of that quantum:

$s = n\, l_I$ (being n an integer)

Now let us write the most simple geometrical relation between the real (L) and the imaginary axes ($iL = T$) in the Intervalic Dimensional Space, that is to say, the classical definition of velocity, v:

$v = L/T = L/iL$
$v = s/t$

Remember that in intervalic dimensions we have the following *identities*:
  1) *velocity* is identical to *energy*: v E
  2) *invertime*, t^{-1}, is identical to *frequorce*: t^{-1} φ
  3) *frequorce* is the unification of *frequency* and *force* in a single physical quantity: φ v F

Therefore the preceding *definition* of velocity can be written in 16 different but equivalent ways when using intervalic dimensions:

$v = s / t = n\, \hbar / t$
$v = s \cdot v = n\, \hbar \cdot v$
$v = s \cdot F = n\, \hbar \cdot F$
$v = s \cdot \varphi = n\, \hbar \cdot \varphi$

$E = s / t = n\, \hbar / t$
$E = s \cdot v = n\, \hbar \cdot v$
$E = s \cdot F = n\, \hbar \cdot F$
$E = s \cdot \varphi = n\, \hbar \cdot \varphi$

All these formulae are distinct manners of expressing an unique and fundamental geometric relation:

v $E = s/t = n\hbar \varphi$

The physical task is precisely to interpret and to connect rightly with natural phenomena the geometric statements deduced by logical means.

In this case we are, although, working *a posteriori*, finding that old physic laws apparently unconnected, are in fact the same equation but expressed in different scales of observation. This is the case, for example, of the equations for macroscopic and microscopic energy, respectively:

$E = s \cdot F$
$E = n\hbar \cdot v$

As can be seen, the classic equation is only an approximation of the last one. Since in the macroscopic world the quantum of real space is too much small to be relevant in any measurement, we simple write 's' instead of the accurate expression '$n\hbar$'. Identity between them probably can be easily seen when they are derived from the same basic equation writing down their intervalic dimensions:

$v(i^{-1})$ $E(i^{-1}) = n\hbar(L) \varphi(i^{-1}L^{-1}) = s(L) F(i^{-1}L^{-1})$
$v(i^{-1})$ $E(i^{-1}) = n\hbar(L) \varphi(i^{-1}L^{-1}) = n\hbar(L) v(i^{-1}L^{-1})$

Besides, the unification of frequency and force in a underlying new physical quantity can be viewed as the merge between the two faces of the physic reality: particles and waves. As Lancelot Whyte affirmed in his quoted book, "it will therefore be necessary to discard entirely the terms and particles in fundamental synthetic theory. Waves and particles are conceptions derived from macroscopic phenomena which provide partial and misleading descriptions of two complementary aspects of one microscopic phenomenon. 'Light and matter are unitary physical phenomena, their apparent duality arises from the essential inadequacy of our language' (Heisenberg). 'The waves and particles should be regarded as two abstractions which are useful in describing the same physical reality' (Dirac)". Now, at least, by means of the introduction of the physical quantity named *frequorce* it is avoided the "essential inadequacy of our language".

Finally, another interpretation of Einstein's equation in intervalic dimensions also explain why appears h instead of ℏ. Really, if the elongation of electromagnetic waves is just the intervalic length as we suppose, the length of the transversal path described by a photon in a cycle is $2\pi l_I$ = $2\pi\hbar$ = h. It is also clear that the fundamental constant is ℏ instead of *h*, which is defined as $h = 2\pi\hbar$. Thus, a more meaningful enunciation of Einstein's equation would be:

$$E = 2\pi \hbar \nu$$

Although the magnitude of Planck's length is near from the magnitude of the intervalic length, it must be pointed out that the first one is not a fundamental quanta but a derived physical quantity, and, what is worse, according to the *intervalic geometry* of Nature, such magnitude is absolutely wrong because it is outside the underlying geometric architecture of Nature defined by the intervalic dimensions.

INTERVALIC ELECTRIC CHARGE
$q_I = \sqrt{-(c^{-1}\hbar)} = 5.9339900 \cdot 10^{-22}$ (C)
Intervalic dimension: ($L^{3/2}M^{1/2}T^{-1} = L^{3/2}i^{1/2}i^{-1}L^{-1} = i^{-1/2}L^{1/2}$)

By means of the intervalic dimensional analysis of the electric charge it is deduced that there must be two types of electric charge, positive and negative, because there are two mathematical solutions for a square root. The almost insulting simplicity of this result should not do us to ignore its importance. It was expected that a faultless theory would explain the existence of the two charge signs; but that was not expected is that it would be reached in a so simple way: from the dimensional analysis of the electric charge.

Sometimes the high specialization reached in Physics make us to forget the real basis of the discipline. I think that the permanent insight on the foundations of any discipline is what distinguish the great minds from the rest. With respect to the dimension of the electric charge, Lancelot Law Whyte advises us that "in fact we have not yet made enough direct experiments to know even whether the dimensional system which is used for electrons is correct" (Lancelot L. Whyte, *Archimedes or The Future of Physics*). I suppose there is not necessary to comment that the charge in the traditional dimensional system is not a fully "consistent" dimension, and in the "geometrized" system is clumsily trivial (as it is the whole system). Besides, if these systems were right, why they do not yield

---

The physical meaning of the **FINE STRUCTURE CONSTANT**, α, is usually interpreted in Physics textbooks as *a measurement of the elementary charge squared, e^2,* **in natural units**, which involves $e \approx 270$ in those unknown *natural units*. Till today no one system of units has been able to derive the elementary charge value through theoretical means (not set by hand). And consequently there has had no system which claimed to be the natural one. However the Intervalic Units are postulated to be just the **natural units**, and its theoretical definition of the *elementary charge* —deduced logical and unavoidably *from the own Intervalic System of Dimensions*— clearly matches with the fine structure constant empirical value. Therefore, we must conclude necessarily that the Intervalic System of Dimensions and Units —which anyway is the only system which is *not equivalent to all the remaining ones*— is just the unique and truthful **Nature's system of units**.

$$q_I = \sqrt{-(c^{-1}\hbar)} = 1 \ (i^{-1/2}L^{1/2}) = 5.93398995 \cdot 10^{-22} \ (C)$$
$$e = 270 \ q_I = 270 \ \sqrt{-(c^{-1}\hbar)} = 1.60217733 \cdot 10^{-19} \ (C)$$
$$\alpha = 270^2 \cdot 10^{-7} = 1/137.1742112 \ (1) \ \text{—exact value—}$$

logically any *fundamental* quanta from its dimensional analysis? Of course, because they are not.

The intervalic quantum of electric charge, q_I, is defined as:

$$q_I = \sqrt{-(c^{-1}\hbar)} = 5.9339900 \cdot 10^{-22} \, (C)$$

Since this is the last and fundamental quantum of this physical quantity, all electric charges in Nature *must be multiple of the intervalic charge,* q_I, in the same way as all lengths in Nature *must be multiple of the intervalic length,* $\hbar$.

Thus, the elementary charge, *e*, is exactly defined as an integer multiple of the intervalic charge:

$$e = 270 \, q_I = 270 \sqrt{-(c^{-1}\hbar)}$$

The extraordinary importance of that equation is to get by first time a theoretical definition of the elementary charge in terms of the two fundamental constants of Nature, c and $\hbar$. Therefore, the elementary charge would already not be a last fundamental constant of Nature but a derived one.

Moreover, it provides an exact theoretical definition of the electromagnetic coupling constant, α, as we have seen in other site, whose exact geometric value is:

$$\alpha = 270^2 \cdot 10^{-7} = 7.2900000 \cdot 10^{-3} = 1/137.1742112$$

According to experimental measurements of *e*, the proposed relation implies a deviation of around 5/10,000 in some of the magnitudes, together considered, of c, $\hbar$ and *e*, with respect to their experimental values. If we suppose that the experimental values of *e* and c are most reliable than the value of $\hbar$, since the last one is related with the elementary charge in measurement experiments, we would have the following set of values taking the experimental values of *e* and c as exact magnitudes:

$$e = 270 \sqrt{-(c^{-1}\hbar)} = 1.6021773 \cdot 10^{-19} \text{ (C)}$$
$$c = 270^2 \, \hbar \, e^{-2} = 2.9979246 \cdot 10^8 \text{ (m/s)}$$
$$\hbar = 270^{-2} \, c \, e^2 = 1.0556363 \cdot 10^{-34} \text{ (m)}$$

On the other hand, if a picturesque prediction of String Theory is right, there would be 16 fractional charges of the elementary charge in Nature. Now, according to the intervalic charge and remembering the high principle of economy that leads all our steps, if *e* is an elementary charge it must be the strongest *state of minimal energy* of the electric charge (instead of the intervalic charge, q_I), and the following states of minimal energy of the electric charge will have to be the set of integers divisors of *e*, which are the following ones:

1, 2, 3, 5, 6, 9, 10, 15, 18, 27, 30, 45, 54, 90, 135, 270.

The number of elements of this set is just 16, according to String Theory. Since the concrete values predicted by both theories are quite different, the determination of those values would be a crucial experiment for the confirmation or falsation of any of the two theories (after viewing the extremely simple and beautiful intervalic structures directly yielded by the intervalic quanta, needless to say which is true and which is false).

Finally, when seeing the surprising result $e = 270 \, q_I$ we can question: what would be the consequences for Physics of that result? To understand that result we need to research and to know the new intervalic symmetries introduced in Physics by the definition $e = 270 \, q_I$. In the worst case those symmetries would add nothing to Physics and we would obtain an equivalent system of dimensions, but in other case they could lead to the discovery of a new Physics, which has been just the case. What is clear is that the values of the elementary charge in the traditional systems of dimensions do not involve and do not introduce any symmetries in

Physics: therefore those systems are surely poorer than the intervalic one as they do not incorporate —nor can do it in any way— that new geometry in the last basis of Physics.

INTERVALIC MAGNETIC CHARGE
$\Phi_I = \sqrt{-(c\hbar)} = 1.7789652 \cdot 10^{-13}$ (Wb)
Intervalic dimension: $(i^{1/2} L^{1/2})$

As we have seen in other site, the magnetic charge can be dimensionally yielded by a -90° rotation —c (i^{-1})— of the electric charge in the Intervalic Space. Therefore, the deduction of the intervalic quantum of magnetic charge is immediate:

$$\Phi_I = c\, q_I = c\sqrt{-(c^{-1}\hbar)} = \sqrt{-(c\hbar)}$$

Due to the curious mathematical properties of the i number, the same result is obtained when rotating the electric charge quantum clockwise, 90° —$c^{-1}(i)$—, in intervalic dimensions:

$$\Phi_I = c^{-1} q_I = c^{-1}\sqrt{-(c^{-1}\hbar)} = \sqrt{-(c^{-3}\hbar)}$$

Since this value is not in the root scale, we shall rotate it ±360° in the Intervalic Space —multiplying it by the dimensionless scalizator $c^{\pm 4}$ (1)— in order to obtain it:

$$\Phi_I = \sqrt{-(c^{-3}\hbar)} \approx \sqrt{-(c^{5}\hbar)} \approx \sqrt{-(c\hbar)}$$

It is obvious that the relation between electric and magnetic intervalic quanta is the speed of light (or the equivalent energy of the intervalic mass):

$$\Phi_I / q_I = \sqrt{(c\hbar)}/\sqrt{(c^{-1}\hbar)} = c$$

Besides, its product makes the intervalic length, $\hbar$:

$$\Phi_I \cdot q_I = \sqrt{(c\hbar)}/\sqrt{(c^{-1}\hbar)} = \hbar$$

This equation is very similar to that postulated by P.A.M. Dirac for the magnetic monopole:

$$e\, m = n\hbar$$

where m is the charge of the magnetic monopole and n an integer number.

The predicted value for the intervalic magnetic charge is, though, very different to that postulated by Dirac. The relation between the intervalic magnetic charge —the intervalic magnetic monopole— and the electron charge is:

$$\Phi_I / e = 1.109783 \cdot 10^6$$

INTERVALIC TEMPERATURE
$\Theta_I = c\, k_B^{-1} = 2.1713738 \cdot 10^{31}\,(K)$
Intervalic dimension: (i^{-1})

As discussed in other site, it is a remarkable fact that the temperature has the same dimensions than the velenergy —velocity-energy— when written in intervalic dimensions. Moreover, Boltzmann constant is *dimensionless* when written in intervalic dimensions, which is a very important fact for the reliability of this constant. Indeed, it is hoped that all reliable fundamental constants of Nature must be dimensionless in intervalic dimensions, since they are simple geometrical constants of proportionality in the Intervalic Dimensional Space. In this sense, it is absolutely meaningful that all fundamental proportionality constants are really dimensionless when written in intervalic dimensions (with the notable exception of the Newton gravitational constant, which has been discussed in other site).

If we write the known classical formula for the conversion between temperature and energy, we can substitute the energy value by its corresponding quantum of energy in the Intervalic Space, that is to say, c (i^{-1}), and directly is obtained the intervalic temperature, Θ_I:

$$T = E / k_B$$
$$\Theta_I = c / k_B$$

The meaning of this geometric limit is similar to the speed of light: it is the maximum temperature allowed in the Universe for any kind of matter or energy, as like as the speed of light is the maximum velocity allowed in a similar way.

For those who feel the deduction of the intervalic temperature too much easy, they can make the following deduction using the intervalic frequorce, $\varphi_I = ch^{-1}$. This frequorce is likewise the highest frequorce allowed in the Universe. Indeed, this is a geometrical limit that can't never be reached, and there are other limits for the gravitational frequorce and also for the electromagnetic frequorce that are both smaller than the value of the intervalic frequorce. These limits are geometrical and meaningfully related.

$$\Theta_I = E\, k_B^{-1} = \hbar\, \varphi_I\, k_B^{-1} = \hbar\, c\, \hbar^{-1}\, k_B^{-1} = c\, k_B^{-1}$$

Of course, to be fully satisfied, we would perhaps be able to get a *geometrical definition* of the Boltzmann constant inside the Intervalic Dimensional Space. In that case we would have a *non arbitrary* formula —unlike the classic one— to convert velocity-energy in its equivalent temperature and vice versa.

It is hard to believe that modern Physics has not been yet aware of the logical impossibility of the existence of unlimited values for the temperature. As Richard Feynman already commented, both energy and temperature can be finally understood in terms of *movement*. But if the movement of any massive body in the Universe is limited by the speed of light, how is possible that its temperature was not limited likewise by a related limit? So, in the mere exposition of the first intervalic quanta which arise immediately from the intervalic dimensions, we already have got a new important geometric limit of Nature: the intervalic temperature, Θ_I, whose behaviour and geometric features are very similar to those of the speed of light.

INTERVALIC QUANTIZATION OR LIMITATION OF PHYSICAL QUANTITIES

The principal quanta and limits described till now are not the unique in the intervalic system of physical quantities and units. Indeed, all physical quantities are quantized because the dimensions that appear in the equation of dimensions of all of them are, precisely, combinations of (L) and (*i*), and both dimensional bases are intervalically quantized by ℏ and c. We could say that the intervalic quantization or limitation is transmitted like a domino effect among physical quantities. No one physical quantity can leave free of quantization or limitation because all of them are dimensionally interconnected and geometrically derived from the same fundamental constants: the quantum ℏ and the limit c.

If we think about it, as philosophers in Ancient Greek already did, there *must be* a limit for the division of Nature. The contrary should end in absurdity. Physics has used —and abused— of the mathematical concept of infinite since it was discovered in the Renaissance. As a result, the magnitudes of physical quantities was erroneously conceived as a mathematical continuum with no ends or limits. But this classic conception has proved to be misleading repeatedly and without doubt, and it is not easy for anybody to understand why it still lies in the unconscious minds of Physics.

In this view, the intervalic quantization or limitation of physical quantities is a logical premise which is totally necessary for the internal consistence in any explanation of the physical world. This conception offers us a new view of the physical quantities like a well organised and finite set, composed by 40 physical quantities —the *intervalic group*—, whose elements are interrelated, and where logical laws are applied in a systematic way, in every case and not only in certain cases for some chosen or privileged physical quantities, like occurs in the traditional system. Moreover, not only they are interrelated, but all of them are *generated* from geometrical transformations of the dimensional base, say (L), in the Intervalic Dimensional Space. These intervalic *transformations* are realized by successive algebric rotations mediated by (i). But the intervalic dimensions of ℏ and c are respectively (L) and (i^{-1}), that is to say, just the dimension of the *generator* and the *transformer* in the dimensional geometry of the Intervalic Dimensional Space. Thus, ℏ and c generates dimensionally all physical quantities, and since they are quantized or limited, all physical quantities must be quantized or limited too.

From systematic definitions (see tables) of intervalic quanta and intervalic limits it is evident that there are geometrical equivalences among them, which are mediated by the transformer in the Intervalic Space, the speed of light —c (i^{-1}) or c^{-1} (i)—, because from now mathematical operations between dimensions are fully algebraically consistent, and therefore, all mathematical and geometrical properties of the i number, etc., must be valid when they appear in the intervalic dimensions of physical quantities.

The wholeness and elegance of that intervalic conception is far from the total absence of an evolving and global organization on physical quantities that poorly shows the rest of known dimensional systems up to date. In the following tables is shown the intervalic quanta and the intervalic limits of all physical quantities of the intervalic group.

GEOMETRIC EQUIVALENCES AMONG INTERVALIC QUANTA AND LIMITS

> We may in fact regard [geometry] as the most ancient branch of Physics.
>
> ALBERT EINSTEIN
> *Sidelights on Relativity*

From the above definitions of intervalic quanta and limits it is evident that there are a set of *equivalences* among them, and that these equivalences are mediated by the speed of light, c. And since the i number is involved in the intervalic dimensions of the speed of light, c (i^{-1}) — or $c^{-1}(i)$ —, those equivalences have a *geometrical* representation as rotations of physical quantities. At first sight, we can see straightforwardly the following geometric equivalences:

$$m_I = c^{-1} p_I$$
$$E_I = c\, p_I$$
$$m_I = c^{\pm 2} E_I$$
$$E_I = c^{\pm 2} m_I$$
$$t_I = c^{-1} l_I$$
$$a_I = c^{\pm 2} k_I$$
$$W_I = c^{\pm 2} k_I$$
$$a_I = c\, \varphi_I$$
$$W_I = c\, \varphi_I$$
$$\varphi_I = c\, k_I$$
$$i_I = c^{-1} \eta_I$$
$$\rho_I = c^{-1} f_I$$
$$u_I = c^{\pm 2} \rho_I$$
$$\theta_I = c^{-1} q_I$$

$$\Phi_I = c^{-1} q_I$$
$$I_I = c^{-1} \Phi^{-1}_I$$
$$V_I = c^{-1} \Phi^{-1}_I$$
$$H_I = c^{-1} P_I$$

All the above relations among intervalic quanta and limits are really simple combinations between the two fundamental constants of the Intervalic Dimensional Space, c and ℏ. Because of it, there is clear that nobody can have any doubt about their strong logical consistency. However, if we make a logical *induction* on those geometrical equivalences, we will obtain some surprising results. Actually, we only have to substitute in those relations the *intervalic quanta and limits* by its corresponding *physical quantities*. The logical step from the first geometrical relation to the second is incontrovertible, and it is hard (or probably impossible) to accept the first equivalences without doing the same with the second ones.

INTERVALIC QUANTA AND INTERVALIC LIMITS OF PHYSICAL QUANTITIES

Representation	Intervalic dimension	Physical quantity	Definition	Geometrical rank	SI value	SIU value
L^0	1	Permitivity	$\varepsilon_I = h^{-1} \hbar$	subatomic limit	1 (F m^{-1})	1 (1)
	1	Momentum (momentum-inertia)	$p_I = c^{-1} c$	subatomic limit	1 (kg m s^{-1})	1 (1)
	1	Boltzmann constant	k_B	-	-	-
	1	Entropy	S	-	-	-
	-1	Permeability	$\mu_I = c^{\pm 2}$	conversion factor	$1.112650 \cdot 10^{-17}$ (H m^{-1})	1 (-1)
	-1	Gravitational potential	Φ	-	-	-
	-1	Antimomentum	$-p_I = c^{\pm 2}$	conversion factor	$8.987552 \cdot 10^{16}$ (-kg m s^{-1})	1 (-1)
L^0	i	Mass (invervelocity-mass)	$m_I = c^{-1}$	subatomic limit	$3.3356409 \cdot 10^{-9}$ (kg)	1 (i)
	i	Invervelocity (invervelocity-mass)	$v^{-1}{}_I = c^{-1}$	subatomic limit	$3.3356409 \cdot 10^{-9}$ (s m^{-1})	1 (i)
	i^{-1}	Velocity (velocity-energy)	$v_I = c$	absolute limit	$2.9979246 \cdot 10^{8}$ (m s^{-1})	1 (i^{-1})
	i^{-1}	Energy (velocity-energy)	$E_I = c$	subatomic limit	$2.9979246 \cdot 10^{8}$ (J) $1.8711569 \cdot 10^{21}$ (MeV)	1 (i^{-1})
	i^{-1}	Temperature	$\Theta_I = c\, k_B^{-1}$	absolute limit	$2.1713738 \cdot 10^{31}$ (K)	1 k_B^{-1} (i^{-1})

Representation	Intervalic dimension	Physical quantity	Definition	Geometrical rank	SI value	SIU value
⬡ L¹	L	Length (real space)	$l_I = \hbar$	quantum	$1.055636 \cdot 10^{-34}$ (m)	1 (L)
		Action	$S_I = \hbar$	quantum	$1.055636 \cdot 10^{-34}$ (J s)	1 (L)
		Capacitance	$C_I = \hbar$	quantum	$1.055636 \cdot 10^{-34}$ (F)	1 (L)
⬡ L⁻¹	-L	Antilength	$-l_I = c^{\pm 2}\hbar$	conversion factor	$9.487583 \cdot 10^{-18}$ (-m)	1 (-L)
		Inductance	$L_I = c^{\pm 2}\hbar$	quantum	$9.487583 \cdot 10^{-18}$ (H)	1 (-L)
⬡ L⁻¹	iL	Time (imaginary space)	$t_I = c^{-1}\hbar$	quantum	$3.5212226 \cdot 10^{-43}$ (s)	1 (iL)
	$i^{-1}L$	Antitime	$-t_I = c\hbar$	conversion factor	$3.164717 \cdot 10^{-26}$ (-s)	1 ($i^{-1}L$)
⬡ L⁻¹	L^{-1}	Wavevector	$k_I = \hbar^{-1}$	absolute limit	$9.472962 \cdot 10^{33}$ (m⁻¹)	1 (L⁻¹)
	-L⁻¹	Acceleration (acceleration-power)	$a_I = c^{\pm 2}\hbar^{-1}$	absolute limit	$8.513874 \cdot 10^{50}$ (m s⁻²)	1 (-L⁻¹)
		Gravitational field	$g_I = c^{\pm 2}\hbar^{-1}$	absolute limit	$8.513874 \cdot 10^{50}$ (m s⁻²)	1 (-L⁻¹)
		Power (acceleration-power)	$W_I = c^{\pm 2}\hbar^{-1}$	absolute limit	$8.513874 \cdot 10^{50}$ (W)	1 (-L⁻¹)
⬡ L⁻¹	iL^{-1}	Antifrequorce	$-\varphi_I = c^{-1}\hbar^{-1}$	conversion factor	$3.163026 \cdot 10^{25}$ (-s⁻¹)	1 (iL^{-1})
		Linear Perdensity	$\Delta^1_I = c^{-1}\hbar^{-1}$	absolute limit	$3.163026 \cdot 10^{25}$ (kg m⁻¹)	1 (iL^{-1})
	$i^{-1}L^{-1}$	Frequorce (frequency-force)	$\varphi_I = c\hbar^{-1}$	absolute limit	$2.8399227 \cdot 10^{42}$ (s⁻¹)	1 ($i^{-1}L^{-1}$)
		Frequency	$\nu_I = c\hbar^{-1}$	absolute limit	$2.8399227 \cdot 10^{42}$ (Hz)	1 ($i^{-1}L^{-1}$)
		Force	$F_I = c\hbar^{-1}$	absolute limit	$2.8399227 \cdot 10^{42}$ (N)	1 ($i^{-1}L^{-1}$)
		Conductivity	$\sigma_I = c\hbar^{-1}$	absolute limit	$2.8399227 \cdot 10^{42}$ (S m⁻¹)	1 ($i^{-1}L^{-1}$)
⬡ L²	L^2	Area	$S_I = \hbar^2$	quantum	$1.114367 \cdot 10^{-68}$ (m²)	1 (L²)
	$-L^2$	Antiarea	$-S_I = c^{\pm 2}\hbar^2$	conversion factor	$1.001544 \cdot 10^{-51}$ (-m²)	1 (-L²)

INTERVALIC THEORY:
The Intervalic Structures of Subatomic Particles and the Last Foundations of Physics

Representation	Intervalic dimension	Physical quantity	Definition	Geometrical rank	SI value	SIU value
L^2	iL^2	Inertia area momentum	$I_I^2 = c^{-1}\hbar^2$	quantum	$3.717129 \cdot 10^{-77}$ (kg m^2)	$1\ (iL^2)$
L^2	$i^{-1}L^2$	?	$c\hbar^2$	–	$3.340789 \cdot 10^{-60}$ (J m^2)	$1\ (i^{-1}L^2)$
L^{-2}	L^{-2}	Viscosity (dynamic)	$\eta_I = \hbar^{-2}$	absolute limit	$8.973701 \cdot 10^{67}$ (Pa s)	$1\ (L^{-2})$
L^{-2}	$-L^{-2}$	Area Power	$W_I^2 = c^{\pm 2}\hbar^{-2}$	absolute limit	$8.065161 \cdot 10^{84}$ (-m^{-2})	$1\ (-L^{-2})$
L^{-2}	iL^{-2}	Area Perdensity	$\Delta_I^2 = c^{-1}\hbar^{-2}$	absolute limit	$2.993305 \cdot 10^{59}$ (kg m^{-2})	$1\ (iL^{-2})$
L^{-2}	iL^{-2}	Inflexion	$i_I = c^{-1}\hbar^{-2}$	absolute limit	$2.993305 \cdot 10^{59}$ (m s^{-3})	$1\ (iL^{-2})$
L^{-2}	$i^{-1}L^{-2}$	Surface Tension	$\sigma_I = c\hbar^{-2}$	absolute limit	$2.690248 \cdot 10^{76}$ (N m^{-1})	$1\ (i^{-1}L^{-2})$
L^3	L^3	Volume	$V_I = \hbar^3$	quantum	$1.176366 \cdot 10^{-102}$ (m^3)	$1\ (L^3)$
L^3	$-L^3$	Antivolume	$-V_I = c^{\pm 2}\hbar^3$	conversion factor	$1.057265 \cdot 10^{-85}$ (-m^3)	$1\ (-L^3)$
L^3	iL^3	Inertia volume momentum	$I_I^3 = c^{-1}\hbar^3$	quantum	$3.923935 \cdot 10^{-111}$ (kg m^3)	$1\ (iL^3)$
L^3	$i^{-1}L^3$	Fermi constant physical quantity	$c\hbar^3$	quantum	$3.526657 \cdot 10^{-94}$ (kg m^3)	$1\ (i^{-1}L^3)$
L^{-3}	L^{-3}	Fluctuation	$f_I = \hbar^{-3}$	absolute limit	$8.500756 \cdot 10^{101}$ (m s^{-4})	$1\ (L^{-3})$
L^{-3}	$-L^{-3}$	Voluminic Power	$W_I^3 = c^{\pm 2}\hbar^{-3}$	absolute limit	$7.640098 \cdot 10^{117}$ (-m^{-3})	$1\ (-L^{-3})$
L^{-3}	$-L^{-3}$	Irradiance	$E_{eI} = c^{\pm 2}\hbar^{-3}$	absolute limit	$7.640098 \cdot 10^{117}$ (W m^{-2})	$1\ (-L^{-3})$
L^{-3}	iL^{-3}	Density	$\rho_I = c^{-1}\hbar^{-3}$	absolute limit	$2.835547 \cdot 10^{93}$ (kg m^{-3})	$1\ (iL^{-3})$
L^{-3}	iL^{-3}	Voluminic Perdensity	$\Delta_I^3 = c^{-1}\hbar^{-3}$	absolute limit	$2.835547 \cdot 10^{93}$ (kg m^{-3})	$1\ (iL^{-3})$
L^{-3}	$i^{-1}L^{-3}$	Pressure	$P_I = c\hbar^{-3}$	absolute limit	$2.548463 \cdot 10^{110}$ (Pa)	$1\ (i^{-1}L^{-3})$
L^{-3}	$i^{-1}L^{-3}$	Energy-tension density	$u_I = c\hbar^{-3}$	absolute limit	$2.548463 \cdot 10^{110}$ (J m^{-3})	$1\ (i^{-1}L^{-3})$

Representation	Intervalic dimension	Physical quantity	Definition	Geometrical rank	SI value	SIU value
L$^{1/2}$	$i^{1/2}$L$^{1/2}$	Magnetic charge	$\theta_I = \sqrt{-(c\hbar)}$	subatomic limit	$1.778965 \cdot 10^{-13}$ (Wb)	$1\,(i^{1/2}$L$^{1/2})$
	$\bar{i}^{1/2}$L$^{1/2}$	Magnetic flux	$\Phi_I = \sqrt{-(c\hbar)}$	subatomic limit	$1.778965 \cdot 10^{-13}$ (Wb)	$1\,(\bar{i}^{1/2}$L$^{1/2})$
L$^{-1/2}$	$i^{1/2}$L$^{-1/2}$	Electric charge	$q_I = \sqrt{-(c^{-1}\hbar)}$	quantum	$5.933989 \cdot 10^{-22}$ (C)	$1\,(i^{1/2}$L$^{-1/2})$
		Current	$I_I = \sqrt{-(c\hbar^{-1})}$	absolute limit	$1.685207 \cdot 10^{21}$ (A)	$1\,(i^{1/2}$L$^{-1/2})$
		Electric potencial	$V_I = \sqrt{-(c\hbar^{-1})}$	absolute limit	$1.685207 \cdot 10^{21}$ (V)	$1\,(i^{1/2}$L$^{-1/2})$
		Magnetic vector potential	$A_I = \sqrt{-(c\hbar^{-1})}$	absolute limit	$1.685207 \cdot 10^{21}$ (Wb m^{-1})	$1\,(i^{1/2}$L$^{-1/2})$
	$\bar{i}^{1/2}$L$^{-1/2}$	Magnetic inverflux	$\Phi^{-1}_I = \sqrt{-(c^{-1}\hbar^{-1})}$	subatomic limit	$5.621246 \cdot 10^{12}$ (Wb^{-1})	$1\,(\bar{i}^{1/2}$L$^{-1/2})$
L$^{3/2}$	$i^{1/2}$L$^{3/2}$	Bohr magneton phys. quantity	$\mu_{BI} = \sqrt{-(c\hbar^3)}$	quantum	$1.877940 \cdot 10^{-47}$ (J T^{-1})	$1\,(i^{1/2}$L$^{3/2})$
	$\bar{i}^{1/2}$L$^{3/2}$	?	$\sqrt{-(c^{-1}\hbar^3)}$	–	$6.264133 \cdot 10^{-56}$ (T^{-1})	$1\,(\bar{i}^{1/2}$L$^{3/2})$
L$^{-3/2}$	$i^{1/2}$L$^{-3/2}$	Electric field strength	$\epsilon_I = \sqrt{-(c\hbar^{-3})}$	absolute limit	$1.596390 \cdot 10^{55}$ (V m^{-1})	$1\,(i^{1/2}$L$^{-3/2})$
		Magnetic field strength	$H_I = \sqrt{-(c\hbar^{-3})}$	absolute limit	$1.596390 \cdot 10^{55}$ (A m^{-1})	$1\,(i^{1/2}$L$^{-3/2})$
		Magnetic flux density	$B_I = \sqrt{-(c\hbar^{-3})}$	absolute limit	$1.596390 \cdot 10^{55}$ (T)	$1\,(i^{1/2}$L$^{-3/2})$
	$\bar{i}^{1/2}$L$^{-3/2}$	Electric polarisation	$P_I = \sqrt{-(c^{-1}\hbar^{-3})}$	absolute limit	$5.324984 \cdot 10^{46}$ (C m^{-2})	$1\,(\bar{i}^{1/2}$L$^{-3/2})$
L$^{5/2}$	$i^{1/2}$L$^{5/2}$	?	$\sqrt{-(c\hbar^5)}$	–	$5.858725 \cdot 10^{-39}$ ( )	$1\,(i^{1/2}$L$^{5/2})$
	$\bar{i}^{1/2}$L$^{5/2}$	?	$\sqrt{-(c^{-1}\hbar^5)}$	–	$1.954260 \cdot 10^{-47}$ (C^{-1} m^3)	$1\,(\bar{i}^{1/2}$L$^{5/2})$
L$^{-5/2}$	$i^{1/2}$L$^{-5/2}$	Charge density	$\rho_I = \sqrt{-(c\hbar^{-5})}$	absolute limit	$5.117026 \cdot 10^{46}$ (C m^{-3})	$1\,(i^{1/2}$L$^{-5/2})$
		Current density	$J_I = \sqrt{-(c\hbar^{-5})}$	absolute limit	$5.117026 \cdot 10^{46}$ (A m^{-2})	$1\,(i^{1/2}$L$^{-5/2})$
	$\bar{i}^{1/2}$L$^{-5/2}$	?	$\sqrt{-(c^{-1}\hbar^{-5})}$	–	$1.706856 \cdot 10^{38}$ ( )	$1\,(\bar{i}^{1/2}$L$^{-5/2})$

Chapter 3

INTERVALIC SYSTEM OF UNITS
The Nature's System of Units

> I do not like it when it can be done this way or that way. It should be: This way or not at all.
>
> ALBERT EINSTEIN
> *Abraham Pais, A Tale of Two Continents*

The intervalic system of dimensions determines unavoidably the existence of the intervalic group of physical quantities. Since all physical quantities of Nature are derived and defined as *combinations* of h and c in intervalic dimensions, they are necessarily *quantized* or *limited* by its own dimensional geometry, and a new set of intervalic quanta and intervalic limits is yielded. The magnitudes and dimensions of these quanta and limits determines logically the allowed values inside a geometrical rank for each physical quantity, being at the same time the most simple, elegant and useful theoretical values conceivable up to date for a reliable and truly universal system of units, which is called the intervalic system of units. There are also treated the underlying geometric meaning of the uncertainty principle and the question about how can affect a change in the values of fundamental constants to the physical world.

LOGICAL AND PHYSICAL FOUNDATIONS OF THE INTERVALIC UNITS

When using intervalic dimensions (L = L, T = iL, M = T/L = i), with $i = \sqrt{-1}$, a new fundamental set of *intervalic quanta* and *intervalic limits* is necessarily yielded from the very own *dimensional basis* of the Intervalic System of Dimensions and Units. We already have pointed out in the preceding chapter (Intervalic Quanta, Intervalic Limits) that this logical fact can be considered as unexpected and almost magic, since no quanta nor any other physic phenomena should be obtained from a mere change in the dimensional system used.

It is clear that if a *quantum* is, by definition, the smallest magnitude allowed for a physical quantity, the most simple magnitude for that quantum is the unity: 1.

On the other hand, the magnitude for a intervalic *limit* is not determinate at first sight like the magnitude for the intervalic quanta. Actually, we need a powerful constraint that determinates logically and necessarily the most logic and simple magnitude for those limits, since in a faultless theory we can't determinate them *by hand*.

Finally, the definition of that system of physical quantities and units must be related only to the last fundamental constants of Nature, that is to say, c and ℏ. Therefore, the units and dimensions of all physical quantities must be defined only in terms of the units and dimensions of these two constants. Perhaps it is difficult to conceive a singular system of units with so extreme simplicity which verifies these powerful logic postulates. But we have found it.

In resume, we have to achieve to join a dimensional system, a set of physical quantities, a set of quanta and limits, and a system of units which can be based, all of them, solely and exclusively on the *single* dimensions and *single* magnitudes of the last fundamental constants of Nature: c and ℏ.

The intervalic system of physical quantities is a system one-proper-dimension based, which only needs to fixing the value of the unit of real space, (L), to be geometrically determinate, since the unit of imaginary space depends on the magnitude of the *i* number, $\sqrt{-1}$, and it is intended to be 1 for obvious geometric reasons. (We ever will write dimensions and physic units between brackets).

Now, the intervalic dimensions of ℏ and c are, just and respectively, (L) and (i^{-1}). This means precisely that all existing physical quantities expressed in intervalic dimensions are *combinations* of these two single base dimensions of the two last fundamental constants of Nature, ℏ and c.

(Please note that if we would choose (i) instead of (i^{-1}) as the intervalic dimension of the speed of light, we will obtain the same dimensional system).

Unlike the other dimensional systems known up to date, the own geometry of the intervalic system of dimensions determinates, surprisingly and necessarily, a *singular* magnitude for *each* physical quantity: they are the *intervalic quanta* and the *intervalic limits*, which we have described previously. If all physical quantities are combinations of ℏ and c —which are a quantum and a limit respectively—, it is logical that all physical quantities must be geometrically quantized or limited too.

Thus, the intervalic quanta and the intervalic limits provide us with a reliable reference for determinate an universal system of units, since the values of c and ℏ are valid in all the Universe and the intervalic quanta and limits are composed only by these two fundamental constants of nature.

Nothing could be more natural than the magnitude of ℏ as the unit of length —or 'real space', (L), in intervalic dimensions—. As Lancelot L. Whyte pointed out usefully in his books **Critique of Physics** (W. W. Norton, New York, 1931) and **Archimedes or The Future of Physics** (E. P. Dutton, New York, 1928), today lamentably forgotten, the essential and truly function of ℏ in all physical equations is not to be an 'action' but to introduce a unit of *length*. It is still more logical in any dimensional system one-proper-dimension-based, whatsoever it is. And the only known dimensional system that verifies those apparently inevitable postulates is just the intervalic one, since any other dimensional system which verifies these postulates will be equivalent to the intervalic system.

Finally, the intuitive meaning of ℏ is now very simple, since it is the transversal path of electromagnetic waves according to one of the various interpretations of Planck's equation of energy when it is written in intervalic dimensions.

Regarding the speed of light, it is not a quantum but a limit, and therefore it is needed a powerful geometrical constraint that sets univocally a singular magnitude for c. In a faultless theory we can't determine it *by hand*, as String Theory have done cheerfully. As we have described in other site, the fact that the intervalic dimension of c is (i^{-1}), involves a *geometrical* relation between the speed of light and all the rest of physical quantities. Since physical quantities are geometrically invariant under rotations of ±360° in the Intervalic Space —that is to say, under multiplications by the dimensionless *transformer* $c^{\pm 4}$—, the magnitude of the speed of light must be 1 in order to conserve the invariance of physical magnitudes under rotations of ±360° in all physical cycles of successive turns.

This means, unexpectedly, that all system of units with a magnitude for the speed of light different from the unit, c ≠ 1, are inevitably *not-singular* systems. Those systems have partial symmetry —*left* or *right*— under rotations in the Intervalic Dimensional Space. But when checking all possible groups of rotations in the Intervalic Dimensional Space it is clear that there is no any rule nor any trace to think that our dimensions and physical quantities have a partial symmetry. Inside those systems of units, rotations of all physical quantities are unconsciously broken for nothing and they only run in one sense, which is not simple, not elegant, not necessary, and finally, absurd. That is to say, their dimensions and physical quantities have only an arbitrary *half* symmetry, which verifies that a turn by +360° —$c^{-4}(1)$— is different from a turn by -360° —$c^{+4}(1)$—, and a rotation by +180° —$c^{-2}(-1)$— is not equal to a rotation by -180° —$c^{+2}(-1)$— (!).

We can hardly understand the radical non sense of that half symmetry, since our traditional system of dimensions is, unfortunately, one of those poor and primitive systems. But when all Physics equations and forces of Nature are transformed into a pure and rich *geometry* as the Intervalic Theory does, we understand that this arbitrary half symmetry showed by not-singular systems of units is a serious lack typical of those primitive systems which is due to a wrong choice in the magnitude and dimension of the speed of light. Starting from here, algebra laws makes its work and some backward creatures can think that their equations do describe correctly the foundations of physics phenomena, but they don't because they are not deduced starting from the underlying fundamental geometry of Nature.

Hence, we are compelled to establish 1 as a *singular* magnitude for the speed of light in intervalic dimensions, since this is the only value that preserves the symmetry of dimensions and physical quantities in the Intervalic Dimensional Space.

In resume, if a dimensional system is physically reliable and fully consistent its measurement units should be determined by the last fundamental constants of **Nature**. This means that the physical quantities that compose the dimensional basis of the system should coincide with the two fundamental constants, c and ℏ. Neither of the known dimensional systems verifies this powerful logical constraint: the definitions of the units of length, time, mass, etc. in the traditional system is almost grotesque, and the same or worst is valid for the degenerate —formerly named "geometrized"— units.

A dimensional system geometrically conformed as the intervalic one, has an additional advantage on the other systems because to the geometrical tools at its disposition. In this case, we can see that all the intervalic physical quantities of the intervalic group can be represented on a sort of Argand Space. In this space we have as *real* axes the dimensionless (± 1) or the dimensionful physical quantities ($\pm L$ raised to any power), and as *imaginary* axes the results derived of their successive rotations by the *i* number.

For the univocal determination of the measurement units of that dimensional geometry we only need to establish the value of the unique dimensional base (L), which is placed along the real axes, and the value of the *conversion constant* between real and imaginary axes, that is to say, the associated physical magnitude for (i). After a look to the physical quantities of the intervalic group we find that the intervalic dimensions of these two values coincide precisely with the intervalic dimensions of the two fundamental constants of nature, c and ℏ. But this *geometrical result* just satisfies the powerful logical constraint described below about the physical quantities that compose the dimensional basis, which must coincide with the two fundamental constants of nature, c and ℏ.

It is interesting to point that we have reached to join the *two* fundamental constants of Nature with the desired *one*-proper-dimension-based system, and it has been possible due to the magic introduction of the imaginary numbers in its dimensional basis —along with its algebraic and geometrical properties—. Thus the imaginary numbers not only appear in the most powerful equations of Physics, but they dwell in the very heart of Physics: in the dimensional basis. We can say that the vast majority of the principal relations and symmetries of Physics can not be yielded without the imaginary numbers, with whom there are composed the last foundations of Physics.

Please note the huge abysm that separates the *intervalic system of units*, SIU:

ℏ = 1 (L)
c = 1 (i^{-1})

with the trivial degenerate system of units. Although the magnitudes for c and ℏ are the same, their respective dimensions are completely different:

ℏ = 1 (1)
c = 1 (1)

The interpretation of physics world are totally different in both systems and therefore they yield necessarily different Physics. Besides, unlike degenerate units, the intervalic units do not need to be converted into SI units to be meaningful at any time of the calculus.

Of course, it is very important to understand that no faultless theory will be reached by the mankind meanwhile Physics will continue to being expressed into wrong systems of dimensions, physical quantities, quanta and units. Perhaps this is the most serious challenge Physics have to confront in modern times, but its future depends on the result of that test. To pass this test implies to reformulate all Physics corpus into intervalic dimensions, physical quantities, quanta and units, but until then we will not go in a Physics truly *universal*. After reading this book, it will be understood that the Intervalic Theory in Physics is the only physics knowledge that can be shared with hypothetical extraterrestrial intelligences for many logical reasons of universality, simplicity, elegance, etc. which we can't enumerate at this moment. I believe that a civilization can considerate itself to be scientifically *advanced* when the Intervalic Theory is discovered because it presuppose a lot of physical achievements — necessarily reached by the theory— and because it opens the doors to a reliable knowledge of the Universe.

ALLOWED RANKS OF PHYSICAL QUANTITIES

Magnitudes for all physical quantities are *geometrically limited* by the Intervalic Theory: each physical quantity has a low height —a intervalic quantum— *or* a high height —a intervalic limit—. Magnitudes of physical quantities are not allowed outside these heights.

In intervalic units, the allowed ranks are the following, being 'x' any magnitude:

Quantum: $x \geq 1$
Limit: $0 \leq x \leq 1$

Measurable magnitudes of any physical quantity which has a *quantum* must be a multiple (integer) of that quantum, since they are *quantized* physical quantities.

Limits can be *subatomic* or *absolute*. In the first case a magnitude only can be greater than its subatomic limit if it is a sum of various subatomic magnitudes, like occurs in macroscopic measurements of some physical quantities at human scale.

Seeing the allowed ranks, it shall be noted that proper *negative magnitudes* (not due to the use of relative coordinates) are not defined in intervalic units. Only *units* can be positive *or* negative (or positive *and* negative, like the electromagnetic units) depending on its intervalic dimension. If we have a negative magnitude, its sign must be passed to its dimension, and then we obtain a different physical quantity. This detail is an unknown and surprising logical property of the intervalic dimensions and units: their signs can *operate algebraically* with the signs of its corresponding magnitude, as we have explained in other site.

HOW CAN AFFECT TO THE PHYSICAL WORLD A CHANGE IN THE VALUES OF FUNDAMENTAL CONSTANTS

Finally, a late comment on the transcendence of the values of the physical constants. It is heard that if the value of only one constant would be slightly different the Universe will not be able to exist. Of course, this would be true if it is possible, but it is not within the intervalic system of dimensions. The underlying conception for that fantastic assumption is the traditional one, which is absolutely misleading, since it conceives a lot of *independent* physical constants and physical quantities. But as we have seen, there are only two reliable fundamental constants, c and $\hbar$ — being this an *affirmation*, not an *assumption* like in String Theory, because their intervalic dimensions just coincide with the intervalic dimensional basis—, and all the rest constants and physical quantities are combinations of these two.

SPECIFIC VALUES OF |c| AND |ℏ| IN NATURAL UNITS

The Avogadro number, N_A, is referred to the SI mass unit of gram. Its value could be meaningless for the intervalic dimensional analysis because the gram is not a natural unit. However if the Avogadro number could be referred to the *intervalic quantum of mass*, $\mathbf{m_I} = c^{-1}$, then such new number —named the *Avogadro intervalic number*, N_{IA}— should be directly related to the number of intervalic strings assembled at the beginning of the primordial Universe, $n(S)$, and hereinafter to the speed of light (since starting from the energy of vacuum we have got the relation $n(S) = c^8$, which is valid exclusively in non-singular units, that is to say, with $c \neq 1$). This is the most logically simple and economic assumption that can be made, and in this way we have got the *Avogadro intervalic number*, N_{IA}:

$$N_{IA} / \mathbf{m_I} = N_A / 10^3$$
$$N_{IA} = c^{-1} \, 10^{-3} \, N_A = 2.008770344 \cdot 10^{12} \quad (1)$$

It is found by exclusion that the unique simple relation available between the Avogadro intervalic number and the number of intervalic strings assembled is: $N_{IA} = n(S)^{1/4} = |c^2|$. (It must be remembered that $c^{\pm 2}$ is the conversion factor *par excellence* in the Intervalic Dimensional Space, which yields the most important geometric equivalences). Thus the specific value of the speed of light in natural units, $|c|$, is:

$$|c| = (1/10) \, N_A^{1/3} = 8.444689071 \cdot 10^7 \, (i^{-1})$$

As natural units verify the following geometric relation: $c^{\pm 4} \hbar = 1$, it may also be derived the specific value of the *intervalic length* in natural units, $|\hbar|$:

$$|\hbar| = 10^4 \, N_A^{-4/3} = c^{-4} = 1.966370477 \cdot 10^{-32} \, (L)$$

These results define specific values of non-singular units of length and time:

$$|l| = 0.005368450717 \, (m)$$
$$|t| = 0.001512209393 \, (s)$$

Therefore, if hypothetically the values of c or ℏ would be fixed in a different magnitude between the Big Crunch and the Big Bang, all the other constants and physical quantities *composed* by c or ℏ would change in the same way. Therefore, a Universe with a bigger value of c would be a Universe *at scale* where all *imaginary* physical quantities (i.e., time) would be greater according to its powers of c which appears in its intervalic dimension. In the same way, a Universe with a bigger value of ℏ would be likewise a Universe *at scale* where all *real* physical quantities (i.e., length) would be greater according to its powers of ℏ which appears in its intervalic dimension. In this view the values acquired by c and ℏ at the beginning of the Universe can be chosen at random. Indeed, they are irrelevant, since the two fundamental constants are incommensurable between them, and their values yield entirely similar models of Universes anyway, where the only difference would be the scale in *real* and/or *imaginary* physical quantities. Certainly, there has no sense to inquiry about those scales, since they could be only compared between different Universes.

Of course, we can make anthropic comments, such as, i.e. regarding the size of the Universe and the human scale, the value of the speed of light is ridiculously small. We could imagine another Universe with a much greater c, where time and all imaginary physical quantities will run more quickly with respect to the present Universe. But this lucubration has little sense because there is no manner anyway to distinguish between both models. Really, we could live in a Universe with *any* value of c and ℏ and we probably could be not capable to appreciate the difference, since dimensionless *ratios* between physical quantities would be exactly the same in all cases, but only in intervalic dimensions! because *real* and *imaginary* dimensions *vanish separately* to leave dimensionless ratios. The dimensional magic arises in intervalic dimensions because we have a *dimensional duality which is just connected with the two fundamental constants*. It does not occur when using physical magnitudes in other systems of dimensions, which lack of that duality and connection, and therefore the vanishing of the dimensional basis has no relation with the dimensions of c and ℏ. Besides, traditional experimental values of constants lack of a geometric basis and their ratios are not defined with the mathematical exactness of a geometric theory.

Moreover, there is no way to determine properly the values of c and ℏ with respect other phenomena, since they are the only *independent* and *last* fundamental constants of nature, those that control the scale of the *real* space and the *imaginary* space in the intervalic geometry of Universe.

Please note the great difference between the intervalic system of dimensions which yields this quiet intervalic conception of paramount simplicity about the fundamental constants and the physical quantities of nature, where all of them are dependent and interrelated, and the clumsy and arranged-by-hand traditional conception of the Universe and the physical laws yielded by the other systems of dimensions, where all phenomena runs almost by an unexplained miracle on the exactness of fundamental constants and the smallest fluctuation on its values will lead inevitably to a universal crash and the ruin of Physics.

This fantastic scene is not allowed within intervalic dimensions, because all constants are derived as combinations of the two fundamental *geometrical* constants, c and ℏ. For example, it is heard that if the value of the elementary charge, e, would change a millionth part, the world could not exist. This is an ordinary deceit, since the intervalic charge, q_I, —and *a fortiori* the elementary charge— is a combination of the fundamental constants, c and ℏ, and therefore a supposed change in e involves necessarily similar changes in the values of c and/or ℏ:

$$e = 270\, q_I = 270\, \sqrt{-(c^{-1}\hbar)}$$

Of course, any change in c or ℏ implies correlative changes in *all* fundamental —and also in all not fundamental— constants, since all of them are derived from c and ℏ, as we have seen in the definitions of intervalic quanta and intervalic limits. To finish with the example, if the value of e would be, say, the double, the values of c or ℏ would *previously* be:

$c = 270^2 \hbar (2e)^{-2}$, if $\hbar$ remains unchanged, or

$\hbar = 270^{-2} c (2e)^2$, if c remains unchanged; and of course, there are possible any other gradation of values comprised between these two limits.

In any case, please note that the pristine relation $e = 270\ \mathbf{q_I}$ does not vary. This means that whatsoever it be the values of c and $\hbar$ supposedly set at the beginning of the Universe, the intervalic symmetries derived from the relation $e = 270\ \mathbf{q_I}$ stay mysteriously unaltered.

Hence, we can affirm that our model of Universe —or other models entirely similar to this one— can be yielded from *any* value of c and $\hbar$ in intervalic dimensions. *"God does not play dice".*

The intervalic dimensional structure of the Universe with its intervalic geometry —which generates all constants of nature— does not depend on the magnitude of the fundamental geometrical constants, c and $\hbar$, like the general properties of a rectangle does not depend on the size of its sides. As a naïve comparison, we can say that the physical quantities and the physical laws of our intervalic Universe come from the geometrical properties of a sort of "rectangle", whose sides c and $\hbar$ —whose magnitudes are dimensionally independent and incommensurable between them— can adopt any value without changing the general geometric properties of the "rectangle". Thus, the ancient question about the unexplained fine adjust in the constants of nature is not solved by an impossible answer, but by the vanishing of the problem, which is revealed as *irrelevant* by means of the intervalic dimensions.

A NEW ERA OF TRUTHFULLY *UNIVERSAL* KNOWLEDGE

As we are going to see along this book, IT opens at last a new era of reliable and truthfully *universal* knowledge in Physics and in Music, as IT does not include any concept or magnitude based on local units or frame of references because all physical quantities are dimensionally expressed as an unique and irreducible combination of the two fundamental constants of Nature: c and ℏ. Therefore, any arbitrariness is avoided in the choosing of a local system of dimensions since our intervalic system of dimensions is actually given by Nature as it is the unique dimensional system in which any physical quantity is expressed in terms of c and ℏ, the two unique fundamental constants. In other words, there is solved the following problem: to make a dimensional system in which the dimensions of all physical quantities are simple combination of the dimensions of c and ℏ. The problem has only one answer: dim c = $i^{\pm 1}$, dim ℏ = X, where X is some physical dimension. To interpret X as some kind of space, X = L, is immediate, and the choosing between c = i or c = i^{-1} is irrelevant as both give the same dimensional system: the intervalic system of dimensions (!). Of course, the setting of *singular* units (c = 1 and ℏ = 1) is the most simple assumption, but it is still irrelevant for the definition of the dimensional system, and we can work perfectly in intervalic dimensions with whatever not-singular units (c ≠ 1, ℏ ≠ 1). To conclude, we arrive to the discovery of the intervalic system of dimensions and units when we are able to allow ourselves to hear without prejudices what Nature is forever quietly whispering.

By this reasons IT in Physics could be communicated and understood across the Universe by any other advanced intelligence. And the same is valid to IT in Music. We can be sure that if an extraterrestrial life make advanced Physics and Music, both will be *intervalic* ones. Since the underlying symmetries of Nature can not be deduced by no means from any other dimensional system, but only from the intervalic system of dimensions, the knowledge of IT is a reliable clue of the scientific degree of development of any civilization. Moreover, the awareness of IT could serve to differentiate an advanced civilization from a primitive one (if this is the case, IT might not be discovered and understood yet by these little human beings, inasmuch as it is doubtful that we really can consider our bellicose civilization as an advanced one).

Surprisingly, the awareness of IT by any civilization can be communicated by the most simple device: it only has to show the knowledge of the geometric equation $e = 270\ \mathbf{q_I}$, which singular value is clearly 270. Therefore, any electromagnetic signal transmitting 270 pulses in any way can be considered as a secure clue for the knowledge of the Intervalic Theory. And to reinforce this message, the frequorce of the signal could be chosen as precisely one of the most important derived constants of Nature (after the two only fundamental ones, c and ℏ): G_I^{-1}, being G_I (iL) the intervalic constant of gravitation, which is derived from c and ℏ in the Intervalic Space. The meaning and geometric relations of G_I, whose magnitude is slightly different from the traditional Newtonian constant G_N, introduces some fascinating new subjects in Physics which can not be fully treated in this book but in other one devoted to Intervalic Cosmology.

By this way we can see once more the deep relation existing between Music, Physics and Mathematics. It was not by chance that the vast majority of the best physicists of the XXth. century were also musicians. The negligence of music only advise for a limited insight to understand the language of Nature, which is just one and the same both for Music and for Physics. As Don Campbell has described in his best-selling books on *The Mozart Effect,* the musical brain uses the two cerebral hemispheres, while the rational brain only uses the left hemisphere. But it is clear that for understanding Nature we need the whole brain. A lot of experiments have proved that what makes the difference between a creative brain and a non-creative one is just musical education.

I am absolutely sure that if Einstein or any other of the best physicists of the XXth. century would have received only a brief note describing the intervalic dimensional basis, any of them would have immediately comprehended the extraordinary importance and transcendence of the theoretical discovery and the implications which would be logical and necessarily derived from the intervalic system of physical quantities.

INTERVALS AND RATIOS

Long time ago physicists were aware that a truthful and universal Physics should perhaps be expressed in *dimensionless ratios* instead of *dimensionful magnitudes*. However, till now only a very few measurements have been able to be expressed in dimensionless ratios, such as velocity (v/c) or some ratios between subatomic particles. Has solved or not this deep problem the intervalic system of units?

The Intervalic Theory in Music has brought about a change in paradigm, as well as in our way of seeing or thinking about music; one no longer works with or thinks in terms of *notes* but in terms of metrical *intervals* instead. Notes relies on the frame of reference chosen, whilst intervals don't, as they are invariant. The organisation of the musical sound according to this new block yields an entirely new musical space: the intervalic space. Really, this is the origin of the term *intervalic* which denotes the theory.

History has demonstrated that the conception and the inner geometry of the physical space and of the musical space in every era are just the same. If we think a little about it, we reach the conclusion that it can't be in other way: mankind has a unique global thought along the epochs, and this unique thinking is expressed through different mediums: sciences, arts, etc. Among them, the most precise in both fields are Physics and Music, and it is a fascinating task to identify the isomorphism between both disciplines in every era. Usually Physics goes chronologically before Music, but sometimes the arts has gone before science, as in the Renaissance, when Filippo Brunelleschi and Leon Battista Alberti invented the perspective geometry and the concept of infinite (cfr. the famous and celebrated essay by Erwin Panofsky, *The Perspective as Symbolic Form* (Berlin, 1927)).

Therefore, since the Intervalic Theory in Music has been postulated, we can deduce that there must exist an isomorphic theory which expresses the same thought in Physics, defining a new physical intervalic space. In such physical theory we only have to substitute notes by intervals. Since a *interval* is, *by definition*, the relation between two notes (being a note just a magnitude of any of the metrical qualities of sound: frequency, time and intensity), it is analogous to a *ratio*, which is, *by definition*, the relation between two magnitudes. Henceforth, a physical theory whose measurements was completely expressed in *ratios* would be isomorphic to a musical theory wholly expressed in metric *intervals*. Of course, such theory is just the Intervalic Theory in Music.

A great difference between the intervalic units and the other systems is that the intervalic physical quantities conforms a *finite* and *ordered* set —the already described intervalic group— and every physical quantity can not acquire any value, as each of them are geometrically closed by its corresponding height: an intervalic quanta or limit. Therefore, a magnitude expressed in intervalic units comprises, apart from the usual meaning of a *dimensionful magnitude*, an additional meaning of a *dimensionless ratio*. Moreover, the value of both ones is just the same. That is to say, if we say, as example, that some magnitude of frequorce is $\varphi = ½ (i^{-1}L^{-1})$, the same magnitude expressed as a dimensionless ratio is: $\varphi = ½ (i^{-1}L^{-1}) / \varphi_I = ½ (1)$. On the contrary a frequency ratio of ½ in any other system of units means nothing because there is no defined any quanta or limit for every physical quantity (with the exception of velocity due to the existence of the limit of the speed of light). Thus, we could say that in intervalic units all physical quantities can be expressed in the same way as we write the velocity as v/c in traditional units. Thus physical quantities become dimensionless and the system of units is transformed into geometry.

The claim of the supposed "geometrized" units to be thus geometric is entirely a fraud because any system of units, including the traditional SI units, can be "geometrized" in that way giving all the magnitudes expressed as dimensionless ratios. However, all those ratios only have got meaning inside every one of those system of units because there is neither natural nor universal magnitudes for every physical quantity (with the only exception of the speed of light). For example, the degenerate units pretends that Planck's length or mass, to say only their stronger physical quantities, are universal magnitudes which serve as a patter of measurement for dimensionless ratios. Far enough. Inasmuch as Planck's length is as universal as a metre or a yard and Planck's mass is as universal as a kilogram or a pound, the so called "geometrized" units are absolutely a fake. If that was the case, *any* system of *singular* units —singular units means c = 1 and ℏ = 1 regardless of the dimensions of c and ℏ— should yield the same dimensionless ratios. However this is simply false as the mere existence of the intervalic units does have demonstrated.

Although modern Physics has lamentably forgotten the importance of the system of *dimensions* and has confounded it with the system of *units*, the fact is that the important one is the dimensional basis, but not the value of the units. It is this way because whatsoever system of singular units or of non-singular units expressed in *intervalic dimensions* — intervalic dimensions means c ($i^{\pm 1}$) and ℏ (L) regardless of the value of c and ℏ— will yield exactly the same dimensionless ratios of universal validity. On the contrary, without intervalic dimensions —c ($i^{\pm 1}$), ℏ (L)— any system of units yield different dimensionless ratios of local (non-universal) validity. By this reason, along this book we may use conventional *units* but with intervalic *dimensions*, because they yield the same universal dimensionless ratios as of singular units —or any other units— but always in intervalic dimensions.

Of course, the units of the intervalic dimensions are unique because it is the only system of units whose units are not arbitrary, like the metre or Planck's length, but they are the genuine geometric heights of Nature: the *intervalic quanta* and the *intervalic limits*. Really this is an essential difference between the intervalic units and all the other systems: it can be said that the intervalic dimensions do not have usual *units* in the usual sense —which by implicit definition are arbitrary and are set by mankind or by any other civilization— but instead *quanta* and *limits* —which are reliable magnitudes of universal validity and are set by Nature—.

We can express completely music and physics through intervals or ratios. The simplicity and elegance introduced by the features of the intervalic units in Physics should already be, in absence of any other further developments, a sufficient reason to adopt them insofar as the aim of science is the understanding of Nature. It is sure that a better understanding of Nature will be followed by practical achievements in technology, but this is not the reason that leads the scientific thinking. Nevertheless, the layperson will foolishly say: so what? It was supposed that neither music nor physics *did not rely* on the dimensional system and units chosen, but against all expected, this traditional supposition has proved to be absolutely *false*, both in music as in physics.

The dramatic fact that the two fundamental constants of Nature define unavoidably a quantum of electric charge which is an exact fraction of the elementary charge, involves that subatomic particles are not *structureless* but *structureful* and *composite*: they must be *states of minimal energy* reached at primordial times by means of a *primordial assembly* of intervalic charges.

THE LOGICAL EXISTENCE OF GEOMETRIC HEIGHTS IN ALL PHYSICAL QUANTITIES

The existence of *infinite* magnitudes is a plague in Physics of XXth. century. In the work **De Rerum Natura** by Titus Lucretius Carus, the disciple of Epicure gave the first demonstration for the *logical necessity* of the existence of a *minimal height* in Nature. As a theoretical thinking it is still completely valid up to present time. However, modern Physics ignores scandalously such logical constraint of physical quantities. Up to the postulation of Special Relativity the magnitude of whatsoever physical quantity could freely be *as great or as small* as desired, *without theoretical limit*. After Einstein postulated the existence of a maximum height, c, for the velocity, the own Einstein postulated the quantization of energy through the introduction of another minimum height: the Planck's action quantum, $\hbar$, which led to the born of Quantum Mechanics. So, these are the two only heights existing in modern Physics, both introduced by Albert Einstein.

Once the intervalic units have been postulated it can be understood that if a magnitude increases near infinite, it can distortion or "go out" from the geometry of intervalic-relativistic space-time. Even worst, there is a logical contradiction in the own existence of an *isolated* limit of the speed of light with respect to the remaining physical quantities. There was thought that the velocity of a body was limited by c, but the remaining physical quantities did not. The contradiction is still more apparent when considering physical quantities closely related with velocity, such as acceleration of gravitational field strength. If the velocity of a body is limited by c, how is possible that its acceleration or its gravitational field strength can be augment infinitely? Meaningfully, both physical quantities have identical intervalic dimensions, ($-L^{-1}$), and are geometrically limited by the intervalic limit of acceleration-power (or poweleration): $c^{\pm 2}\hbar^{-1}$ = 1 ($-L^{-1}$) = $8.513874 \cdot 10^{50}$ (m s^{-2}).

In a similar way, if the action is quantized by the Planck's constant, its related physical quantities such as space or time should also be quantized. But as all physical quantities are dimensionally interrelated, it is absolutely impossible to admit that one, two or three alone physical quantities have got a geometrical height and the remaining ones have not got it yet.

Since in intervalic dimensions every physical quantity is a combination of the dimensions of the limit c and the quantum ℏ, we gave got a precise definition of such corresponding quantum or limit for all physical quantities. Therefore, every physical quantity must logically have a geometrical height: a maximum one —*limit*— or a minimum one —*quantum*—. The capital difference between the intervalic units and the remaining systems of units is that the first one give us a precise definition of such *geometric heights* and the others can't make such heights.

The measurement of all *limited* physical quantities (see tables to identify them marked with the legend "absolute limit") must have, say, a "geometric distortion" because any measurement of such magnitudes can't be greater than their corresponding geometric limits. This "geometric distortion" is a consequence of the existence of those limits, and it would not exist if there was not such limits. They affect not only to *velocity*, as postulated by Special Relativity, but to all physical quantities limited by the intervalic geometry. The Lorentz-Einstein transformations with respect to velocity is the traditional name of only *one* of those 18 intervalic physical quantities (that one of *velocity*), but there are seventeen ones more; but as some intervalic dimensions gather more than one traditional physical quantities, the number of limits according to the traditional physical quantities is greater, as it can be seen in detail in the tables of intervalic units at the final of the chapter. Writing down only one physical quantity per every limited intervalic dimension we have:

- Velocity: $\mathbf{v}_I = c = 2.9979246 \cdot 10^8$ (m s^{-1})
- Temperature: $\mathbf{\Theta}_I = c\, k_B^{-1} = 2.1713738 \cdot 10^{31}$ (K)
- Wavevector: $\mathbf{k}_I = \hbar^{-1} = 9.472962 \cdot 10^{33}$ (m^{-1})
- Poweleration: $\mathbf{a}_I = c^{\pm 2}\hbar^{-1} = 8.513874 \cdot 10^{50}$ (m s^{-2})
- Linear perdensity: $\mathbf{\Delta}^1_I = c^{-1}\hbar^{-1} = 3.163026 \cdot 10^{25}$ (kg m^{-1})
- Frequorce: $\mathbf{\varphi}_I = c\, \hbar^{-1} = 2.8399227 \cdot 10^{42}$ (s^{-1})
- Viscosity (dynamic): $\mathbf{\eta}_I = \hbar^{-2} = 8.973701 \cdot 10^{67}$ (Pa s)
- Area power: $\mathbf{W}^2_I = c^{\pm 2}\hbar^{-2} = 8.065161 \cdot 10^{84}$ (-m^{-2})
- Inflexion: $\mathbf{i}_I = c^{-1}\hbar^{-2} = 2.993305 \cdot 10^{59}$ (m s^{-3})
- Surface tension: $\mathbf{\sigma}_I = c\,\hbar^{-2} = 2.690248 \cdot 10^{76}$ (N m^{-1})
- Fluctuation: $\mathbf{f}_I = \hbar^{-3} = 8.500756 \cdot 10^{101}$ (m s^{-4})
- Voluminic power: $\mathbf{W}^3_I = c^{\pm 2}\hbar^{-3} = 7.640098 \cdot 10^{117}$ (-m^{-3})
- Density: $\mathbf{\rho}_I = c^{-1}\hbar^{-3} = 2.835547 \cdot 10^{93}$ (kg m^{-3})
- Pressure: $\mathbf{P}_I = c\,\hbar^{-3} = 2.548463 \cdot 10^{110}$ (Pa)
- Electric potential: $\mathbf{V}_I = \sqrt{-(c\hbar^{-1})} = 1.685207 \cdot 10^{21}$ (V)
- Electric field strength: $\mathbf{\mathcal{E}}_I = \sqrt{-(c\hbar^{-3})} = 1.596390 \cdot 10^{55}$ (V m^{-1})
- Electric polarisation: $\mathbf{P}_I = \sqrt{-(c^{-1}\hbar^{-3})} = 5.324984 \cdot 10^{46}$ (C m^{-2})
- Charge density: $\mathbf{\rho}_I = \sqrt{-(c\hbar^{-5})} = 5.117026 \cdot 10^{46}$ (C m^{-3})

All of them are logically deduced and physically interpreted as a simple and characteristic feature of the Intervalic Geometry in IT. As desired by heuristic rules of Epistemology, the former Lorentz-Einstein transformations remain as a special case —the most simple one— subsumed into a more general and powerful geometry of Physics.

MACROSCOPIC CONTINUITY *VERSUS* MICROSCOPIC DISCONTINUITY (OR WHY THE UNCERTAINTY PRINCIPLE IS UNCERTAIN)

At this paragraph we can't avoid to be very critical because we are dealing with one of the most misleading and captious principles of Quantum Mechanics which has blocked a lot of possible developments. It can be said that from the postulation of such deceitful principle, all Quantum Mechanics was unfortunately infected by the philosophy which is derived without any mental rigour from such idea. By no means and under no circumstances, we regret to say that this can not be called 'good science'.

The nauseating relations with nazis of Werner Heisenberg, showed in the impressive and well documented book by Paul Lawrence Rose **Heisenberg and the Nazi Atomic Bomb Project. A study in German Culture** (University of California Press, 1998) is not relevant for our purpose now.

As Relativity, the uncertainty principle was historically born as a consequence of the troubles which arose in Physics when going into the study of macroscopic and microscopic scales respectively. The first one was symbolized by Einstein's limit, the speed of light, c, and the second by Planck's quantum of action, $\hbar$. However, today we can see that these troubles appear necessarily when we are going to approximate near anyone of the intervalic *limits* or *quanta* of Nature. This means that there is a astonishing simple explanation for such troubles in terms of the intervalic *geometry* of nature. Let us begin with the speed of light.

The links with the postulation of Special Relativity, such as the Michelson-Morley experiment, etc., are very interesting but we have to suppose that they are already known by the reader. *Massful* particles can not travel quicker than *massless* particles which move at the speed of light, *ergo* c must be a fundamental limit of Nature. The trouble appears when we have a massive particle approaching to the velocity c. If we supply more and more energy to such particle, why it can't finally move faster that light? Hypothetically the particle could shock abruptly with such barrier, but this naïve and coarse possibility is not the way of Nature, but to approaching asymptotically to those limits without touching them never —what we usually know as the Lorentz-Einstein transformations—. This was one of the great contributions of Albert Einstein to Physics, which now we can place and interpret in a wider theory of Nature.

As soon as the experimental devices will be available, we will see how the temperature of a particle follows a behaviour similar to its velocity when approaching to its intervalic limit, the intervalic temperature, $\Theta_I = c\, k_B^{-1} = 2.1713738 \cdot 10^{31}\,(K)$. And in an analogous way, we will see that regardless the energy supplied to a particle, it can never surpass the geometric intervalic limits, like the intervalic acceleration, $\mathbf{a}_I = c^{\pm 2} \hbar^{-1} = 8.513874 \cdot 10^{50}\,(m\ s^{-2})$, or the intervalic frequency, $\varphi_I = c\, \hbar^{-1} = 2.8399227 \cdot 10^{42}\,(s^{-1})$, and so on. Please let us note that this could not be other way because all physical quantities are made from dimensional combinations of c and ℏ, and these two ones are respectively limited and quantized.

Since the Lorentz-Einstein transformations —and the enhanced intervalic transformation of physical quantities in the Intervalic Space— give us the new way of *measurement* of any physical quantity when approaching to the intervalic limits, we have not found too many troubles with the maximum heights (limits).

On the contrary, when dealing with the minimal heights (quanta), we have not got a new mode of measurement at microscopic scale for such physical quantities, and therefore troubles will surely arise... if we do not know the underlying geometry of Nature. According to the deluding Heisenberg interpretation, there is an essential uncertainty in Nature at microscopic scale. It does not say that we are not able to measure at such scale, but that Nature is *essentially undetermined* at that level of reality. As we are going to see straightaway, this is a gross mistake. The only thing which is essentially undetermined is the Copenhagen interpretation of the physical world.

It would be very recommended to know the reasoning of ancient Greeks about the *logic necessity* of the existence of a final limit — whatever it be— in the division of matter. To anyone with a peer reason it becomes evident that there must exist some kind of limit to the *continuity* of matter as we progress towards the microscopic scale. And this must be certain for a lot of features of Nature. In other way, the process of division of matter would not be endless and infinite, as Greeks were already aware. This means that, at some level, we have to find some kind of discontinuity. The stuff of which is made such *discontinuity* will be the elemental *blocks* of Nature corresponding *to that level* of the measurement scale. And of course, there must be a final level with its corresponding final block of Nature.

To a better understanding we can illustrate it with a suitable comparison with plastic arts. Lets imagine the poster of a movie in a wall. When we are watching it at large or medium distance, we see an almost infinite tones of colours, that is to say, we see *continuity* in colour. However, if we watch the poster at 20 cm. or less, we will discover that there are only three or four colours: cyan, magenta and yellow (and possibly black depending on the printing system). In other words, we have crossed some threshold about "measurement" and have seen the underlying *discontinuity* of colour at the psycho physic level. Now Heisenberger would say that the nature of colour is "undetermined" at small scale; and he would explain it affirming that there is an "essential uncertainty" of colour at small scale in Nature. However, as anyone can easily see, the colour is neither 'undetermined' nor there is an 'uncertainty', as we see clearly three colours —CMY— fully determined. What we have got is simply a *discontinuity* in the colour when the observation becomes closer to the poster, a fact that a smart child may understand. Instead the almost infinite tones of colour that can be seen at longer distance, we discover that all those colours are only made from three fundamental ones, the corresponding blocks at this level, the *primary colours*: red, yellow and blue.

To affirm that there is an "essential uncertainty" of *colour* is just the same as Heisenberg did with *action*, which intervalic dimensions is that of *length*. At macroscopic scale space is obviously *continuous*. However, when approaching progressively to the scale of the intervalic length, $l_I = \hbar = 1.0556363 \cdot 10^{-34}$ (m), the space becomes granulated, and finally arises the fundamental block of length at such microscopic scale. What was continuous at great scales becomes *discontinuous* at this level, in the same way as colour. There is no 'uncertainty', but simple 'discontinuity' when we move near the scale of every one intervalic quanta.

The above reasoning can surely be understood by a kid of five years. Please note that to hold the existence of a continuous space at any scale, that is to say, *ad infinitum*, is just the same that to affirm the existence of the medieval *ether*, in vogue up to the postulation of Special Relativity. Thus, a *continuous* space involves necessarily the existence of a *universal ether* (this is no other than the *absolute space* of Newton). Really, both concepts are interchangeable. Someone might say that Heisenberg postulated uncertainty with the aim to saving the *continuity* of space, which was a concept hardly rooted in Classic Physics. Once more we can see that an unconscious philosophy may damage seriously the Physics research. In order to maintain a *continuous space*, Heisenberg paid the price to introduce 'uncertainty' in the heart of Physics, where it only was discreteness or discontinuity, but by no means indeterminacy.

Another comparison: this is just the same as if we have a ruler divided in millimetres to measure the distances of a picture drawn on a sheet paper. Lets suppose that the millimetre of the ruler is like the intervalic length, $\hbar$. It is clear that if we try to make a measurement with a precision smaller than the millimetre, we will go into a trouble. To this trouble the cute Heisenberg answered that the own picture was undetermined, instead to conclude that the ruler, composed by millimetres, can not make any measurement smaller that the last block of the ruler, the millimetre.

Unfortunately, Physics followed such absurdity. In Music there happened a similar scene, although it was discarded in a few years: the undetermined and random music was in vogue at the middle of the XXth. century by means of some unscrupulous composers. However, soon after the common sense come back to Music and everybody comprehended and agreed that *the composer does not play dice*. And if the composers — the little creators of the artificial branch of Music (the musical works)— do not play dice, how can we pretend that the supreme creator of the natural branch of Music (that is to say, the Universe) does play dice?

In the Intervalic Space the discreteness of physical quantities affects to all physical quantities whose intervalic unit is a quantum. As with the intervalic limits, a single intervalic dimension may gather more than one traditional physical quantity, as can be seen in the tables. Some of these quanta are:

- Length, $l_I = \hbar = 1.0556363 \cdot 10^{-34}$ (m),
- Inductance: $L_I = c^{\pm 2}\hbar = 9.487583 \cdot 10^{-18}$ (H)
- Time: $t_I = c^{-1}\hbar = 3.5212226 \cdot 10^{-43}$ (s)
- Area: $S_I = \hbar^2 = 1.114367 \cdot 10^{-68}$ (m²)
- Area inertia momentum: $I^2_I = c^{-1}\hbar^2 = 3.717129 \cdot 10^{-77}$ (kg m²)
- Volume: $V_I = \hbar^3 = 1.176366 \cdot 10^{-102}$ (m³)
- Inertia volume momentum: $I^3_I = c^{-1}\hbar^3 = 3.923935 \cdot 10^{-111}$ (kg m³)
- Fermi constant physical quantity: $c\hbar^3 = 3.526657 \cdot 10^{-94}$ (kg m³)
- Electric charge: $q_I = \sqrt{-(c^{-1}\hbar)} = 5.933989 \cdot 10^{-22}$ (C)
- Bohr magneton: $\mu_{BI} = \sqrt{-(c\hbar^3)} = 1.877940 \cdot 10^{-47}$ (J T⁻¹)

Please be aware that any measurement involving any of those intervalic quanta (and not only the length) is geometrically subdued to discreteness when approaching to the scale of its corresponding intervalic quantum. Thus, for example, discontinuity will arise immediately in everyone of the following measurements at microscopic scale (the intervalic dimensions are showed between brackets):

$\Delta l\ (L) \cdot \Delta p\ (1) \geq l_I = \hbar\ (L)$
$\Delta t\ (iL) \cdot \Delta p\ (1) \geq t_I = c^{-1}\hbar\ (iL)$

$\Delta q\ (i^{-1/2}L^{1/2}) \cdot \Delta p\ (1) \geq \mathbf{q_I} = \sqrt{-(c^{-1}\hbar)}$

$\Delta l\ (L) \cdot \Delta m\ (i) \geq \mathbf{t_I} = c^{-1}\hbar\ (iL)$

$\Delta S\ (L) \cdot \Delta m\ (i) \geq \mathbf{t_I} = c^{-1}\hbar\ (iL)$

$\Delta t\ (iL) \cdot \Delta v\ (i^{-1}) \geq \mathbf{l_I} = \hbar\ (L)$

$\Delta t\ (iL) \cdot \Delta E\ (i^{-1}) \geq \mathbf{l_I} = \hbar\ (L)$

$\Delta p\ (1)\ /\ \Delta \varphi\ (i^{-1}L^{-1}) \geq \mathbf{t_I} = c^{-1}\hbar\ (iL)$

$\Delta l\ (L)\ /\ \Delta \varphi\ (i^{-1}L^{-1}) \geq \mathbf{l^2_I} = c^{-1}\hbar^2\ (iL^2)$

$\Delta t\ (iL)\ /\ \Delta a\ (-L^{-1}) \geq c\hbar^2\ (i^{-1}L^2)$

$\Delta t\ (iL)\ /\ \Delta W\ (-L^{-1}) \geq c\hbar^2\ (i^{-1}L^2)$

$\Delta m\ (i)\ /\ \Delta a\ (-L^{-1}) \geq -\mathbf{t_I} = c\hbar\ (i^{-1}L)$

$\Delta m\ (i)\ /\ \Delta g\ (-L^{-1}) \geq -\mathbf{t_I} = c\hbar\ (i^{-1}L)$

$\Delta \theta\ (i^{1/2}L^{1/2}) \cdot \Delta \Phi\ (-1) \geq \mathbf{q_I} = \sqrt{-(c^{-1}\hbar)}$

$\Delta p\ (1)\ /\ \Delta A\ (i^{1/2}L^{-1/2}) \geq \mathbf{q_I} = \sqrt{-(c^{-1}\hbar)}$

$\Delta V\ (i^{1/2}L^{-1/2}) \cdot \Delta t\ (iL) \geq \mathbf{q_I} = \sqrt{-(c^{-1}\hbar)}$

...

The first relation of the above ones is no other than the infamous *Heisenberg uncertainty principle*, which is only one among many others of the intervalic quanta which appears inasmuch as we make some measurements near the microscopic scale. Quantum philosophy did not hesitate to sacrifice 'determinacy' to save 'continuity' in Physics, postulating that the world was *undetermined* and *continuous* at Planck scale, while the truth is absolutely just the opposite: Nature is *determined* and *discrete* at the scale of her last blocks —the intervalic quanta—.

To conclude, please let us note that all bodies of the Universe are composed in the same manner: with single blocks. Why all physical bodies of Nature would not be composed in the same way? A continuous matter or stuff divisible *ad infinitum* simply does not exist, and probably can not exist. Continuity is only a macroscopic appearance, a very useful one for human perception. At the last geometric foundations of Nature there is no universal ether, there is no absolute space, there is no a doubtful uncertainty... but only a clever discreteness.

THE GROSS MISTAKE OF THE UNCERTAINTY PRINCIPLE AND OF THE COPENHAGEN INTERPRETATION

Among all gross errors along the history of Physics there is one which stands out among the remaining ones. It is perhaps the greatest hoax passed off in science: the so called *uncertainty principle*, postulated by Werner Heisenberg, greatly acclaimed by all sorts of Relativisms. It is said that such misleading idea was proposed to avoid the granular conception of space, unacceptable to the physics community by that time. But the Copenhagen interpretation made things even worst: according to it, there is an essential uncertainty in Nature at microscopic scale. It does not say that we are not able to measure at such scale, which is a simple fact, but that Nature is unavoidably undetermined at that level of reality. This affirmation is completely a gross mistake, mistake which besides blocked and prevented the physics research on the underlying geometry of Nature during decades.

To a better understanding we can illustrate it with a suitable comparison with plastic arts. Lets imagine the poster of a film in a wall. When we are watching it from a large distance, we see an almost infinite tones of colours, that is to say, we see *continuity* in colour. However, if we watch the poster at 20 cm. or less, we will discover that there are only four colours: cyan, magenta, yellow and black. In other words, we have cross the threshold of a scale of "measurement" and have seen the underlying *discontinuity* of colour at a lesser scale. Now Heisenberg would say that the nature of colour is "uncertain" at small scale; and he would explain it affirming that there is an "essential uncertainty" of colour at small scale in Nature. However, as anyone can easily see, the colour is neither 'undetermined' nor there is any 'uncertainty', as we see clearly four colours.

To affirm that there is an essential uncertainty of *colour* is just the same as Heisenberg did with *action*, which intervalic dimensions is that one of *length*. At macroscopic scale, space is obviously *continuous*. However, when approaching progressively to the scale of the intervalic length, $l_I = \hbar = 1.0556363 \cdot 10^{-34}$ (m), the space becomes granulated, and finally arises the fundamental block of length. What was continuous at great scales becomes *discontinuous* at this level, in the same way as colour. There is no 'uncertainty', but simple 'discontinuity' when we move near the scale of *each intervalic quanta*, and not only the intervalic quantum of *length* but all of them, since the intervalic dimension of every intervalic quanta is made from a combination of the intervalic dimensional basis (L, *i*). For example:

$\Delta l\,(L) \cdot \Delta p\,(1) \geq l_I = \hbar\,(L)$ $\quad\quad\quad$ $\Delta l\,(L) / \Delta\varphi\,(i^{-1}L^{-1}) \geq l^2_I = c^{-1}\hbar^2\,(iL^2)$

$\Delta t\,(iL) \cdot \Delta v\,(i^{-1}) \geq l_I = \hbar\,(L)$ $\quad\quad\quad$ $\Delta q\,(i^{-1/2}L^{1/2}) \cdot \Delta p\,(1) \geq q_I = \sqrt{-(c^{-1}\hbar)}$

$\Delta l\,(L) \cdot \Delta m\,(i) \geq t_I = c^{-1}\hbar\,(iL)$ $\quad\quad\quad$ $\Delta p\,(1) / \Delta\varphi\,(i^{-1}L^{-1}) \geq t_I = c^{-1}\hbar\,(iL)$

$\Delta t\,(iL) / \Delta a\,(-L^{-1}) \geq c\hbar^2\,(i^{-1}L^2)$ $\quad\quad\quad$ $\Delta m\,(i) / \Delta a\,(-L^{-1}) \geq -t_I = c\hbar\,(i^{-1}L)$

The uncertainty principle is absolutely useless because of the disparate scales of measuring involved. It affirms a simple irrelevance: we can not measure the position of a subatomic particle —which radius is around 10^{-15} (m)— with a precision greater than $\hbar = 1.0556363 \cdot 10^{-34}$ (m). To make this would be applicable in practice we should need to know the values of the electron or nucleon radii with a precision of 20 to 28 non zero digits. This only would be possible if the values of the intervalic dimensional basis, c and $\hbar$ (the speed of light and the Planck's constant) would have equally got 20 to 28 significant digits.

But the uncertainty principle is wrong in theory too. The *intervalic state* of any subatomic particle —determined by the allowed intervalic symmetries of the constituent particles of the exchange intervalic structure— is always in perpetual exchange interaction; and besides each intervalic structure, in every level, is spinning with a linear velocity on surface whose value is in the order of the speed of light. It is a plain misunderstanding to talk about the size of an object out of the value of its scale of measurement. This elemental reasoning can surely be understood by a child of five years, but surprisingly it couldn't and wasn't by the Copenhagen physicists.

Finally and by the way, Copenhagen fans should read perhaps the book by Paul Lawrence Rose, *Heisenberg and the Nazi Atomic Bomb Project*, UCP, 1998.

Main differences between the INTERVALIC SYSTEM of PHYSICAL QUANTITIES and all other systems of units and dimensions

- The Intervalic System is the unique dimensional system whose **dimensional basis** —(**L**, ***i***)— is just composed by the single intervalic dimensions of the **last fundamental constants of Nature, ℏ and c**: dim ℏ = (L), dim c (i^{-1}). Of course, the system has two formulations: dim c = (i^{-1}) or dim c = (i), which are absolutely equivalent.
- Existing physical quantities are *generated* by all algebraic combinations between the two dimensional basis —L and *i*— which makes a *finite* and *ordered* set of **40 physical quantities**. The *number of physical quantities* is given by the formula: $4 + 12n$, being *n* the number of actual dimensions of space.
- There are no physical quantities whose equation of dimensions have more than the *actual dimensions* of space (3) and time (1), as in other systems, which is absolutely a nonsense or, at least, an inconsistency.
- There are neither different physical quantities with the same equation of dimensions (as in traditional units), nor different dimensions with the same physical quantity (as in the misleading called "geometrized" units, which is the poorest system of units ever made, not even being consistent).
- The intervalic dimensions of all physical quantities can *operate algebraically* with the signs of its corresponding magnitudes in any equation.
- The own definition of all existing physical quantities as an algebraic combination of c (i^{-1}) and ℏ (L) yields unavoidably a *geometric height* for every physical quantity, making the full set of INTERVALIC QUANTA and INTERVALIC LIMITS, which form not only the Intervalic Units, but above all, the foundations of the **underlying fundamental geometry of Nature**, from which are derived the genuine **intervalic symmetries of Nature**, long time searched by Physics.
- Therefore, every physical quantity can not acquire any value, as each of them is *geometrically* closed by its corresponding height: an intervalic quanta or limit, which is another great difference between the intervalic system of units and the remaining systems, which like to play with infinites and singularities.
- Being the intervalic dimensions the truthful units of Nature, the intervalic physical quantities have got an *epistemological rank* by means

of their equations of dimensions have got an *heuristic value*, which is lacked in other systems. The most important example is the merging of two traditional physical quantities into a new one because their intervalic dimensions are identical. That is the case of: velenergy (velocity-energy), frequorce (frequency-force) and poweration (power-acceleration)-gravitational field strength. The merged physical quantities means that they are really the same underlying physical quantity, although in phenomenology there may appear as different in incomplete or false dimensional systems. All this allow to unify intervalic dimensions, physical quantities and units in a unique concept, if desired.

- When applying basic geometry to the intervalic dimensions inside the Argand-like Intervalic Dimensional Space, a full set of invariant **Intervalic Transformations** of physical quantities is *geometrically* derived. The Intervalic Transformations comprise the former Lorentz-Einstein transformations of Special Relativity, which stays as a specific case inside a much wider geometry.
- Contrarily to supposed, the Intervalic System of Dimensions is the unique system which is **not equivalent** to all the remaining dimensional systems of units (which are, from now on, irrelevant in Physics research).
- All results yielded in the Intervalic Theory of Particle Physics are *geometric statements* logical and unavoidably deduced from the *intervalic quanta and limits* of the **Intervalic System of Units** *without using any mathematical formalism.*
- The Intervalic Theory is the unique Physics theory ever postulated which has *no one arbitrary constant*.
- Inasmuch as c and ℏ are universal constants, the *intervalic quanta and limits* are reliable physical quantities of *universal validity*. The Intervalic System of Units is not an arbitrary one but the genuine **system of units of Nature**. It must be noted that the *intervalic symmetries of Nature* can not be deduced by means of any other dimensional system, but only from the intervalic one. Thus its knowledge might be viewed as a clue of the scientific degree of development of a civilization, and so it is also apparent that the Intervalic Units are the unique which could be shared with hypothetical advanced extraterrestrial intelligences.
- Any value expressed in *intervalic units* can be interpreted as a *dimensionless interval or ratio* and Physics really becomes truthful Geometry. Hence the name of the Intervalic Theory.

INTERVALIC QUANTA, INTERVALIC LIMITS, INTERVALIC UNITS

Intervalic dimens.	Physical quantity	Definition	Geometrical rank	Equivalence: SIU magnit.(unit) = SI magnit.(unit)
1	Permitivity	$\varepsilon_I = \hbar^{-1}\hbar$	subatomic limit	$1\,(1) = 1\,(F\,m^{-1})$
	Momentum	$\mathbf{p}_I = c^{-1}c$	subatomic limit	$1\,(1) = 1\,(kg\,m\,s^{-1})$
	Boltzmann constant	k_B	-	-
	Entropy	S	-	-
-1	Permeability	$\mu_I = c^{\pm 2}$	conversion factor	$1\,(-1) = 1.112650 \cdot 10^{-17}\,(H\,m^{-1})$
	Gravitational potential	Φ	-	-
	Antimomentum	$-\mathbf{p}_I = c^{\pm 2}$	conversion factor	$1\,(-1) = 8.987552 \cdot 10^{16}\,(-kg\,m\,s^{-1})$
i	Mass (invervelocity-mass)	$\mathbf{m}_I = c^{-1}$	subatomic limit	$1\,(i) = 3.3356409 \cdot 10^{-9}\,(kg)$
	Invervelocity-mass	$\mathbf{v}^{-1}{}_I = c^{-1}$	subatomic limit	$1\,(i) = 3.3356409 \cdot 10^{-9}\,(s\,m^{-1})$
i^{-1}	Velocity (velenergy)	$\mathbf{v}_I = c$	absolute limit	$1\,(i^{-1}) = 2.9979246 \cdot 10^{8}\,(m\,s^{-1})$
	Energy (velenergy)	$\mathbf{E}_I = c$	subatomic limit	$1\,(i^{-1}) = 2.9979246 \cdot 10^{8}\,(J)$ $1\,(i^{-1}) = 1.8711569 \cdot 10^{21}\,(MeV)$
	Temperature	$\Theta_I = c\,k_B^{-1}$	absolute limit	$1\,k_B^{-1}\,(i^{-1}) = 2.1713738 \cdot 10^{31}\,(K)$
L	Length (real space)	$l_I = \hbar$	quantum	$1\,(L) = 1.055636 \cdot 10^{-34}\,(m)$
	Action	$S_I = \hbar$	quantum	$1\,(L) = 1.055636 \cdot 10^{-34}\,(J\,s)$
	Capacitance	$C_I = \hbar$	quantum	$1\,(L) = 1.055636 \cdot 10^{-34}\,(F)$
-L	Antilength	$-l_I = c^{\pm 2}\hbar$	conversion factor	$1\,(-L) = 9.487583 \cdot 10^{-18}\,(-m)$
	Inductance	$L_I = c^{\pm 2}\hbar$	quantum	$1\,(-L) = 9.487583 \cdot 10^{-18}\,(H)$
$i\,L$	Time (imaginary space)	$t_I = c^{-1}\hbar$	quantum	$1\,(iL) = 3.5212226 \cdot 10^{-43}\,(s)$
$i^{-1}\,L$	Antitime	$-t_I = c\,\hbar$	conversion factor	$1\,(i^{-1}L) = 3.164717 \cdot 10^{-26}\,(-s)$
L^{-1}	Wavevector	$\mathbf{k}_I = \hbar^{-1}$	absolute limit	$1\,(L^{-1}) = 9.472962 \cdot 10^{33}\,(m^{-1})$

Intervalic dimens.	Physical quantity	Definition	Geometrical rank	Equivalence: SIU magnit.(unit) = SI magnit.(unit)
$-L^{-1}$	Acceleration (poweleration)	$\mathbf{a}_I = c^{\pm 2}\hbar^{-1}$	absolute limit	$1\,(-L^{-1}) = 8.513874 \cdot 10^{50}\,(m\,s^{-2})$
	Gravitational field	$\mathbf{g}_I = c^{\pm 2}\hbar^{-1}$	absolute limit	$1\,(-L^{-1}) = 8.513874 \cdot 10^{50}\,(m\,s^{-2})$
	Power (poweleration)	$\mathbf{W}_I = c^{\pm 2}\hbar^{-1}$	absolute limit	$1\,(-L^{-1}) = 8.513874 \cdot 10^{50}\,(W)$
$i\,L^{-1}$	Antifrequorce	$-\varphi_I = c^{-1}\hbar^{-1}$	conversion factor	$1\,(i\,L^{-1}) = 3.163026 \cdot 10^{25}\,(-s^{-1})$
	Linear Perdensity	$\Delta^1_I = c^{-1}\hbar^{-1}$	absolute limit	$1\,(i\,L^{-1}) = 3.163026 \cdot 10^{25}\,(kg\,m^{-1})$
$i^{-1}L^{-1}$	Frequorce (frequency-force)	$\varphi_I = c\hbar^{-1}$	absolute limit	$1\,(i^{-1}L^{-1}) = 2.8399227 \cdot 10^{42}\,(s^{-1})$
	Frequency (frequorce)	$\nu_I = c\hbar^{-1}$	absolute limit	$1\,(i^{-1}L^{-1}) = 2.8399227 \cdot 10^{42}\,(Hz)$
	Force (frequorce)	$F_I = c\hbar^{-1}$	absolute limit	$1\,(i^{-1}L^{-1}) = 2.8399227 \cdot 10^{42}\,(N)$
	Conductivity	$\sigma_I = c\hbar^{-1}$	absolute limit	$1\,(i^{-1}L^{-1}) = 2.8399227 \cdot 10^{42}\,(S\,m^{-1})$
L^2	Area	$S_I = \hbar^2$	quantum	$1\,(L^2) = 1.114367 \cdot 10^{-68}\,(m^2)$
$-L^2$	Antiarea	$-S_I = c^{\pm 2}\hbar^2$	conversion factor	$1\,(-L^2) = 1.001544 \cdot 10^{-51}\,(-m^2)$
$i\,L^2$	Inertia area momentum	$\mathbf{I}^2_I = c^{-1}\hbar^2$	quantum	$1\,(i\,L^2) = 3.717129 \cdot 10^{-77}\,(kg\,m^2)$
$i^{-1}L^2$	?	$c\,\hbar^2$	-	$1\,(i^{-1}L^2) = 3.340789 \cdot 10^{-60}\,(J\,m^2)$
L^{-2}	Viscosity (dynamic)	$\eta_I = \hbar^{-2}$	absolute limit	$1\,(L^{-2}) = 8.973701 \cdot 10^{67}\,(Pa\,s)$
$-L^{-2}$	Area Power	$\mathbf{W}^2_I = c^{\pm 2}\hbar^{-2}$	absolute limit	$1\,(-L^{-2}) = 8.065161 \cdot 10^{84}\,(-m^{-2})$
$i\,L^{-2}$	Area Perdensity	$\Delta^2_I = c^{-1}\hbar^{-2}$	absolute limit	$1\,(i\,L^{-2}) = 2.993305 \cdot 10^{59}\,(kg\,m^{-2})$
	Inflexion	$\mathbf{i}_I = c^{-1}\hbar^{-2}$	absolute limit	$1\,(i\,L^{-2}) = 2.993305 \cdot 10^{59}\,(m\,s^{-3})$
$i^{-1}L^{-2}$	Surface Tension	$\sigma_I = c\,\hbar^{-2}$	absolute limit	$1\,(i^{-1}L^{-2}) = 2.690248 \cdot 10^{76}\,(N\,m^{-1})$
L^3	Volume	$V_I = \hbar^3$	quantum	$1\,(L^3) = 1.176366 \cdot 10^{-102}\,(m^3)$
$-L^3$	Antivolume	$-V_I = c^{\pm 2}\hbar^3$	conversion factor	$1\,(-L^3) = 1.057265 \cdot 10^{-85}\,(-m^3)$
$i\,L^3$	Inertia vol. momentum	$\mathbf{I}^3_I = c^{-1}\hbar^3$	quantum	$1\,(iL^3) = 3.923935 \cdot 10^{-111}\,(kg\,m^3)$
$i^{-1}L^3$	Fermi const. phys. qty.	$c\,\hbar^3$	quantum	$1\,(i^{-1}L^3) = 3.526657 \cdot 10^{-94}\,(kg\,m^3)$
L^{-3}	Fluctuation	$\mathbf{f}_I = \hbar^{-3}$	absolute limit	$1\,(L^{-3}) = 8.500756 \cdot 10^{101}\,(m\,s^{-4})$

Intervalic dimens.	Physical quantity	Definition	Geometrical rank	Equivalence: SIU magnit. (unit) = SI magnit. (unit)
$-L^{-3}$	Voluminic Power	$W^3_I = c^{\pm 2}\hbar^{-3}$	absolute limit	$1\ (-L^{-3}) = 7.640098 \cdot 10^{117}\ (-m^{-3})$
	Irradiance	$E_{el} = c^{\pm 2}\hbar^{-3}$	absolute limit	$1\ (-L^{-3}) = 7.640098 \cdot 10^{117}\ (W\ m^{-2})$
$i\,L^{-3}$	Density	$\rho_I = c^{-1}\hbar^{-3}$	absolute limit	$1\ (i\,L^{-3}) = 2.835547 \cdot 10^{93}\ (kg\ m^{-3})$
	Voluminic Perdensity	$\Delta^3_I = c^{-1}\hbar^{-3}$	absolute limit	$1\ (i\,L^{-3}) = 2.835547 \cdot 10^{93}\ (kg\ m^{-3})$
$i^{-1}L^{-3}$	Pressure	$P_I = c\,\hbar^{-3}$	absolute limit	$1\ (i^{-1}L^{-3}) = 2.548463 \cdot 10^{110}\ (Pa)$
	Energy-tension density	$u_I = c\,\hbar^{-3}$	absolute limit	$1\ (i^{-1}L^{-3}) = 2.548463 \cdot 10^{110}\ (J\ m^{-3})$
$i^{1/2}L^{1/2}$	Magnetic charge	$\theta_I = \sqrt{-(c\hbar)}$	subatomic limit	$1\ (i^{1/2}L^{1/2}) = 1.778965 \cdot 10^{-13}\ (Wb)$
	Magnetic flux	$\Phi_I = \sqrt{-(c\hbar)}$	subatomic limit	$1\ (i^{1/2}L^{1/2}) = 1.778965 \cdot 10^{-13}\ (Wb)$
$i^{-1/2}L^{1/2}$	Electric charge	$q_I = \sqrt{-(c^{-1}\hbar)}$	quantum	$1\ (i^{-1/2}L^{1/2}) = 5.933989 \cdot 10^{-22}\ (C)$
$i^{1/2}L^{-1/2}$	Current	$I_I = \sqrt{-(c\hbar^{-1})}$	absolute limit	$1\ (i^{1/2}L^{-1/2}) = 1.685207 \cdot 10^{21}\ (A)$
	Electric potential	$V_I = \sqrt{-(c\hbar^{-1})}$	absolute limit	$1\ (i^{1/2}L^{-1/2}) = 1.685207 \cdot 10^{21}\ (V)$
	Magnetic vector potential	$A_I = \sqrt{-(c\hbar^{-1})}$	absolute limit	$1\ (i^{1/2}L^{-1/2}) = 1.685207 \cdot 10^{21}\ (Wb\ m^{-1})$
$i^{-1/2}L^{-1/2}$	Magnetic influx	$\Phi^{-1}_I = \sqrt{-(c^{-1}\hbar^{-1})}$	subatomic limit	$1\ (i^{-1/2}L^{-1/2}) = 5.621246 \cdot 10^{12}\ (Wb^{-1})$
$i^{1/2}L^{3/2}$	Bohr magneton ph. qty.	$\mu_{BI} = \sqrt{-(c\hbar^3)}$	quantum	$1\ (i^{1/2}L^{3/2}) = 1.877940 \cdot 10^{-47}\ (J\ T^{-1})$
$i^{-1/2}L^{3/2}$	?	$\sqrt{-(c^{-1}\hbar^3)}$	-	$1\ (i^{-1/2}L^{3/2}) = 6.264133 \cdot 10^{-56}\ (T^{-1})$
$i^{1/2}L^{-3/2}$	Electric field strength	$\mathcal{E}_I = \sqrt{-(c\hbar^{-3})}$	absolute limit	$1\ (i^{1/2}L^{-3/2}) = 1.596390 \cdot 10^{55}\ (V\ m^{-1})$
	Magnetic field strength	$H_I = \sqrt{-(c\hbar^{-3})}$	absolute limit	$1\ (i^{1/2}L^{-3/2}) = 1.596390 \cdot 10^{55}\ (A\ m^{-1})$
	Magnetic flux density	$B_I = \sqrt{-(c\hbar^{-3})}$	absolute limit	$1\ (i^{1/2}L^{-3/2}) = 1.596390 \cdot 10^{55}\ (T)$
$i^{-1/2}L^{-3/2}$	Electric polarisation	$P_I = \sqrt{-(c^{-1}\hbar^{-3})}$	absolute limit	$1\ (i^{-1/2}L^{-3/2}) = 5.324984 \cdot 10^{46}\ (C\ m^{-2})$
$i^{1/2}L^{5/2}$	?	$\sqrt{-(c\hbar^5)}$	-	$1\ (i^{1/2}L^{5/2}) = 5.858725 \cdot 10^{-39}\ (\ )$
$i^{-1/2}L^{5/2}$	?	$\sqrt{-(c^{-1}\hbar^5)}$	-	$1\ (i^{-1/2}L^{5/2}) = 1.954260 \cdot 10^{-47}\ (C^{-1}\ m^3)$
$i^{1/2}L^{-5/2}$	Charge density	$\rho_I = \sqrt{-(c\hbar^{-5})}$	absolute limit	$1\ (i^{1/2}L^{-5/2}) = 5.117026 \cdot 10^{46}\ (C\ m^{-3})$
	Current density	$J_I = \sqrt{-(c\hbar^{-5})}$	absolute limit	$1\ (i^{1/2}L^{-5/2}) = 5.117026 \cdot 10^{46}\ (A\ m^{-2})$
$i^{-1/2}L^{-5/2}$	?	$\sqrt{-(c^{-1}\hbar^{-5})}$	-	$1\ (i^{-1/2}L^{-5/2}) = 1.706856 \cdot 10^{38}\ (\ )$

Chapter 4

INTERVALIC GEOMETRY OF THE SPEED OF LIGHT

GEOMETRICAL PROPERTIES OF THE SPEED OF LIGHT IN THE INTERVALIC SPACE

As we know, the speed of light or *intervalic velenergy*, $v_I = c\ (i^{-1})$, is one of the two fundamental constants of nature, being the other one the *intervalic length*, $l_I = \hbar\ (L)$ —the former Planck constant—. The first geometrical constant is *the intervalic limit* and the second one the *intervalic quantum* of the Intervalic Space.

The speed of light establishes the ratio of conversion between *real* and *imaginary* dimensions —and therefore, between physical quantities— in the Intervalic Dimensional Space (for now on 'Intervalic Space' for short). Hence, the speed of light determines a systematic set of *geometrical relations* among physical quantities.

These geometrical properties are due to the extraordinary fact that the intervalic dimension of the speed of light is precisely the i number, that is to say, $c\ (i^{-1})$ —or $c^{-1}\ (i)$—. Unlike traditional dimensions, the intervalic dimensions are *algebraically consistent* within themselves as well as with the magnitudes of physical quantities. This remarkable fact involves that we can make use of the well known Argand space to represent the axes of the real and the imaginary intervalic dimensions in that frame of reference, which we will name the *Intervalic Space*. However, in the Argand space it is only meaningful the real axe, being meaningless the imaginary axe, which only serves for projecting their measurements on the real axe. On the contrary, in the Intervalic Space both real and imaginary axes are always totally meaningful. Really, it is hard to believe that after having used the Argand space the discovery of the Intervalic Space has not been postulated till our days. But this way is the tortuous way of the human progress in Physics.

To say it in other manner. The great difference between the Intervalic Space and the traditional frames of reference of Physics is that the

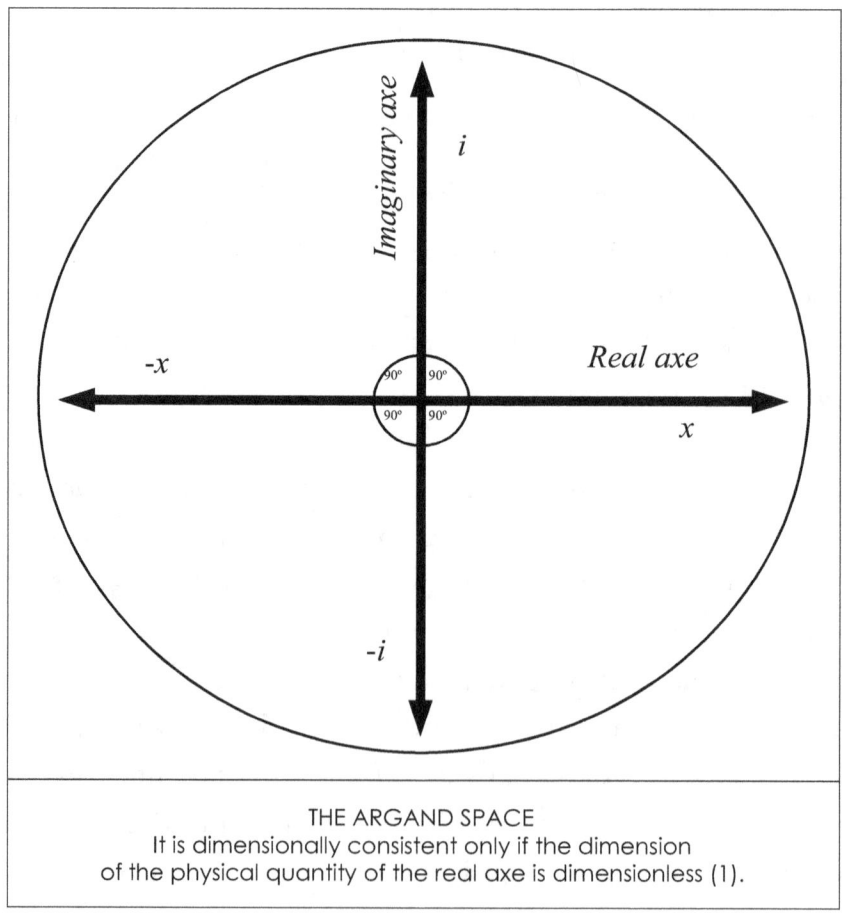

THE ARGAND SPACE
It is dimensionally consistent only if the dimension
of the physical quantity of the real axe is dimensionless (1).

frames of reference in the Intervalic Space have an additional meaning because they are *dimensionally* coherent, whilst the traditional ones are not. It is like to have got two frames of reference in one: the measurement system plus the dimensional system. The axes of the Intervalic Space are dimensionally alive and have new *dimensional* geometric relations between them. In traditional or modern Physics this can never be possible because the *dimensions* of the physical quantities do no have *geometric relations* between themselves. For example, we can represent space and time as usually, in two orthogonal axes, but there is no geometric relations of any kind between their *dimensions* of space and time. On the contrary, space and time in the Intervalic Space are geometrically related by their intervalic dimensions, which are respectively (L) and (*i*L).

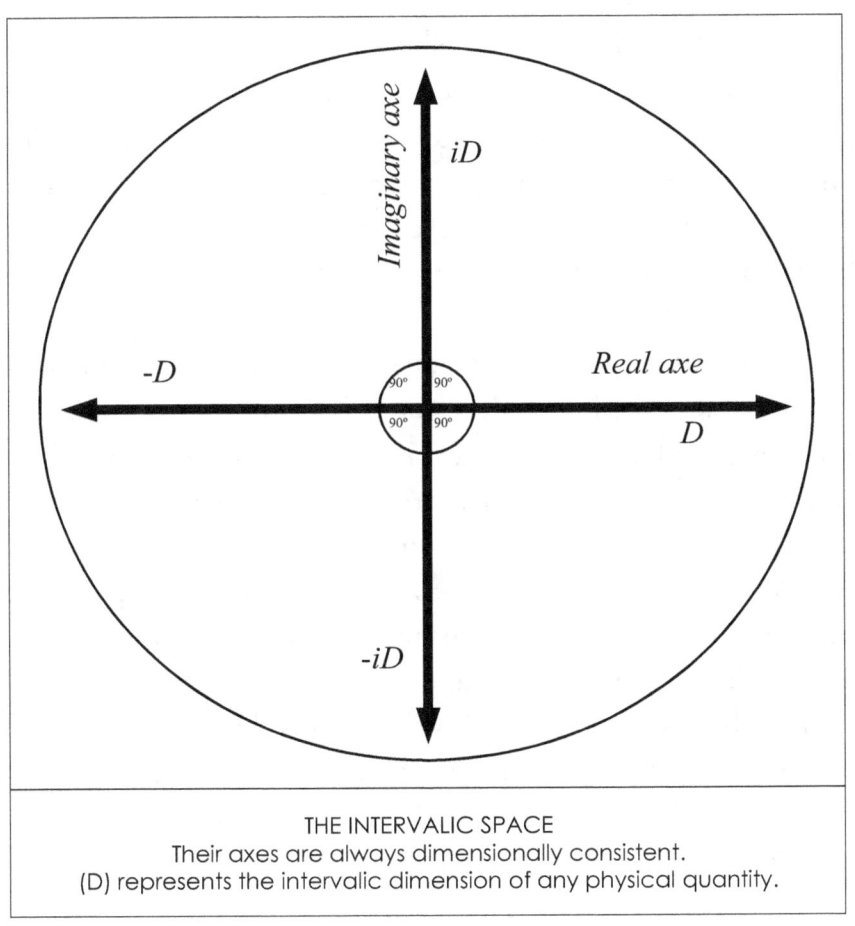

THE INTERVALIC SPACE
Their axes are always dimensionally consistent.
(D) represents the intervalic dimension of any physical quantity.

Actually, we can operate algebraically with the dimensions of the physical quantities in any equation. This is a totally new field of *dimensional algebra* of the physical quantities which is absolutely unthinkable in modern Physics. It may be viewed as a strange fact of the destiny that modern Physics has reached to uncover some hidden relations of Nature but has not found one of the most important and simple of it, those ones of the own dimensional system. The irony is that the so called "geometrized" units claims to be geometric but they have not seen the inner and fundamental geometry of the own physical dimensions.

We usually take the dimension (i) as a rotation of +90° in the Intervalic Space and its inverse ($-i$) as a rotation of -90°. Obviously, their magnitudes are respectively: $c^{-1}(i)$ and c (i^{-1}). Therefore we will call the speed of light as the geometrical *transformer* in the Intervalic Space. The rising of the i number to the successive integers powers produce other intervalic transformers: $c^{-2}(-1)$ and c^2 (-1), which rotates respectively +180° and -180°; $c^{-3}(i^{-1})$ and c^3 (i), which rotates respectively +270° and -270°; and finally $c^{-4}(1)$ and $c^4(1)$, which rotates respectively +360° and -360°.

As can be easily understood, the four last transformers are redundant because they are equivalent respectively to -90°, +90°, -0° and +0°; besides, the rotation by $c^{-2}(-1)$ or by c^2 (-1) makes the same geometrical final result.

These groups of rotations are governed by the curious mathematical properties of the i number, which still appear more strange, in a classical Physics context, when applying to physical quantities. Nevertheless, we don't know any phenomena which breaks the invariance of the geometrical rotations, and therefore the Intervalic Space must conserve such invariance. Moreover, as we will see suitably, invariance of physical quantities and physical equations can be interpreted as a simple consequence derived and due to the geometrical properties inside the Intervalic Space. Geometric relations are invariant in themselves. Hence, the laws of Physics must be geometrically invariant under rotations by +360° —$c^{-4}(1)$— or by -360° —$c^4(1)$—. Of course, this only can be valid in a *singular* system of units, that is to say, a system where the magnitude of c is set to 1.

INTERVALIC THEORY:
The Intervalic Structures of Subatomic Particles and the Last Foundations of Physics

The singular system of units in intervalic dimensions is named the Intervalic System of Units, which we have also described in other site.

Now, lets take any physical quantity, 'x', with intervalic dimension (D). Its rotations in the Intervalic Space are the following:

$-0°$: x (D)
$-90°$: cx (i^{-1}D)
$-180°$: c^2x (-D) $\approx$ $+180°$: $c^{-2}x$ (-D)
$-270°$: c^3x (iD) $\approx$ $+90°$: $c^{-1}x$ (iD)
$-360°$: c^4x (D) $\approx$ $+0°$: x (D)

$+0°$: x (D)
$+90°$: $c^{-1}x$ (iD)
$+180°$: $c^{-2}x$ (-D) $\approx$ $-180°$: c^2x (-D)
$+270°$: $c^{-3}x$ (i^{-1}D) $\approx$ $-90°$: cx (i^{-1}D)
$+360°$: $c^{-4}x$ (D) $\approx$ $-0°$: x (D)

The physical quantity 'x' is therefore called a *generator* in the Intervalic Space. It is very important to notice that all existing physical quantities known up to date (plus five more ones unknown yet) are logically generated by a unique generator in the Intervalic Space: the *intervalic length*, ℏ —the former Planck constant—. The solely properties of the *dimensional product* between the intervalic length, ℏ and the transformers of the Intervalic Space, $c^{\pm n}$, yields all the *derived physical quantities*. In this way it can be generated by means of simple rotations all the physical quantities existing in the Intervalic Space, which is a remarkable characteristic that shows the logical simplicity and elegance of mediums in the Intervalic Space.

The rotations produced by a generator reveal unavoidably a *geometrical equivalence* between physical quantities. Some of these equivalences were unknown till this moment, and other now acquire a new physical meaning. Many of these equivalences were expressed in Classic Physics as *equations*. It can be said that now these physical equations leave to be so,

\multicolumn{5}{c}{**THE GROUP OF GEOMETRICAL *TRANSFORMERS* IN THE INTERVALIC SPACE**}
System of Units	*Name*	*Rotation*	*Magnitude*	*Intervalic dimension*
Intervalic system of units and *singular* systems of units ($c = 1$)	Direct Ortizator	$+90°$	c^{-1}	i
	Inverse Ortizator	$-90°$	c^{1}	i^{-1}
	Antizator	$\pm 180°$	$c^{\pm 2}$	$i^{\pm 2} = -1$
	Direct Pseudortizator	$+270° = -90°$	c^{-3}	$i^{3} = i^{-1}$
	Inverse Pseudortizator	$-270° = +90°$	c^{3}	$i^{-3} = i$
	Scalizator	$\pm 360°$	$c^{\pm 4}$	$i^{\pm 4} = 1$
Traditional systems of units and *not-singular* systems of units ($c \neq 1$)	Direct Ortizator	$+90°$	c^{-1}	i
	Inverse Ortizator	$-90°$	c^{1}	i^{-1}
	Direct Antizator	$+180°$	c^{-2}	$i^{2} = -1$
	Inverse Antizator	$-180°$	c^{2}	$i^{-2} = -1$
	Direct Pseudortizator	$+270° = -90°$	c^{-3}	$i^{3} = i^{-1}$
	Inverse Pseudortizator	$-270° = +90°$	c^{3}	$i^{-3} = i$
	Direct Scalizator	$+360°$	c^{-4}	$i^{4} = 1$
	Inverse Scalizator	$-360°$	c^{4}	$i^{-4} = 1$

and acquire the epistemological rank of simple *geometrical statements*. The study of those geometrical equivalences is a fascinating task which conforms a new branch of Physics: the *Intervalic Geometry,* which we can not treat in this book.

When translate those geometrical statements to traditional dimensions, *only a half* of the symmetries are conserved. Moreover, there is no known rule to determine which way of a intervalic rotation will be conserved when passing to traditional dimensions: if clockwise or the contrary. This arbitrariness reinforce our affirmation about the not-basic rank and the epistemological foundationlessness of the traditional system of units. Broken half symmetry showed by physical quantities in traditional dimensions is not a truthful property of Nature but a serious lack of the ancient theory.

GEOMETRICAL EQUIVALENCES AMONG PHYSICAL QUANTITIES IN THE INTERVALIC SPACE

Some *geometrical equivalences* between diverse physical quantities will probably be understand after a look to the following pictures. It is very important to remark that these equivalences are totally *independent* from the traditional Physics equations known up to date. Therefore, we have find an unexpected independent way to *deduce* geometrically some physical relations which has nothing in common in its origin with the traditional physical work. Nevertheless, its validity is absolutely reliable due to the impressive simplicity involved in the intervalic geometry, where there no exists any assumption that has other origin than the most elemental logical postulates.

The problem is that we have a huge amount of information, and perhaps we have too much new physical relations to be processed at once. And don't forget that above we have not written down all the possible rotations of physical quantities in the Intervalic Space. In example, no one of the rotations of electromagnetic physical quantities —those with fraction powers in its intervalic equation of dimensions— have been drawn up. In the following tables we have written the most familiar expressions for the rotations by ±180°, choosing the half symmetry which is experimentally conserved —known or predicted— in traditional units. This provisional facility don't have to make us to forget that the reliable geometrical symmetries are the intervalic ones, but not the traditional ones.

Some of the geometrical equivalences derived from rotations in the Intervalic Space are unknown, and others coincide with already known equations. For the last ones we now have a powerful and independent way to verify its truthfulness. Moreover, a big part of Physics —as a minimum, all Physics equations involving the speed of light, and probably many equations involving the intervalic length (the former Planck's quantum)— has been immediately converted into pure *geometry*, an unexpected achievement which is widely considered as one of the most reliable *advances* which can be realized in Physics, in a similar way as Einstein did partially General Relativity.

Regarding the unknown geometrical equivalences, it is important to notice the systematic work from which they have been born of. Although at first sight some of them might seem wrong, the logical —and systematic— deduction involved have been applied in the same manner in all physical quantities. Therefore, we are compelled to think that what is wrong, if any, shall be any step in our traditional definition and conception of physical quantities (some of them clearly obsolete and inadequate), but not the logical deductions yielded from the intervalic physical quantities. Really, all scientific knowledge is based on the Pythagorean assumption about the reliability of the *number* and its *geometrical* relations. If the Universe is composed by numbers, as Pythagoras said, there is no logical reason to think that a part of Physics verifies that postulate, and other part —such as the dimensional analysis and the dimensional relations among physical quantities— does not do it. In fact, there is hard —and surely wrong— to affirm that physical quantities are composed by numbers but physical dimensions are not, since the concept of 'dimension' is logically *before* the concept of 'magnitude'.

As P.A.M. Dirac said with great *finezza*, equations are ever more intelligent than their authors. In my case, equations are ever much more intelligent than his author. Really, the true work of the physicist is *to interpret* suitably the results yielded trough mathematical processes. And they must try ever to minimize the logical assumptions underlying in physical concepts and equations. Since the logical basis underlying the intervalic dimensions are by far more simple than those shivering ones which are holding the traditional dimensions, I think that physicists have now an exceptionally good opportunity to make a significant advance in the most important —and forgotten— theme of Physics: its foundations.

In order to finish this epigraph we can list some of the physical quantities born from ortizations, that is to say, from a rotation of $\pm 90°$ in the Intervalic Space:

Time: $T = c^{-1} L$
Antitime: $T_a = c L$
c-velocity: $v_c = c^{-1} v$
c-mass: $m_c = c v^{-1}$
c-momentum: $p_c = c^{-1} p$
Energy: $E = c p$
Wavevector: $\lambda^{-1} = c^{-1} \varphi$
Power: $W = c \varphi$
Antifrequorce: $\varphi_a = c^{-1} \lambda^{-1}$

In the same way, we have the following physical quantities generated by antizations —or rotations of $\pm 180°$— in the Intervalic Space:

Antispace: $L_a = c^{\pm 2} L$
Mass: $m = c^{\pm 2} E$
Energy: $E = c^{\pm 2} m$
Antimomentum: $p_a = c^{\pm 2} p$
Antifrequorce: $\varphi_a = c^{\pm 2} \varphi$
Acceleration: $a = c^{\pm 2} \lambda^{-1}$

Here we can't explain no one of these statements, but only to point out that the logical deduction of the following known formulas as mere geometrical statements yielded from the simplest rotations of physical quantities in the Intervalic Space. Written in traditional units:

Time: $T = c^{-1} L$
Energy: $E = c\, p$
Wavevector: $\lambda^{-1} = c^{-1}\, \varphi$
Mass: $m = c^{-2} E$
Energy: $E = c^2 m$

Among equations unknown to date, the following are interesting at first sight because they join traditional macrocosmic physical magnitudes (W, a) with the microcosmic ones (v, λ). Besides, the first of them make trivial the photoelectric effect since introduces a relation between power and frequorce:

Power: $W = c\, \varphi$
Acceleration: $a = c^{\pm 2}\, \lambda^{-1}$

Among other dimensional symmetries, we have the following between basic physical quantities in the Intervalic Space:

Symmetries of 0° —these ones are indeed *unifications* between physical quantities—
Velocity—Energy (= velenergy)
Invervelocity—Mass
Acceleration—Power (= poweleration)
Frequency—Force (= frequorce)
Momentum—Inertia

Symmetries of 90°
Space—Time
Velenergy—Momentum

Frequorce—Wavevector
Frequorce—Poweleration
Wavevector—Antifrequorce

Symmetries of 180°
Invervelocity-Mass—Velenergy
Poweleration—Wavevector

For the sake of illustration I give some pictures with the groups of intervalic transformations of some physical quantities, a subject pertaining (as really all this chapter) to the Intervalic Geometry, but not to Particle Physics. There is no place in this book to explain them, so if you do not understand something, feel free to take a quick look at them and skip to the next paragraph where we will describe briefly the basics of the dimensional geometry in the Intervalic Space, a theme of great importance which is linked with all fields of Physics because the whole Special Relativity is completely deduced in an independent way as a limit case of the group of intervalic transformations in the Intervalic Space.

	THE GROUP OF INTERVALIC TRANSFORMATIONS OF MASS				
Generator	Transformer	Rotation	Value and dimension	Derived physical quantity and dimension	Geometrical equivalence in intervalic units / traditional units
Mass $m\ (i)$	Direct Ortizator	$+90°$	$c^{-1}\ (i)$	Ortomass- Direct antizator $m_{ort}\ (-1)$	$m_{ort} = c^{-1}\ v^{-1}$ $Ant_{dir} = c^{-2}$ (for $m = c^{-1}$)
	Inverse Ortizator	$-90°$	$c^1\ (i^{-1})$	Light-mass $m_c\ (1)$	$m_c = c\ v^{-1}$
	Antizator	$\pm 180°$	$c^{\pm 2}\ (-1)$	Velenergy $v\ (i^{-1})$	$E = c^{\pm 2}\ m$ $v = c^2\ m$ By definition: $v \equiv E$ $E = c^2\ m$

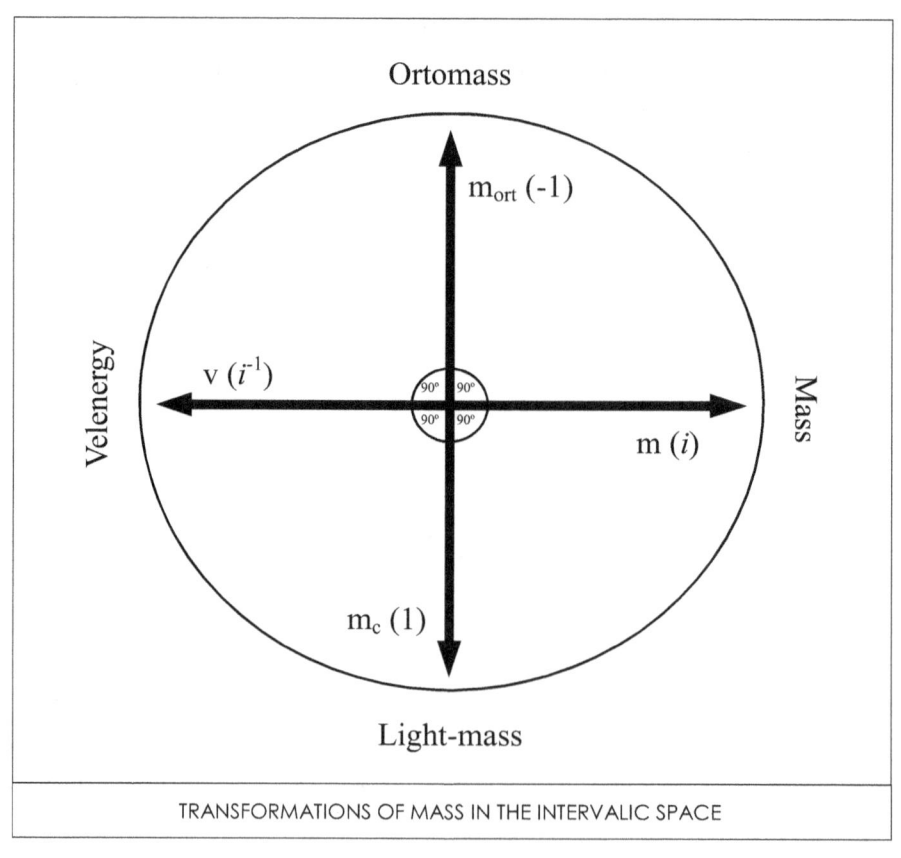

TRANSFORMATIONS OF MASS IN THE INTERVALIC SPACE

INTERVALIC THEORY:
The Intervalic Structures of Subatomic Particles and the Last Foundations of Physics

THE GROUP OF INTERVALIC TRANSFORMATIONS OF VELENERGY

Generator	Transformer	Rotation	Value and dimension	Derived physical quantity and dimension	Geometrical equivalence in intervalic units / traditional units
Velenergy $v\,(i^{-1})$ $E\,(i^{-1})$	Direct Ortizator	$+90°$	$c^{-1}\,(i)$	Light-velocity $v_c\,(1)$	$v_c = c^{-1}\,v$
	Inverse Ortizator	$-90°$	$c^{1}\,(i^{-1})$	Ortovelocity-Inverse antizator $v_{ort}\,(-1)$	$v_{ort} = c\,v$ For: $v = c$: $Ant_{inv} = c^2$
	Antizator	$\pm 180°$	$c^{\pm 2}\,(-1)$	Mass $m\,(i)$	$m = c^{\pm 2}\,E$ $m = c^{-2}\,v$ By definition: $v \equiv E$ $m = c^{-2}\,E$

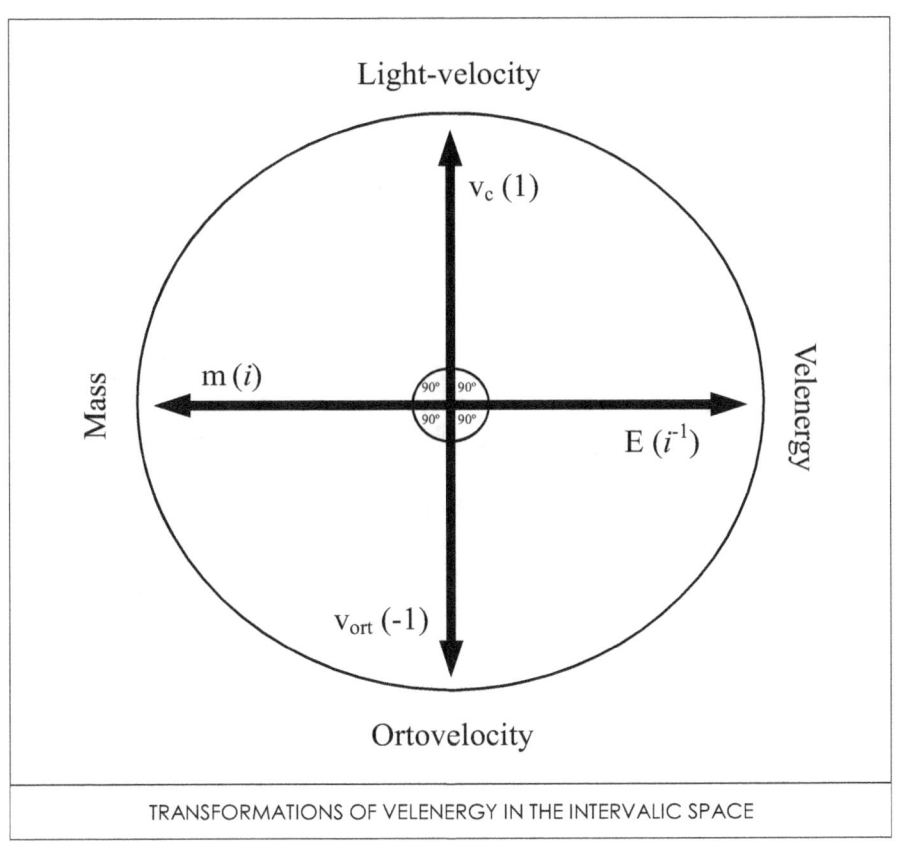

TRANSFORMATIONS OF VELENERGY IN THE INTERVALIC SPACE

\multicolumn{6}{c}{**THE GROUP OF INTERVALIC TRANSFORMATIONS OF LENGTH (REAL SPACE)**}
Generator	Transformer	Rotation	Value and dimension	Derived physical quantity and dimension	Geometrical equivalence in intervalic units / traditional units
Real space $L\,(L)$	Direct Ortizator	$+90°$	$c^{-1}\,(i)$	Time $T\,(iL)$	$T = c^{-1} L$
	Inverse Ortizator	$-90°$	$c^{1}\,(i^{-1})$	Antitime $T_a\,(i^{-1}L)$	$T_a = c\,L$
	Antizator	$\pm 180°$	$c^{\pm 2}\,(-1)$	Antilength $L_a\,(-L)$	$L_a = c^{\pm 2} L$
					$L_a = c^{2} L$

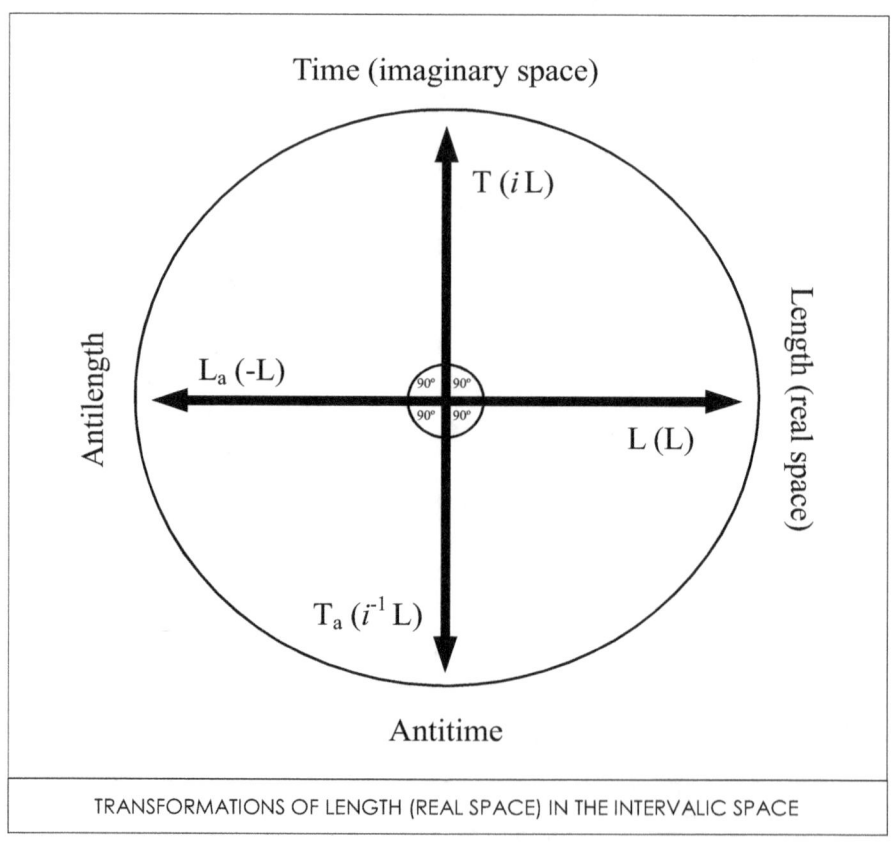

TRANSFORMATIONS OF LENGTH (REAL SPACE) IN THE INTERVALIC SPACE

INTERVALIC THEORY:
The Intervalic Structures of Subatomic Particles and the Last Foundations of Physics

THE GROUP OF INTERVALIC TRANSFORMATIONS OF WAVEVECTOR

Generator	Transformer	Rotation	Value and dimension	Derived physical quantity and dimension	Geometrical equivalence in intervalic units / traditional units
Wavevector λ^{-1} (L^{-1})	Direct Ortizator	$+90°$	c^{-1} (i)	Anti-frequorce φ_a ($i\,L^{-1}$)	$\varphi_a = c^{-1} \lambda^{-1}$
	Inverse Ortizator	$-90°$	c^{1} (i^{-1})	Frequorce φ ($i^{-1}L^{-1}$)	$\varphi = c\,\lambda^{-1}$
	Antizator	$\pm 180°$	$c^{\pm 2}$ (-1)	Acceleration-power a-W ($-L^{-1}$)	$a = c^{\pm 2} \lambda^{-1}$
					$a = c^{2} \lambda^{-1}$

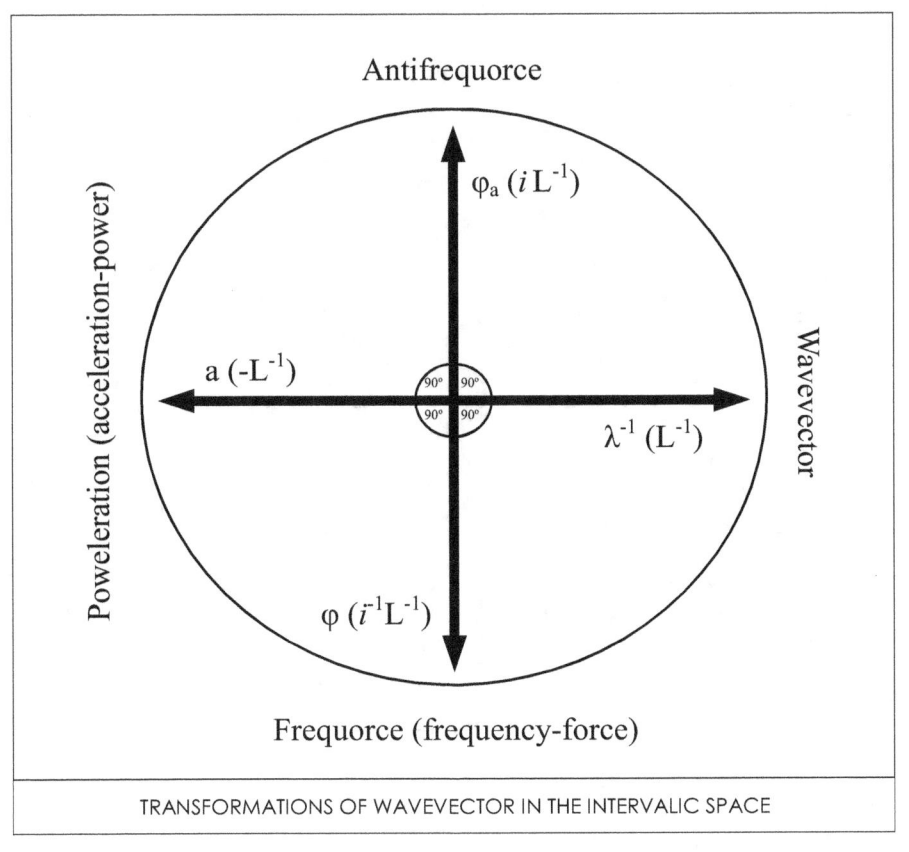

TRANSFORMATIONS OF WAVEVECTOR IN THE INTERVALIC SPACE

THE GROUP OF INTERVALIC TRANSFORMATIONS OF MOMENTUM

Generator	Transformer	Rotation	Value and dimension	Derived physical quantity and dimension	Geometrical equivalence in intervalic units / traditional units
Momentum p (1)	Direct Ortizator	$+90°$	c^{-1} (i)	Light-momentum p_c (i)	$p_c = c^{-1} p$
	Inverse Ortizator	$-90°$	c^1 (i^{-1})	Velenergy E (i^{-1})	$E = c\, p$
	Antizator	$\pm 180°$	$c^{\pm 2}$ (-1)	Antimomentum p_a (-1)	$p_a = c^{\pm 2} p$
					$p_a = c^{-2} p$

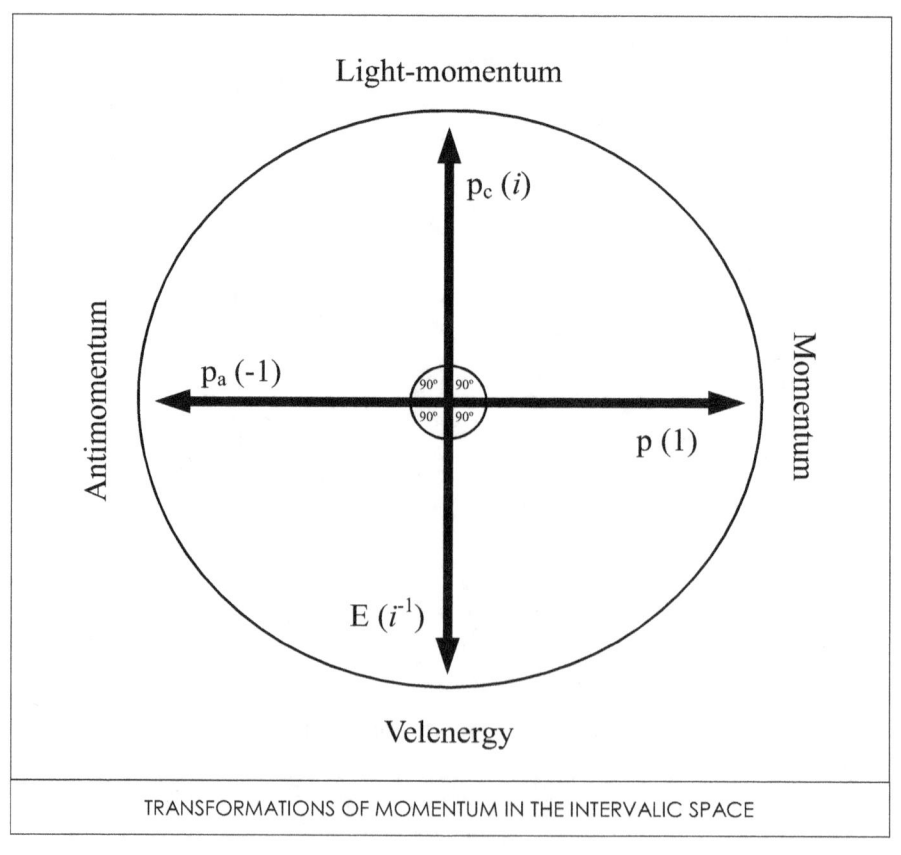

TRANSFORMATIONS OF MOMENTUM IN THE INTERVALIC SPACE

THE GROUP OF INTERVALIC TRANSFORMATIONS OF PERMITIVITY

Generator	Transformer	Rotation	Value and dimension	Derived physical quantity and dimension	Geometrical equivalence in intervalic units / traditional units
Permitivity ε (1)	Direct Ortizator	+90°	c^{-1} (i)	Mass m (i)	$m = c^{-1} \varepsilon$
	Inverse Ortizator	-90°	c^{1} (i^{-1})	Velenergy E (i^{-1})	$E = c \varepsilon$
	Antizator	±180°	$c^{\pm 2}$ (-1)	Inverpermeability μ^{-1} (-1)	$\mu^{-1} = c^{\pm 2} \varepsilon$
					$\mu^{-1} = c^{2} \varepsilon$

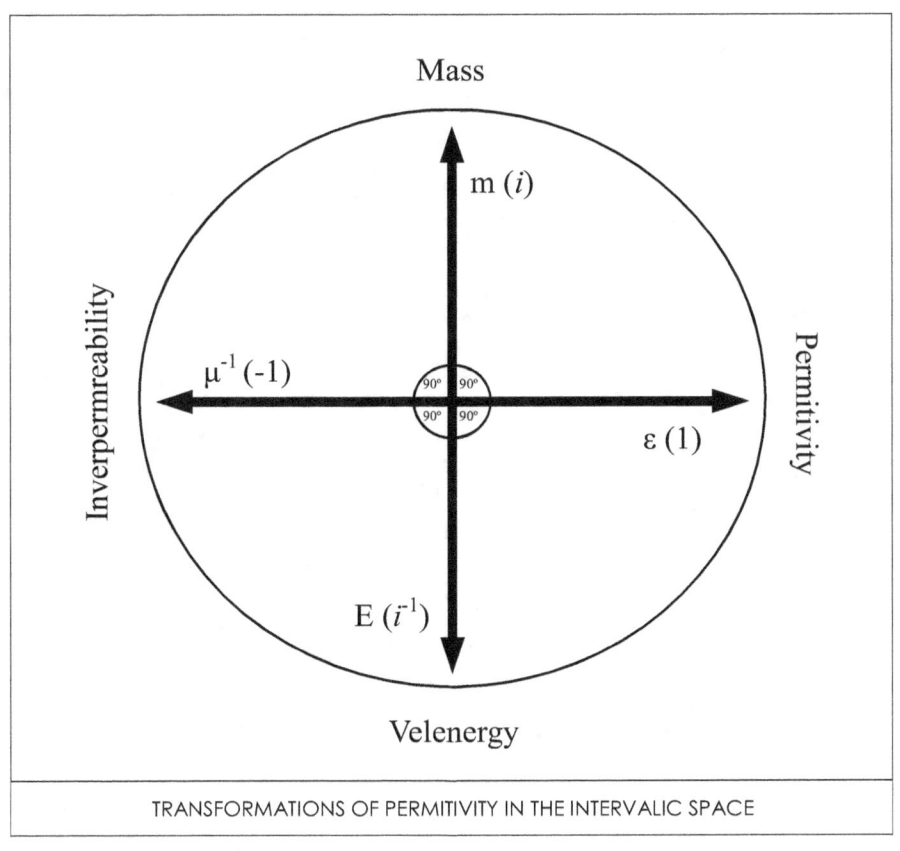

TRANSFORMATIONS OF PERMITIVITY IN THE INTERVALIC SPACE

THE GROUP OF INTERVALIC TRANSFORMATIONS OF FREQUORCE (FREQUENCY-FORCE)

Generator	Transformer	Rotation	Value and dimension	Derived physical quantity and dimension	Geometrical equivalence in intervalic units / traditional units
Frequorce φ ($T^{-1} = i^{-1}L^{-1}$)	Direct Ortizator	+90°	c^{-1} (i)	Wavevector λ^{-1} (L^{-1})	$\lambda^{-1} = c^{-1} \varphi$
	Inverse Ortizator	-90°	c^{1} (i^{-1})	Acceleration-power a-W ($-L^{-1}$)	$W = c \varphi$
	Antizator	±180°	$c^{\pm 2}$ (-1)	Anti-frequorce φ_a (iL^{-1})	$\varphi_a = c^{\pm 2} \varphi$
					$\varphi_a = c^{-2} \varphi$

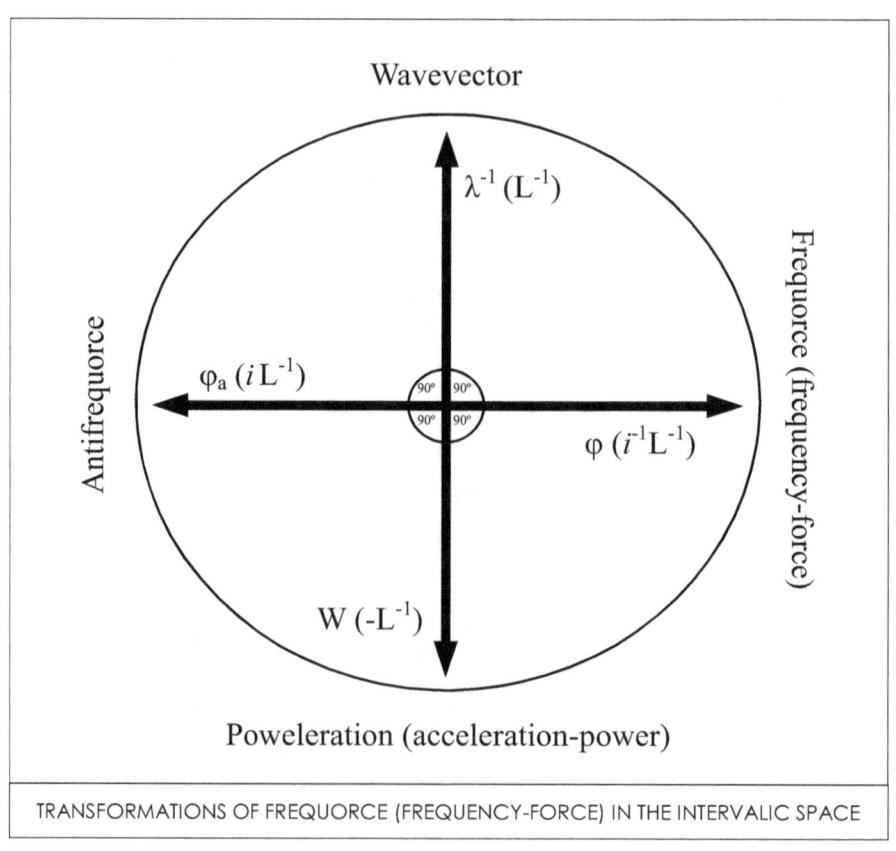

TRANSFORMATIONS OF FREQUORCE (FREQUENCY-FORCE) IN THE INTERVALIC SPACE

THE GEOMETRY OF SPACE-TIME IN THE INTERVALIC SPACE

> Henceforth space by itself, and time by itself, are doomed to fade away into mere shadows, and only a kind of union of the two will preserve an independent reality.
>
> HERMANN MINKOWSKI
> *Space and Time* (1908)

All physical quantities exist in precise dimensional geometrical relations to other physical quantities in the Intervalic Space. This dimensional geometry affects not only to the pure equation of dimensions of the involved physical quantities, but to the whole relation comprising the magnitudes of the measurement and its intervalic dimensions insofar as the dimensions of the physical quantities operate algebraically along with the magnitudes. This is unthinkable in traditional and modern Physics where the dimensions pertain to an abstract, closed and endogamy realm, isolated from any kind of measurement, which never merges with any magnitude under no one circumstance. But in IT the situation of the physical quantities is very different: they "live" in the geometry of the Intervalic Space. They live in dimensional relation with other physical quantities, governed by the geometry of the speed of light. We already have seen some sites where they dwell: the dimensional frames of reference of the intervalic transformation of physical quantities described hereinbefore. For example, in the case of real space, it is always in geometric relation with time (90º), antilength (180º) or antitime (270º). Disregarding virtual states, we will also find the physical bodies existing in the quadrant delimited by real space (L) and imaginary space (iL), shadowed in the picture. Thus, any body may be measured with respect of space and time (space-time). The astonishing fact is that in this dimensional frame of reference it can be applied all statements of traditional Geometry. For example, we can determine the position of any body applying the usual tools of the theorem of Pythagoras operating algebraically with the involved intervalic dimensions. Moreover, the geometric results obtained in this way are *invariant* because they are derived from geometric statements —abstract statements— which do not rely on any physical frame of reference. Clearly any magnitude derived mathematically through the Pythagoras' theorem is, *a fortiori*, invariant because Geometry is an abstract science. Thus we do not have to make any probe to know whether any magnitude is or not invariant in the Intervalic Space: all of them are invariant insofar as all of them are purely geometric statements inside the Intervalic Space.

Being x_1, x_4 and s any dimensionless magnitudes in abscissas, ordinates and hypotenuse, we have:

$$[x_1(L)]^2 + [x_4(iL)]^2 = [s(X)]^2$$

The X is the intervalic dimension of the hypotenuse and represent some dimension composed from (L) and (iL). Obviously this is not admissible for traditional Physics where all physical quantities have got a determined dimension, but it is unimaginable to have got a dimension which is not predetermined but determined by an algebraic relation from others, and which by this reason, that dimension can take itself several values.

For more generality we can add the other two real spatial dimensions to the above geometric statement, where x_2 and x_3 are their corresponding magnitudes:

$$[x_1(L)]^2 + [x_2(L)]^2 + [x_3(L)]^2 + [x_4(iL)]^2 = [s(X)]^2$$

If we repeat the measurement for that body after a differential of time, the magnitudes x_1, x_2, x_3, x_4 and s will now be interpreted as differential intervals. Any physical body must be in that quadrant limited by the axis of real and imaginary spaces, so it will be comprised between the two following geometric limits:

- If $x_1 = x_2 = x_3 = 0$, the body is at rest in real space and moves only along the imaginary space —that is to say, along time—. Real space does not elapse for that event. Therefore its velocity, v, is zero: v = 0. And the dimension of X is time:

$$s(X) = x_4(iL)$$

- If $x_4 = 0$, the body is at rest in imaginary space and moves only along the real space. Imaginary space —time— does not elapse for that

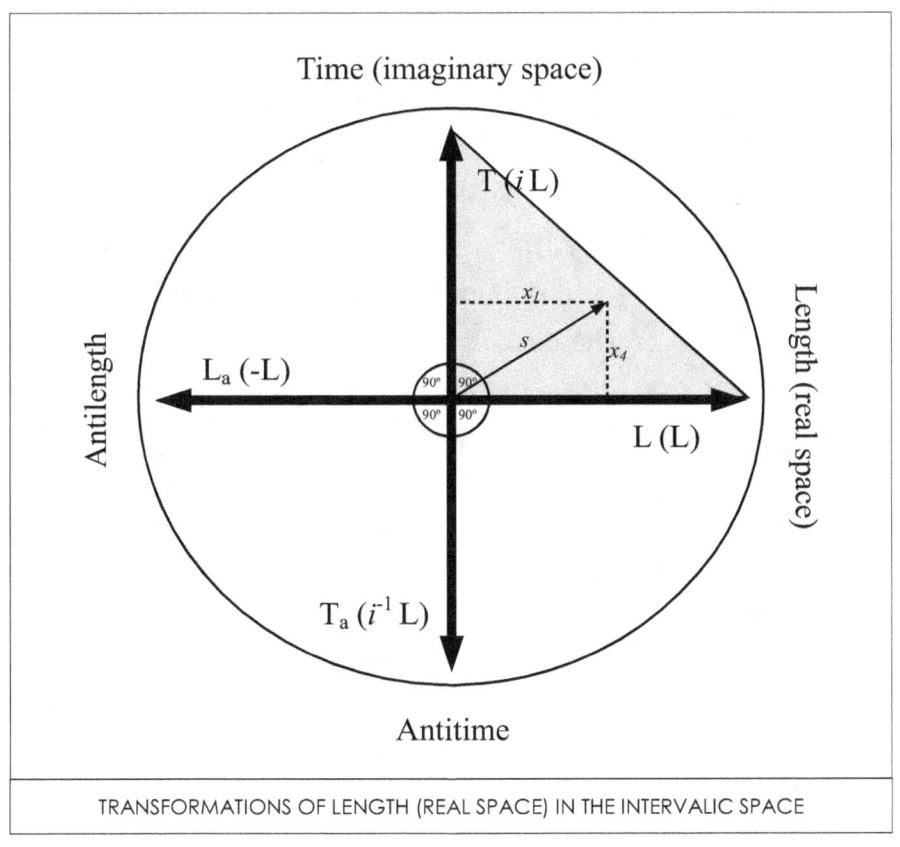

TRANSFORMATIONS OF LENGTH (REAL SPACE) IN THE INTERVALIC SPACE

event. Therefore its velocity, v, is infinite: v = ∞. And the dimension of X is real space:

$$s(X) = \sqrt{[x_1^2 + x_2^2 + x_3^2]}\,(L)$$

Between these two limits cases we have, applying elemental algebra for operating with the intervalic dimensions:

$$[x_1(L)]^2 + [x_2(L)]^2 + [x_3(L)]^2 - [x_4(L)]^2 = [s(X)]^2$$
$$[x_1^2 + x_2^2 + x_3^2 - x_4^2]\,(L)^2 = s^2(X)^2$$
$$s(X) = \sqrt{[x_1^2 + x_2^2 + x_3^2 - x_4^2]}\,(L)$$

From here it can be yielded three cases:

- If $x_1^2 + x_2^2 + x_3^2 > x_4^2$, the dimension of X is (L) —that is to say: $s(X) = \sqrt{x^2}\,(L) = x\,(L)$— and the interval s is *spatial*. It means the space distance between two events.

- If $x_1^2 + x_2^2 + x_3^2 < x_4^2$, the dimension of X is (iL) —that is to say: $s(X) = \sqrt{-x^2}\,(L) = i\,x\,(L) = x\,(i\mathrm{L})$— and the interval s is *temporal*. It means the time distance between two events.

- If $x_1^2 + x_2^2 + x_3^2 = x_4^2$, the dimension of X is not defined —that is to say: $s(X) = \sqrt{0}\,(L)$— and the interval s is *zero*. It means that the undefined distance between these two events is zero, or in other words, such distance does not exist. It is obvious that it involves necessarily that the velocity of the body is just the speed of light, v = c. This case represent the so named *cone light* of Special Relativity. To be "at rest" in the Intervalic Space means to be moving at the speed of light. Thus the singular result $s = 0$ means strangely that neither space nor time elapse for light — or for the frame of reference of a photon—.

By the way, we can comment a curious detail in relation with the preceding result. If a photon is "at rest" in the Intervalic Space it means that the perception of the arrow of time is merely a limitation of our senses. In the same way as we do not see the complete dimension of the real space of the Universe, but only a diminutive part of it, we only see a diminutive part of the imaginary space. One may wonder about why the first perceptual limitation does not trouble our mind, but the second one certainly does. We do not say that real space is *unfolding* because we do not see it in its full size. Analogously we should not say that imaginary space —time— is unfolding because we can not see it in full. Both real and imaginary space are at once completely unfolded. There are only our perceptual limitations which impede to realize that fact in an oscillating Universe. In this way the speed of light would be better interpreted and measured as *energy* instead of *velocity*, just as Lancelot L. Whyte advised cleverly in his **Critique of Physics**. Please remember that in the Intervalic Space energy and velocity have just the same underlying dimension: *velenergy*, what is meaningful due to the epistemological rank of the intervalic dimensions. Nevertheless you do not need to realize this detail to understand the intervalic geometry of light.

In our Universe there are only allowed events which occur inside the light cone, which means that there are allowed only *temporal* and *zero* intervals (for events involving *energy*). As we will see later, the existence of time in the Intervalic Space always involves necessarily the existence of energy, and vice versa. Nevertheless, *spatial* intervals are allowed for events which do not involve energy (that is to say, for events involving only *information*).

What we have done here with the intervalic geometry of space-time can be done in a similar way for every one of the physical quantities of the Intervalic Group, and specially with the 16 physical quantities containing integers powers of the *i* number in their equations of dimensions (the two dimensionless ones —(1) and (-1)— are generated by the powers of the *i* number). Thus we will obtain the intervalic geometry of, for example, frequorce-powelaration, etc. although the case of space-time is perhaps one of the most interesting ones due to its remarkable simplicity.

THE MINKOWSKI ANTECEDENT

After viewing this geometry it is obvious that the tetradimensional reformulation of Special Relativity by Minkowski in only a particular case of the intervalic geometry of physical quantities. The famous idea of Minkowski of substituting *by hand* the temporal variable by the expression *ict*, involves clearly the intervalic dimension for time. However, since imaginary dimensions were not allowed in the dimensional basis, Physics has not interpreted correctly this fact, but only as a good tool suitable for a better formulation of General Relativity, which probably had not existed without this reformulation. It is clear that the aim of Minkowski was to express the relation between space and time of Special Relativity in a manner which seems the Euclidian geometry. Then his substitution *ad hoc* of the temporal component by *ict*. Really the own Minkowski said this in his famous *Address* delivered at the 80th Assembly of German Natural Scientist and Physicians, at Cologne, 21 September, 1908:

We can determine the ratio of units of length and time beforehand in such a way that the natural limit of velocity becomes $c = 1$. If we then introduce, further, $\sqrt{-1}\, t = s$ in place of t, the quadratic differential expression

$$d\tau^2 = -dx^2 - dy^2 - dz^2 - ds^2$$

Thus becomes perfectly symmetrical in x, y, z, s; and this symmetry is communicated to any law which does not contradict the world-postulate. Thus the essence of this postulate may be clothed mathematically in a very pregnant manner in the mystic formula

$$3 \cdot 10^5 \text{ km} = \sqrt{-1} \text{ sec.}$$

It can be said louder but not clearer: $c = i$. Please remember that the intervalic system of dimensions has two identical formulations: one with c (i) and the other with c (i^{-1}), which is what we have used in this book by no special reason. They both are absolutely equivalent. So it could be said that the seed of the intervalic system of dimensions was already planted by Minkowski in 1908, although the panic fear to the imaginary dimensions prevented the logic —and logically unavoidable— development of the germinal idea towards the conformation of a new physical system.

THE GEOMETRY OF VELOCITY IN THE INTERVALIC SPACE

We have seen the basic enouncement of the geometry of space-time in the Intervalic Space. Now let us go the describe the basics of the intervalic geometry of velocity. We already have make a picture with the dimensional transformation of velenergy in the Intervalic Space. The main difference of the intervalic geometry of velocity with respect to that one of space is that the intervalic unit of velocity is a *limit*, c (i^{-1}), whilst the intervalic unit of length is a *quantum*, ℏ (L). Therefore we must incorporate such limit which is inherent to the own geometry of the Intervalic Space to any measurement made in the Intervalic space. When dealing with space there was no such problem because it is implicitly understood in the Intervalic Space that no one measurement smaller than the intervalic length, ℏ, is allowed (and anyway is out of any possibility of practical measurement). So we have drawn this way in the picture near this lines, and applying straightforwardly the Pythagorean theorem —as if we would making any measurement in this intervalic frame of reference—, we have got the following dimensional relation, where v_c is the *light-velocity*, $v_c = c^{-1}v$, a dimensionless physical quantity which is usually written in relativistic Physics as $v_c = v/c$:

$$s^2 = c^2 + v_c^2$$

The magnitude of s^2 is an *interval*, and it is once more *invariant* as commented, since it states and comes from a geometrical theorem. Obviously the above relation can be extended to the other two remaining axes of the real space and we would have:

$$s^2 = c^2 + v_{c(x)}^2 + v_{c(y)}^2 + v_{c(z)}^2$$

But as without lacking generality we can always choose a frame of reference where $v_{c(y)} = v_{c(z)} = 0$, we can maintain the first simpler formulation:

$$s^2 = c^2 + v_c^2 \text{, that is to say, } s = \sqrt{(c^2 + v_c^2)}$$

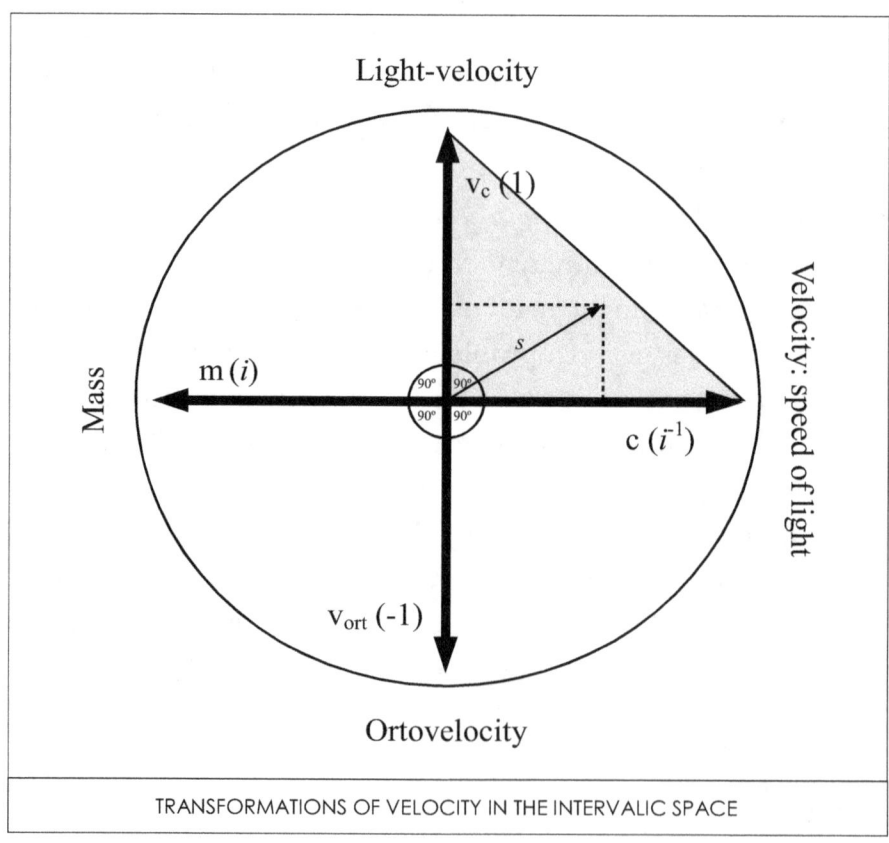

TRANSFORMATIONS OF VELOCITY IN THE INTERVALIC SPACE

With the only aim to make a formulation whose symbols were similar to those of Special Relativity —which is an insignificant detail without any relevance—, it can be introduced the *intervalic factor xi*, ξ, defined simply as the inverse of the invariant interval, s:

$$\xi \quad 1/s = 1/\sqrt{(c^2 + v_c^2)}$$
$$\xi^{-1} \quad s = \sqrt{(c^2 + v_c^2)}$$

The intervalic dimension of ξ is obviously $1/\sqrt{(-1)} = (i^{-1})$.

It must be noted that we have to introduce necessarily intervalic units in this expression in order to preserve its consistency, since both c and the *light-velocity* v_c appear inside the formula. Therefore, we are going to substitute c^2 by its corresponding magnitude in intervalic units, 1, and its dimension, (-1):

$$\xi \quad 1/s = 1/\sqrt{[1(-1) + v_c^2]}$$

A before-mentioned property of the intervalic dimensions is that they operate algebraically with its corresponding magnitudes. That is to say, if 'x' is a magnitude and 'D' an intervalic dimension, we have:

$$-x\,(D) = x\,(-D)$$

Of course, the minus sign of a magnitude can never be interpreted as a value due to incompleteness of the system, as in Quantum Mechanics, but as a minus sign intrinsic value. For example, it is the same to have in the Intervalic Space -7 seconds or 7 antiseconds. Thus, operating algebraically with magnitudes and dimensions in the preceding formulation of the intervalic xi factor, we get the canonical —relativistic-like— expression of ξ (i^{-1}):

$$\xi = 1/\sqrt{(-1 + v_c^2)}$$

Since the intervalic dimension of ξ is (i^{-1}), this means that ξ is an *ortizator* in the Intervalic Space, just like c (i^{-1}). And since its expression is a simple theorem that refers to a geometrical relation, it can be used an ortizator instead of c (i^{-1}) to obtain a set of invariant transformations in the Intervalic Space. Among the most simple of them, we have the rotations the five basic magnitudes: real space (L), imaginary space —that is to say, time— (iL), velenergy (i^{-1}), mass (i) and momentum (1):

- $L \cdot \xi^{-1}$ (i) = T_a ($i^{-1}L$), antitime
- $L \cdot \xi$ (i^{-1}) = T (i L), time
- $T \cdot \xi^{-1}$ (i) = L (L), space
- $T \cdot \xi$ (i^{-1}) = L_a (-L), antispace
- $E \cdot \xi^{-1}$ (i) = p_a (-1), antimomentum
- $E \cdot \xi$ (i^{-1}) = p (1), momentum
- $m \cdot \xi^{-1}$ (i) = p (1), momentum
- $m \cdot \xi$ (i^{-1}) = p_a (-1), antimomentum
- $p \cdot \xi^{-1}$ (i) = E (i^{-1}), velenergy
- $p \cdot \xi$ (i^{-1}) = m (i), mass

Now to get a most suitable expressions we can operate algebraically with the intervalic dimensions. If we multiply dimensionally the odd lines by (i) and the even lines by (i^{-1}), we get the full set of the so named Lorentz-Einstein transformations, which are only the below group of the intervalic invariant transformations with the *dimensionless* factor $\gamma = 1 / \sqrt{1 - v_c^2}$ instead of the dimensional one ξ (i^{-1}). The conversion of ξ (i^{-1}) into γ (1) is immediate when multiplying dimensionally by (i):

$$1/\sqrt{-1 + v_c^2} \cdot (\sqrt{-1}) = 1 / \sqrt{1 - v_c^2}$$
$$\xi (i^{-1}) \cdot (i) = \gamma (1)$$

Therefore, drawing the mentioned operations in odd and even lines, we get directly the Lorentz-Einstein transformations of Special Relativity as a simple invariant intervalic rotation in the Intervalic Space. It can be noted that which is logical and automatically obtained is, besides, the full basic set of relativistic transformations, that is to say, just the relativistic equations for the longitude, time, mass, energy and momentum, five relativistic physical quantities that coincide exactly with the five basic physical quantities of the Intervalic Space —logically born of the simplest geometrical relations among the intervalic dimensions—:

Longitude
$L_0 \cdot \xi^{-1} = $ antitime, $T_a \, (i^{-1}L)$
$L_0 \cdot \sqrt{(-1 + v_c^2)} \cdot (i) = (i^{-1}L) \cdot (i) \quad L$
$L_0 \sqrt{(1 - v_c^2)} = L$
$L_0 \, \gamma^{-1} = L$

Longitude
$L \cdot \xi = $ time, $T_0 \, (i \, L_0)$
$L \cdot [1 / \sqrt{(-1 + v_c^2)}] \cdot (i^{-1}) = (i \, L_0) \cdot (i^{-1}) \quad L_0$
$L \cdot 1 / \sqrt{(1 - v_c^2)} = L_0$
$L = \sqrt{(1 - v_c^2)} \, L_0$
$L = \gamma^{-1} L_0$

Time
$T \cdot \xi^{-1} = $ space, $L_0 \, (L)$
$T \cdot \sqrt{(-1 + v_c^2)} \cdot (i) = L_0 \cdot (i) \quad T_0$
$T \cdot \sqrt{(1 - v_c^2)} = T_0$
$T = T_0 / \sqrt{(1 - v_c^2)}$
$T = \gamma \, T_0$

Time
$T_0 \cdot \xi = $ antispace, $L_a \, (-L)$
$T_0 \cdot [1 / \sqrt{(-1 + v_c^2)}] \cdot (i^{-1}) = L_a \cdot (i^{-1}) = (-L) \cdot (i^{-1}) = i \, L \quad T$
$T_0 \cdot 1 / \sqrt{(1 - v_c^2)} = T$
$T_0 \cdot \gamma = T$

Energy

$E \cdot \xi^{-1} = $ antimomentum, p_a (-1)
$E \cdot \sqrt{(-1 + v_c^2)} \cdot (i) = (-1) \cdot (i) = (-i) = (i^{-1}) \quad E_0$
$E \cdot \sqrt{(1 - v_c^2)} = E_0$
$E = E_0 / \sqrt{(-1 + v_c^2)}$
$E = \gamma E_0$

Energy

$E_0 \cdot \xi = $ momentum, p (1)
$E_0 \cdot [1 / \sqrt{(-1 + v_c^2)}] \cdot (i^{-1}) = (1) \cdot (i^{-1}) = (i^{-1}) \quad E$
$E_0 \cdot 1 / \sqrt{(1 - v_c^2)} = E$
$E_0 \cdot \gamma = E$

Mass

$m \cdot \xi^{-1} = $ momentum, p (1)
$m \cdot \sqrt{(-1 + v_c^2)} \cdot (i) = (1) \cdot (i) = (i) \quad m_0$
$m \cdot \sqrt{(1 - v_c^2)} = m_0$
$m = m_0 / \sqrt{(-1 + v_c^2)}$
$m = \gamma m_0$

Mass

$m \cdot \xi = $ antimomentum, p_a (-1)
$m_0 \cdot [1 / \sqrt{(-1 + v_c^2)}] \cdot (i^{-1}) = (-1) \cdot (i^{-1}) = -(-i) \quad m$
$m_0 \cdot 1 / \sqrt{(1 - v_c^2)} = m$
$m_0 \cdot \gamma = m$

Momentum

$p \cdot \xi^{-1} = $ velenergy, E (i^{-1})
$p \cdot \sqrt{(-1 + v_c^2)} \cdot (i) = (i^{-1}) \cdot (i) = (1) \quad p_0$
$p \cdot \sqrt{(1 - v_c^2)} = p_0$
$p = p_0 / \sqrt{(-1 + v_c^2)}$
$p = \gamma p_0$

Momentum

$p \cdot \xi =$ mass, m (i)

$p_0 \cdot [1 / \sqrt{(-1 + v_c^2)}] \cdot (i^{-1}) = (i) \cdot (i^{-1}) = (1)$ p

$p_0 \cdot 1 / \sqrt{(1 - v_c^2)} = p$

$p_0 \cdot \gamma = p$

Thus, the Lorentz-Einstein transformations of Special Relativity are less more than the enouncement of the Pythagoras' theorem in the Intervalic Space regarding velocity. And the famous relation $E = mc^2$ is merely an almost trivial geometric statement: the dimensional rotation of velenergy by 180º in the Intervalic Space: $E = c^{\pm 2} m$. It must be noted once more that all *geometric relations* in the Intervalic Space are *invariant*. In fact it can't be in other way inasmuch as those geometric relations are mathematical theorems which do not depend on a system of coordinates. Troubles and pseudo problems arise often in modern Physics when using a dimensional system which is simply wrong or incomplete to describe the physical world, but never when using the unique right one which is according with the nature and the underlying geometry of our Universe.

THE GEOMETRY OF TEMPERATURE IN THE INTERVALIC SPACE

As we know, temperature has got the same dimensions than velenergy in intervalic dimensions, (i^{-1}). And Boltzmann constant is dimensionless, (1). If we write the known classical formula for the conversion between temperature and energy, we can substitute the energy value by its corresponding quantum of energy in the Intervalic Space, that is to say, c (i^{-1}), and directly is obtained the intervalic temperature, Θ_I:

$\Theta_I = c \, k_B^{-1} = 2.171374 \cdot 10^{31}$ (K)

The meaning of this quantum is similar to the speed of light: it is the maximum temperature allowed in the Universe for any kind of matter or energy, as like as the speed of light is the maximum velocity allowed in a similar way.

For those who feel the deduction of the intervalic temperature too much easy, they can make the following deduction using the intervalic frequorce, $\varphi_I = ch^{-1}$. This frequorce is likewise the highest frequorce allowed in the Universe. Indeed, this is a geometrical quantum limit that can't never be reached, and there are other limits for the gravitational frequorce and also for the electromagnetic frequorce that are both smaller that the intervalic frequorce. These limits are geometrical and meaningfully related.

$$\Theta_I = E / k_B = \hbar \, \varphi_I / k_B = \hbar \, c \, \hbar^{-1} / k_B = c / k_B$$

Due to the epistemological rank of the intervalic dimensions, we can assure that if two physical quantities have the same dimensions, they are solely one underlying physical quantity and have also the same physical properties, although they may have apparently different behaviors at micro and macroscales of observation. In short, we can predict that the measurement of *heat* bodies in the Intervalic Space —that is to say, its intervalic-relativistic behavior— is similar to the electrodynamics of *moving* bodies. The geometry of the *temperature* in the Intervalic Space will be very similar to the geometry of *velenergy* in the Intervalic Space. In this case, the roles of *velocity* and *speed of light* in the classical formulation are substituted respectively by *temperature* and *intervalic temperature*. In traditional terminology, it could be said that there is a Special Relativity regarding *temperature*, while the classic one would be the Special Relativity regarding *velocity*. The relativistic gamma factor regarding the temperature is:

$$\gamma(\Theta) = 1 / \sqrt{(1 - \Theta_c^2)}$$

where Θ_c is the *light-temperature*, defined as:

$$\Theta_c = \Theta/\Theta_I$$

Therefore, the corresponding intervalic transformations regarding temperature are, written in the relativistic classic mode:

$$L = L_0\, \gamma^{-1}(\Theta)$$
$$T = \gamma(\Theta)\, T_0$$
$$E = \gamma(\Theta)\, E_0$$
$$m = \gamma(\Theta)\, m_0$$
$$p = \gamma(\Theta)\, p_0$$

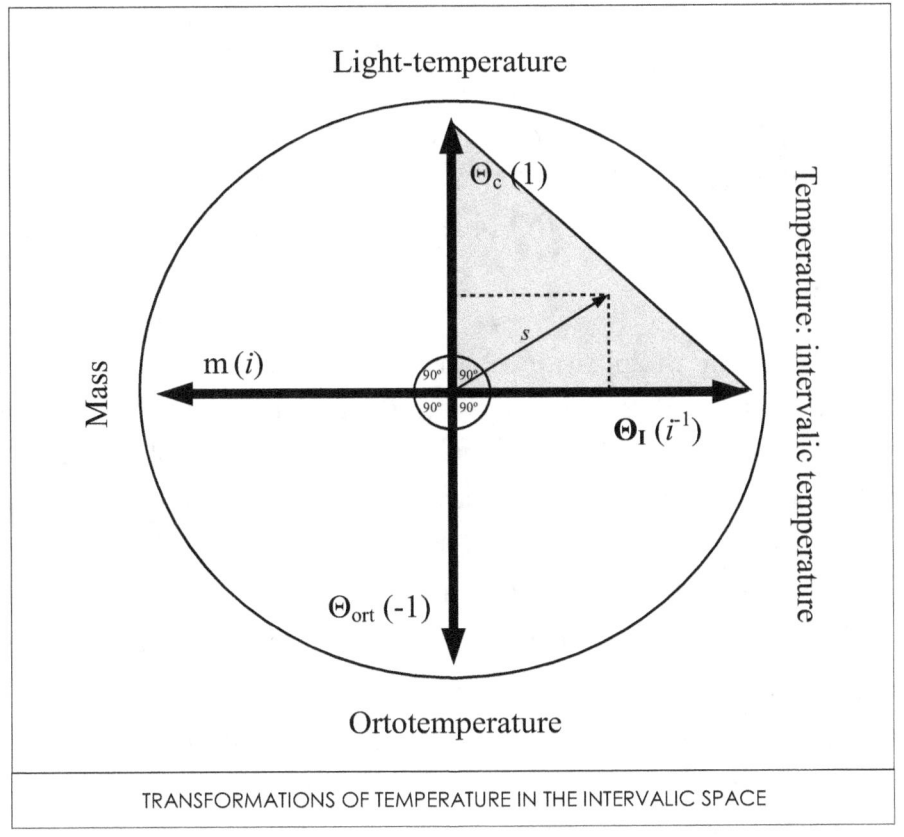

TRANSFORMATIONS OF TEMPERATURE IN THE INTERVALIC SPACE

Since the intervalic temperature is terribly high, these transformations are hardly measurable in our world. Nevertheless, in the primitive Universe they could have played an important role. For example, if we think that in the beginning of the Universe the temperature of the primordial light egg reached the intervalic temperature, the measurement of the first seconds of the Big Bang would not be seconds, but years. With more precision, at the zero time, with just Θ_I temperature, the dilatation of time would be infinite, that is to say, the Universe would be timeless at this point. Of course, this is a fascinating problem of mathematical limits which we can not treat here.

Finally, a crucial experiment could be realized in order to determine a possible falsation or verification of these intervalic transformations regarding temperature. The high precision of the atomic clocks could measure the following dilation of time due to temperature:

$$T = \gamma(\Theta) T_0$$

INTERVALIC GEOMETRY OF SPACE:
SPACE GEOMETRY IN THE INTERVALIC DIMENSIONAL SPACE

To determine the position of any body we can apply the usual tools of the Theorem of Pythagoras operating algebraically with the involved intervalic dimensions. Moreover, the geometric results obtained in this way will be *invariant* because they are derived from geometric statements which do not rely on the frame of reference.

$$[x_1(L)]^2 + [x_2(L)]^2 + [x_3(L)]^2 + [x_4(iL)]^2 = [s(X)]^2$$

where x_1, x_2, x_3, x_4 and s are any dimensionless magnitudes and (X) some dimension composed from real space (L) and imaginary space (iL).

If we repeat the measurement for that body after a differential of time, the magnitudes x_1, x_2, x_3, x_4 and s will now be interpreted as differential intervals. Any physical body must be in that quadrant limited by the axis of real and imaginary spaces, so it will be comprised between the two following geometric limits:

- If $x_1 = x_2 = x_3 = 0$, the body is at rest in real space and moves only along the imaginary space —that is to say, along time—. Real space does not elapse for that event. Therefore its velocity, v, is zero: v = 0. And the dimension of (X) is time:

$$s(X) = x_4 (iL)$$

- If $x_4 = 0$, the body is at rest in imaginary space and moves only along the real space. Imaginary space —time— does not elapse for that event. Therefore its velocity, v, is infinite: v = ∞. And the dimension of (X) is real space:

$$s(X) = \sqrt{[x_1^2 + x_2^2 + x_3^2]}\,(L)$$

Between these two extreme cases we have, operating with intervalic dimensions:

$$[x_1^2 + x_2^2 + x_3^2 - x_4^2]\,(L)^2 = s^2(X)^2$$
$$s(X) = \sqrt{[x_1^2 + x_2^2 + x_3^2 - x_4^2]}\,(L)$$

From here it can be yielded three cases:

- If $x_1^2 + x_2^2 + x_3^2 > x_4^2$, the dimension of (X) is (L) —that is to say: $s(X) = \sqrt{x^2}\,(L) = x\,(L)$— and the interval s is *spatial*. It means the space distance between two events.

- If $x_1^2 + x_2^2 + x_3^2 < x_4^2$, the dimension of (X) is (iL) —that is to say: $s(X) = \sqrt{-x^2}\,(L) = i\,x\,(L) = x\,(iL)$— and the interval s is *temporal*. It means the time distance between two events.

- If $x_1^2 + x_2^2 + x_3^2 = x_4^2$, the dimension of (X) is not defined — that is to say: $s(X) = \sqrt{0}\,(L)$ — and the interval s is *zero*. It means that the distance between these two events is zero, or in other words, such distance does not exist. It is obvious that it involves necessarily that the velocity of the body is just the speed of light, $v = c$. This case represent the so named *cone light* of Special Relativity. To be "at rest" in the Intervalic Space means to be moving at the speed of light. Thus the singular result $s = 0$ means that neither space nor time elapse for light, or for the frame of reference of a photon.

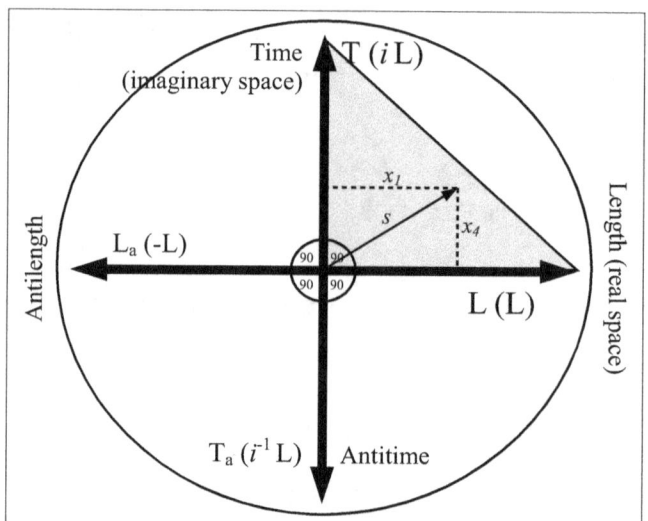

Since a photon is "at rest" in the Intervalic Space it means that the perception of the arrow of time is merely an illusion of our senses. In the same way as we do not see the complete dimension of the real space of the Universe, but only a diminutive part of it, we only see a diminutive part of the imaginary space. One may wonder about why the first perceptual limitation does not trouble our mind, but the second one certainly does. We do not say that real space is *unfolding* because we do not see it in its full size. Analogously we should not say that imaginary space —

time— is unfolding because we can not see it in full. Both real and imaginary space are at once completely unfolded. There are only our perceptual limitations —that work immersed inside spacetime— which impede to realize that fact. And not only time but all physical quantities whose intervalic dimension contains some (i) factor make the illusion of being time-arrowed.

In our Universe there are only allowed *temporal* and *zero* intervals for events involving *energy*. The existence of time in the Intervalic Space always involves necessarily the existence of energy, and vice versa. Nevertheless, *spatial* intervals are allowed for events which do not involve energy, that is to say, for events involving only physical quantities whose intervalic dimension contains only some (L) factor, such as *information*, *space* or *consciousness*.

Finally, what has been done here for the intervalic geometry of *space*, can be done in a similar way for every one of the 40 physical quantities of the Intervalic Group.

Chapter 5

$I = c^{\pm 2} \hbar Q^{-2}$, INTERVALIC ENERGY
Intervalic Principles of Equivalence between Electric Charge, Energy and Matter

INTERVALIC PRINCIPLE OF EQUIVALENCE BETWEEN MASS AND ENERGY

> It follows from the theory of relativity that mass and energy are both different manifestations of the same thing —a somewhat unfamiliar conception for the average man—. Furthermore, $E = mc^2$, in which energy is put equal to mass multiplied with the square of the velocity of light, showed that a very small account of mass may be converted into a very large amount of energy... the mass and energy in fact were equivalent.
>
> ALBERT EINSTEIN
> *Alice Calaprice, The Expanded Quotable Einstein*

It can be said that there are three kinds of "substances" in the physical world: matter, energy and electric charge. Up to date Physics only knows the equivalence between two of them, matter and energy, through the famous Einstein equation: $E = mc^2$. It remains unknown the equivalence between matter and electric charge and between energy and electric charge. According to Intervalic Theory there are an extraordinary simple *geometrical equivalence* between all of them. Moreover, the equivalence between matter and energy is now defined as a mere *geometrical statement* derived from a rotation in the Intervalic Space. The former Einstein equation acquires a new meaning in intervalic dimensions and it is named inside the Intervalic Theory as the *intervalic principle of equivalence between matter and energy*:

$$E = c^{\pm 2} m$$

As usually, we use intervalic dimensions ($L = L$, $T = iL$, $M = T/L = i$). In traditional dimensions we have that the rotation of mass in the Intervalic Space by $-180°$ —$E = c^2 m$— is different to its rotation by $+180°$ —$E = c^{-2} m$—; but this algebraic fact is due only to the partial symmetries of our traditional system of dimensions, and not to a *real* half symmetry showed by dimensions in Nature. In IT the famous Einstein equation is a mere *geometrical statement* about a $\pm 180°$ rotation of mass in the Intervalic Space, and that formula is written as: $E = c^{\pm 2} m$, since c^{+2} (-1) c^{-2}(-1) in intervalic units. Please note that the speed of light is ever placed in front of the other physical quantities in all equations because $c^{\pm n}$ is a mere *geometrical transformer* in the Intervalic Space. Showing its intervalic dimensions between brackets:

$$E\,(i^1) = c^{\pm 2}\,(-1)\,m\,(i)$$

On the contrary, in traditional dimensions we have: $-180° \neq +180°$. Obviously or not, this is due to the fact that SI or traditional systems are not *singular* systems of units —$c \neq 1$—, and of course, because the dimension of c^2 is different from the dimension of c^{-2} in traditional dimensions, since their physical quantities and dimensions don't have any *geometrical* properties.

When quantizing through substituting in that formula the mass, m, by the intervalic quantum of mass, $\mathbf{m_I} = c^{-1}(i)$, we have in the traditional half symmetry:

$$E(i^{-1}) = c^2(-1)\,\mathbf{m_I}(i) = c^2(-1)\,c^{-1}(i) = c(i)$$

which is, by definition, the intervalic quantum of energy: $\mathbf{E_I} = c(i)$

If we take the contrary rotation, which does not exist in traditional dimensions, we will got justly the same result (in a different *cycle* of powers of c —$c^{\pm 4n}(1)$—, which is geometrical and dimensionally equivalent):

$$E(i^{-1}) = c^{-2}(-1)\,\mathbf{m_I}(i) = c^{-2}(-1)\,c^{-1}(i) = c^{-3}(i) \approx c(i)$$

Thus, in intervalic dimensions we have anyway a *geometrical statement*, which is, *a fortiori*, *invariant*, since it is previously a geometrical property that does not depend on the system of coordinates chosen:

$$E(i^{-1}) = c^{\pm 2}(-1)\,\mathbf{m_I}(i) = c^{\pm 2}(-1)\,c^{-1}(i) = c(i)$$

Deductions made until here are unavoidable and without any possibility of logical choice. Nevertheless, when using intervalic dimensions arise up to three possible new sets of *intervalic symmetries* for the gravitational interaction between mass and energy. Since these three sets are equally *consistent* from a logical point of view we arrive to a crossroads where we have to choose one option and to discard the other two. Although a set of symmetries appears to be much more elegant than the others, the final choice only can be leaded by empirical results. Unfortunately, experimental Physics seems to be not capable to carry out the needed experiments yet.

INTERVALIC PRINCIPLE OF EQUIVALENCE BETWEEN ENERGY AND ELECTRIC CHARGE

Starting from the intervalic dimensional analysis of the electric charge, $Q = i^{-1/2}L^{1/2}$, it is straightforwardly and necessarily deduced a new relations between charge and the other intervalic physical quantities. Actually, we have:

$$Q = i^{-1/2}L^{1/2} = \sqrt{-T}$$
$$Q^2 = -T \; T_a$$

When antitime, T_a, is rotated $\pm 180°$ —$c^{\pm 2}(-1)$— in the Intervalic Space, we have:

$$Q^2 = c^{\pm 2} T$$
$$\varphi = c^{\pm 2} Q^{-2}$$

where φ is the frequorce —frequency-force—.

And finally, substituting the terms of the last equation in the intervalic definition of quantum velenergy, being J the spin and φ the frequorce: $E = 2\pi J \varphi$ —from which derives the known Einstein's formula, in intervalic dimensions: $E\,(i^{-1}) = h\,(L)\,\nu\,(i^{-1}L^{-1})$—, we directly obtain the desired intervalic geometrical equivalence:

$$I = c^{\pm 2} \hbar \, Q^{-2}$$

This is the equation of the *intervalic principle of equivalence between energy and electric charge*, perhaps the most important equation in Physics. Showing its intervalic dimensions:

INTERVALIC THEORY:
The Intervalic Structures of Subatomic Particles and the Last Foundations of Physics

$$I(i^{-1}) = c^{\pm 2}(-1) \hbar (L) Q^{-2} (iL^{-1})$$

In not-singular units, as the traditional ones, the half symmetry which is conserved is, obviously, the following:

$$I = c^{-2} \hbar Q^{-2}$$

which is the expression of the same principle written in any system of units with $c \neq 1$.

Now we can know which is, for example, the intervalic energy of the intervalic quantum of electric charge, $q_I = \sqrt{-(c^{-1}\hbar)}$:

$$I(q_I) = c^{\pm 2} \hbar \, q_I^{-2} = c^{\pm 2} \hbar \, [\sqrt{-(c^{-1}\hbar)}]^{-2} = c^{-1} = 1 \, (i^{-1})$$

The same magnitude in SI units is:

$$I(q_I) = c^{-2} \hbar \, q_I^{-2} = c^{-1} = 3.3356409 \cdot 10^{-9} \, (J) = 3.7114011 \cdot 10^{-26} \, (kg)$$

Remembering that the elementary charge, e, is 270 times the intervalic charge, q_I, the intervalic energy of the elementary charge is, in SIU units:

$$I(e) = c^{\pm 2} \hbar \, e^{-2} = c^{\pm 2} \hbar \, (270 \, q_I)^{-2} = c^{\pm 2} \hbar \, [270 \sqrt{(c^{-1}\hbar)}]^{-2} = 270^{-2} \, c^{-1}(i^{-1})$$

In traditional SI units this magnitude is:

$$I(e) = c^{-2} \hbar \, e^{-2} = 270^{-2} c^{-1} = 4.5756390 \cdot 10^{-14} (J) = 5.0910850 \cdot 10^{-31} \, (kg)$$

INTERVALIC PRINCIPLE OF EQUIVALENCE BETWEEN MATTER AND ELECTRIC CHARGE

Combining this principle with the equation of the intervalic geometrical equivalence between matter and energy, $E(i^{-1}) = c^{\pm 2}(-1)\, m(i)$, we obtain immediately the intervalic principle of equivalence between matter and electric charge:

$$m = c^{\pm 4}\, \hbar\, Q^{-2}$$
$$Q = c^{\pm 2}\, \sqrt{(\hbar/m)}$$

The same equivalence written in not-singular units is converted in:

$$m = c^{-4}\, \hbar\, Q^{-2}$$
$$Q = c^{-2}\, \sqrt{(\hbar/m)}$$

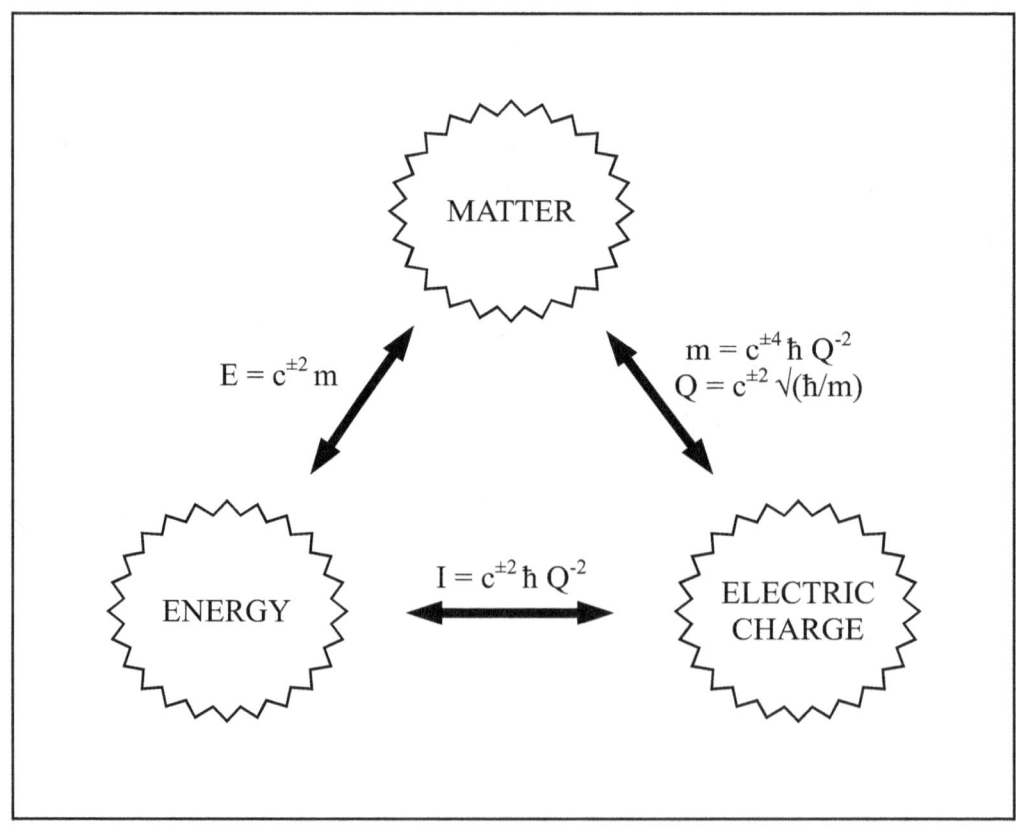

INTERVALIC ENERGY AND ELECTROMAGNETIC ENERGY

The difference between *intervalic* energy and *electromagnetic* energy is similar to that between *material* energy and *gravitational* energy. Intervalic and material energies are the equivalent energies of the two physical "substances" *at macroscopic scale*: electric charge and matter. Electric charge and matter, by themselves, can be viewed like huge "condensations" of energy (or "solidified" energy, as some authors have called them). On the contrary, electromagnetic and gravitational energies are like the "emanation" of those condensations (or "gaseous" energies, if it can be said, in order to follow the comparison). These energies are properly the energies of the *field* —electromagnetic or gravitational— created respectively by the electric charge and the matter around themselves.

Someone will be thinking that these commentaries are as evident as superfluous. But the truth is that traditional and modern Physics have confounded and mistaken these elemental concepts. Since the intervalic principle of equivalence between energy and electric charge was not known, the electromagnetic energy has been taken many times as it was intervalic energy (i.e., in the definition of the classical electron radius). Moreover, it is surprising how it can be usually managed the principle of equivalence between matter and energy, $E(m) = c^{\pm 2} m$ —formerly Einstein equation $E = mc^2$— without appreciating that there *must* exists an analogous equivalence between electric charge and energy. If somebody thinks that the *electromagnetic* energy covers the total energy of the electric charge, he also should think that *gravitational* energy recovers likewise the total energy of the matter. Of course, these energies are only the energies of the *field*, but not the *equivalent energies* of the own electric charge and matter, which are respectively the intervalic equations of equivalence: $I(Q) = c^{\pm 2} \hbar\, Q^{-2}$ and $E(m) = c^{\pm 2} m$.

	MATTER	ELECTRIC CHARGE
Equivalent energy	$E(m) = c^{\pm 2} m$	$I(Q) = c^{\pm 2} \hbar Q^{-2}$
Classic field energy	$U(m, r) = G m^2 / r$	$U(Q, r) = (1/4\pi\varepsilon) Q^2 / r$

INTERVALIC ENERGY AND MATERIAL ENERGY

The mass energy of a body or particle at rest is the sum of its constituent masses due: first, to electric charge, which is named *intervalic* energy; and second, the remaining energy —which will be named *material* energy— that has an *electromagnetic* origin.

$$E = c^{\pm 2} \hbar Q^{-2} + c^{\pm 2} m$$

Or in not-singular units:

$$E = c^{-2} \hbar Q^{-2} + c^{2} m$$

Of course, it is understood that the actual mass of subatomic particles is already that sum. Therefore, those different kinds of masses can be separated according to its different origins. Actual mass will be named *total* mass, m_{tot}; mass due to the electric charge will be named *intervalic* mass, m_{in}; and the remaining mass will be named *material* mass, m_{mat} —

or *electromagnetic* mass, m_{em}, if it comes completely from electromagnetic sources—.

The *intervalic* mass is obviously the commonly named *intrinsic* mass of a subatomic particle. This is a very important result because we can explain that the 'intrinsic mass' of a particle is a concept that really does not exists, since there is no a kind of *mass* with an *intrinsic* origin in that sense. *Mass* appears to don't be a primordial physical quantity, since every mass comes from an *intervalic* and/or *electromagnetic* energy, and both ones relies on the *electric charge*.

INTERVALIC ENERGY AT MICROSCOPIC AND MACROSCOPIC SCALES

It must be noted that contrarily to the equivalence between matter and energy, and against all expected, *electric charge is inversely proportional to energy* at quantum scale. At macroscopic scale we do not have got neither electric charge nor its equivalent energy, but the *sum* of the electric charges of subatomic particles and the *electromagnetic energy*, which is not by no means the *equivalent energy* of the charges themselves. This equivalence explains why the minimum state of energy of the 16 allowed subatomic charges predicted by the Intervalic Theory (but with very different values than those of the String Theory) is that of the elementary charge —which is the greater single charge: 270 q_I—, instead of the intervalic charge —the smaller one, the intervalic quantum of electric charge: q_I—.

The question may arise when passing through microscopic scale to macroscopic scale: until when is valid the quantum measurement of the intervalic energy? Really there is no contradiction because we have *never* measured *intervalic* energy at macroscopic scale but only *electromagnetic* energy. Really, the intervalic energy is only applicable to particles *assembled* by its corresponding *intervalic interaction*; since atoms and molecules of any macroscopic body are not assembled in this way, the intervalic energy of such a body as a whole is irrelevant, that is to say, it is zero.

Nevertheless and although it is completely useless, we can calculate the hypothetical intervalic energy of, for example, a macroscopic ball with radius = 0.01 (m), charged with 0.01 (C), their involved energies are, in SI units:

$$I(Q) = c^{\pm 2}\hbar\, Q^{-2} = 1.1745538 \cdot 10^{-47} \, (J)$$

Comparing it with the electromagnetic potential energy:

$$U(Q, r) = (3/5)(1/4\pi\varepsilon_0)\, Q^2 / r = 5.3925312 \cdot 10^7 \, (J)$$

it is clear that this hypothetical magnitude of the intervalic energy at macroscopic scale is totally non measurable at laboratories.

Someone can be thinking that it is not intuitive that the intervalic energy decreases as the electric charge of a particle increases. Of course, intuition is not a reason in quantum Physics, but I can comment what I think about it. All physical idea must have an intuitive meaning; if it have not, it does not mean that the physical idea may be wrong, but our intuition has not grown to understand that new physical behaviour. As we progressively educate our intuition, the next generation of physicists will feel as intuitive concepts which were felt as awkward by earlier generations. Perhaps when we remember that the intervalic energy relies on *spin*, its behaviour could be better understood. Spin is a quantum number which becomes very small as soon as we go up to the structure levels of matter: subatomic particles, atoms, molecules,... and spin is already a despicable feature. It can be said that a macroscopic body has its spin, but it is practically despicable in the behaviour of the body at macroscopic scale. Well, something similar occurs with the intervalic energy, and I think this simple explanation is very intuitive.

RELATION BETWEEN INTERVALIC ENERGY AND THE STRUCTURE OF MATTER

Intervalic energy shows surprising features according to the usual manner of thinking Physics. Although it is known that the *form* of bodies can have influence on some of its physical properties —such as moment of inertia, angular momentum, etc.—, this is a not important detail that is not studied by the core of Physics. However the intervalic energy, responsible of almost all mass in our Universe, has an important relation with the *structure* of physical particles.

The intervalic energy of any body is a wide concept which has no sufficient precision to be rightly understood without further specifications. First of all, we have to distinguish the intervalic energy of the body as a whole from the sum of the individual intervalic energies of its constituent parts. This rule is applicable to all scales and levels, form the microscopic to the macroscopic world. Thus, the *total inner* intervalic energy of a body is the total sum of the partial intervalic energies corresponding to *all structure levels* that compounds the body, making a complex cascade which can be compared to a pyramid of energies.

Now it can be easily understood that a body can be uncharged at macroscopic scale but can enclose a great amount of intervalic energy in its inferior constituent levels. Really, by this simple way is produced the magic of the creation of *mass* in nature.

Let us examine the making of sense of this process. It involves by first time the *structure* of matter as a principal agent in Physics. Up to date, it was supposed that the *form* of bodies could have a sort of "translation" into the physical quantities as mere *information* —which anyway does not pertain to the core of Physics yet—. But now we are seeing that this form, this structure of matter is not only information but *energy*.

Really, when we say that the mass of a body is 'x', we are meaning that this mass is the total *sum* of all massive particles compounding the body. But this total sum does not run with the electric charge because it can be positive or negative. Thus the vast majority of bodies have null charge at macroscopic scale, but this does not mean in any manner that they have null charge at microscopic levels. This reasoning looks obvious, but it don't be so since it has not been questioned until now. It is evident that there is a fabulous organization along atomic and subatomic structures which involve a huge amount of energy. Where is and from where comes this energy? To say that it is the "intrinsic mass" does not solve the problem; and it is not *gravitational* nor *electromagnetic* energy (although the last one has been intended for). The answer is no other than the *intervalic energy*: $I = c^{\pm 2} \hbar\, Q^{-2}$.

Although it is not the subject of this chapter —which seem almost to be an essay of "classic physics"—, it can be pointed out that contrarily to the formula $E = c^{\pm 2} m$, the equation $I = c^{\pm 2} \hbar\, Q^{-2}$ is not only a principle of equivalence, as we have seen, but it is also the equation of the intervalic energy corresponding to its related interaction: the *intervalic interaction*, from which derives all the four supposed frequorces of Nature, as we will demonstrate circumstanciately along this book.

Needless to say that I hope there will be nobody that thinks about the *intervalic structure* of subatomic particles as rigid structures in a classical sense, as they are dynamic structures which shows perhaps one of the most clear phenomena of quantum behaviour. However, that quantum features are derived unexpectedly from the own *intervalic geometry*, the same one which yields the intervalic quanta and the intervalic symmetries of compositeness, leading to a new structurefulness Particle Physics, and *a fortiori*, to a new Physics.

DERIVATION OF COULOMB'S ELECTROMAGNETIC ENERGY FROM THE INTERVALIC ENERGY

As can be clearly viewed in the Intervalic System of Units, the economy is one of the greatest achievements in IT. All the 40 existing physical quantities and units recovers all possible logical combinations between the two fundamental constants of Nature, c and ħ. This means that the inverse of any dimensions, of any physical quantity and of any unit is also a meaningful dimension, physical quantity or unit. We can hope that this logical economy will appear at any place in the last foundations of a faultless theory. Therefore, we could check if the intervalic principle of equivalence between energy and electric charge, which seems to be to most important equation in Physics since all forces can be derived from it, verifies or not that logical economy reserved only for the last foundations of the theory. Let us write its inverse formulation: $I^{-1} = c^{\pm 2} ħ^{-1} Q^2$. Remembering that at macroscopic scale we have the identity $nħ \; r$, we have:

$$I^{-1} = c^{\pm 2} (nħ)^{-1} Q^2 = c^{\pm 2} Q^2 / r$$

This is just the Coulomb's law written in intervalic units, where in place of the traditional electric constant $1/4\pi\varepsilon_0$ appears the dimensionless factor $c^{\pm 2}$ (-1). Remembering that in SI units is *set by definition* the value of the magnetic constant as $\mu_0/4\pi = 10^{-7}$ instead of 1, as would be more logical, we have to add finally this factor, which intervalic dimension is (-1). Since the intervalic dimensions are fully consistent and they operate algebraically, we have: I^{-1} (i) · 1 (-1) = I^{-1} (i^{-1}). Therefore the dimension of I^{-1} is now energy in intervalic units, and the same in SI units:

$$E = c^2 \, 10^{-7} \, Q^2 / r \quad (1/4\pi\varepsilon_0) \, Q^2 / r$$

Thus, Coulomb's law and the electromagnetic interaction is also derived from the intervalic principle of equivalence between energy and electric charge, $I = c^{\pm 2} ħ \, Q^{-2}$.

INTERVALIC PRINCIPLE OF ENERGY BALANCE FOR SUBATOMIC PARTICLES

Following the above results, the total *mass energy*, E_{mass}, of any subatomic particle can be written in a surprisingly simple equation plenty of elegance. It is the sum of the *intervalic energy* plus its *inverse* —which is the electromagnetic potential energy—. In intervalic units:

$$E_{mass} = I + I^{-1} = c^{\pm 2} \hbar\, Q^{-2} + 1/(c^{\pm 2} \hbar\, Q^{-2})$$

Apart from them, a subatomic particle at rest only have the energy due to its intrinsic angular momentum: the *spin energy*, $E(J)$. Thus adding the spin energy to the preceding mass energy we will obtain the total *structural energy*, E_{struc}, of a subatomic particle:

$$E_{struc} = I + I^{-1} + E(J)$$

And in a splendorous achievement of logical economy, IT has postulated that this structural energy is the same for all subatomic particles. Of course, that value is *zero*, which means that the sum of all structural energy of the Universe is zero:

$$I + I^{-1} + E(J) = 0$$

While the intervalic energy is always attractive, the spin energy has always a repulsive effect. For any intervalic structure with like charges the electromagnetic potential energy is opposite to the intervalic energy. Thus we can write the classic formulation of the *intervalic principle of energy balance for subatomic particles*:

$$I - I^{-1} - E(J) = 0$$

It must be noted that the above intervalic and electromagnetic energies are manifested as *mass energy*, while spin energy is not manifested as mass but as *kinetic energy*. In the Intervalic Theory of Particle Physics this principle is applied to deducting the features of subatomic particles, and astonishing results are obtained with great simplicity. It has to be noted that the Standard Model (SM) has a lack of underlying principles and it only has deduced some correct results adjusting by hand some constant and, at last, due to the great generality of Lagrangian formalism. To obtain the same —and more— results than SM, but in a *right* and suitable way, it is not necessary to use the Lagrangian formalism in IT. On the contrary, all known results yielded onerously from SM, now appears as a naïve set of results in IT. In a nutshell, the Lagrangian formalism allow to avoid the infinite quantum paths postulated by Quantum Theory between two states of energy. The great idea in IT is that these quantum paths are not introduced in the formalism, and the difference between two states of energy, called *energy interval*, is deduced by means of *geometrical relations* in the Intervalic Space. Really, IT convert Physics equations in *geometrical statements*, which are of course and *a fortiori*, *invariant* since they determine a *geometrical relation* in the Intervalic Space. By these reasons, it is surprising that IT in Physics has not been postulated half a century ago, but this is already history (a similar question could be commented about IT in Music, although its simple formalism is still hardly understood by some contemporary musicians without a minimal mathematical baggage).

DEFINITIONS OF POTENTIAL, ACTUAL AND PRIMORDIAL INTERVALIC ENERGIES

Potential intervalic energy is defined as the sum of all intervalino's intervalic energies accumulated into a body:

$$E_{pot} = \Sigma\, n(\mathbf{q_I}) \cdot E(\mathbf{q_I})_{in} = c^{-1} \Sigma\, n(\mathbf{q_I})$$

Actual intervalic energy is the no other than the usual intervalic energy of a body:

$$E_{act} = c^{\pm 2} \hbar Q^{-2}$$

And *primordial* intervalic energy is the subtract between both energies, which was liberated when the primordial particle was assembled:

$$E_{prim} = E_{pot} - E_{act} = c^{-1} \Sigma\, n(\mathbf{q_I}) - c^{\pm 2} \hbar Q^{-2}$$

THE INTERVALIC DUALITIES OF ENERGY: MASS-ENERGY AND ELECTRIC CHARGE-ENERGY

The principle of equivalence between matter and energy has several physical interpretations, and the most of them can not be derived from the former Einstein equation $E = mc^2$, but only from the new one using intervalic dimensions, $E = c^{\pm 2} m$. One of these interpretations is shared with the intervalic principle of equivalence between energy and electric charge, $I = c^{\pm 2} \hbar Q^{-2}$, and refers to the dualities of energy: mass-energy and electric charge-energy.

Traditional Physics timidly suggests that the mass of a body could be balanced with the corresponding energy of its gravitational field. And since the gravitational energy is attractive, it has a minus sign; therefore both energies could be exactly compensated, being the total sum zero.

It is hard to conceive such lack of intellectual ambition in any discipline of human knowledge as this poor example showed by Physics. The principal problem from the traditional view is the us sign in gravitational energy. Can we deduce such serious affirmation from only a *conventional* minus sign? Surely we can't, but when we look at these physical quantities in within the intervalic dimensions a radical change is operated. Actually, intervalic dimensions of mass (i) and energy (i^{-1}) are *opposite* —and *inverse* at once—. Therefore, its sum is not one of the intervalic dimensionless (± 1) —which at last are the dimensions of some physical quantities—, but zero:

$$(i) + (i^{-1}) = (0)$$

Since (0) dimension is not defined in intervalic dimensions, it means the absence of any dimension, that is to say, dimensions of mass (i) and energy (i^{-1}) vanish when they are summed. So now the previous assumption is not based on a conventional sign, but on dimensional foundations.

If this assumption is naively developed in a systematic manner, we have we following balances, where Q is the electric charge, I is intervalic energy, M mass, U field energy, and K kinetic energy; all of them referred to the whole Universe:

$\Sigma Q^{\pm} = 0$
$Q \ U(Q) + K(Q)$
$Q = I(Q)$
$I(Q) = M(Q)$
$M(Q) \ U(M) + K(M)$

In English: first, the sum of positive and negative charges is zero.

Second, there is an exact balance between a part of the electric charge of the Universe and its corresponding electromagnetic and kinetic energies generated by the electric charge.

Third, there is an equality between the remaining part of the electric charge of the Universe and its corresponding energy determined by the intervalic principle of equivalence between energy and electric charge: $I = c^{\pm 2} h \, Q^{-2}$.

Fourth, this intervalic energy has been converted into mass according to the intervalic principles of equivalence.

Fifth, there is an exact balance between the mass of the Universe and its corresponding gravitational and kinetic energies generated by the mass.

In the case of subatomic particles it is clear that a state of lowest energy must be already reached when it was created. Therefore some meaningful *balance* had to arise at any time of its involved structures. The intervalic postulate is that the intervalic energy has to be exactly balanced with the remaining ones —electromagnetic and kinetic—. In this case, the second and third equations of above take the following form:

½Q U(Q) + K(spin)
½Q = I(Q)

The application of this balance —which coincides exactly with the *intervalic principle of energy balance for subatomic particles* (as macrocosmos, so microcosmos; "As above, so below")—, developed for all particles, can be thoroughly viewed in IT, which substitutes the formalism of SM —whose complete name perhaps should preferably be the Standard Model of Structureless Particles, or simply the Structureless Model (abbreviated equally as SM)—.

According to these balances it is clear that all masses and energies of the Universe derive only from the electric charge. And since the sum of all electric charges is zero, the total sum of matter, energy and electric charge of the Universe is zero. This result is in agreement with the named *intervalic zero assumption* of IT: the sum of all magnitudes of physical quantities of the Universe is zero. And as I have already demonstrated in other site, the dimensional sum of all the 40 physical quantities —which conform the intervalic group— existing in the Universe is equally zero.

Intervalic Principles of Equivalence between Electric Charge, Energy and Mass

$$E = c^{\pm 2} m$$
$$I = c^{\pm 2} \hbar Q^{-2}$$
$$Q = c^{\pm 2} \sqrt{(\hbar/m)}$$

The most fundamental equations in Physics are merely *dimensional equivalences* —or simple *geometric statements*— in the Intervalic Dimensional Space. It should be noted that those intervalic equations are always *invariant* since they just describe a dimensional equivalence between physical quantities that does not depend on the frame of reference chosen.

For example, the famous Einstein equation is a mere *geometrical statement* which describes a ±180° rotation of mass in the Intervalic Dimensional Space, and that formula is written as: $E = c^{\pm 2}m$, since $c^{+2}(-1) \equiv c^{-2}(-1)$ in intervalic units. The equation just describes the *geometric equivalence* between energy and mass, a relation which is almost trivial inside the Intervalic Dimensional Space: $E(i^{-1}) = c^{\pm 2}(-1) m(i)$.

To deduce whether it is a intervalic equivalence between electric charge and energy, we start from the intervalic dimensional analysis of the electric charge, $Q = i^{-1/2} L^{1/2} = \sqrt{-T}$. Then: $Q^2 = -T \equiv T_a$. When antitime, T_a, is rotated ±180° —$c^{\pm 2}(-1)$— in the Intervalic Dimensional Space, we have: $Q^2 = c^{\pm 2} T$. Inverting the equation (since intervalic dimensions can operate algebraically) we have got: $\varphi = c^{\pm 2} Q^{-2}$, where φ is the frequorce. The geometric equivalence of the frequorce is: $\varphi(i^{-1}L^{-1}) = \hbar^{-1}(L^{-1}) E(i^{-1})$, a simple statement which at macroscopic scale becomes the traditional equation: $E = n\hbar$ $F = s F$, where s is *space* and F *force*. When it is applied to dalinoless particles s means the length of the string: $s = 2\pi J$ (being J the *spin* —or dynamic radius— of the particle) and the same equation becomes: $E = 2\pi J \varphi$, from which is derived the traditional Planck's equation —a special case for particles with spin 1—. As charged particles are dalinoful, substituting the geometric equivalence of the frequorce in the previous dimensional equation, it is obtained the geometric equivalence: $I = c^{\pm 2} \hbar Q^{-2}$. This is the **Intervalic Principle of Equivalence between Energy and Electric Charge**, perhaps the most important equation in Physics. Showing its intervalic dimensions:

$$I(i^{-1}) = c^{\pm 2}(-1) \hbar(L) Q^{-2}(i L^{-1})$$

In the remaining not-geometric or incomplete systems of units the half symmetry which is conserved is only the following one: $I = c^{-2} \hbar Q^{-2}$.

EQUIVALENT AND FIELD ENERGIES	Mass	Electric Charge
Equivalent energy	$E = c^{\pm 2} m$	$I = c^{\pm 2} \hbar Q^{-2}$
Classic field energy	$U = G m^2 / r$	$U = (1/4\pi\varepsilon) Q^2 / r$

Chapter 6

INTERVALINO

INTERVALIC ENERGY OF THE INTERVALIC CHARGE

Let us remember the equation of the *intervalic principle of equivalence between energy and electric charge*: $I = c^{\pm 2} \hbar Q^{-2}$. Showing its intervalic dimensions between brackets:

$$I(i^{-1}) = c^{\pm 2}(-1) \hbar (L) Q^{-2}(iL^{-1})$$

In not-singular units, as the traditional ones, only a half symmetry is conserved. It is obviously the following:

$$I = c^{-2} \hbar Q^{-2}$$

which is the expression of the same principle written in any system of units with $c \neq 1$.

Now we can know how is the intervalic energy of the intervalic quantum of electric charge, $q_I = \sqrt{-(c^{-1}\hbar)}$:

$$I(q_I) = c^{\pm 2} \hbar \, q_I^{-2} = c^{\pm 2} \hbar \, [\sqrt{-(c^{-1}\hbar)}]^{-2} = c^{-1} (i^{-1})$$

The same magnitude in SI units is:

$$I(q_I) = c^{-2} \hbar \, q_I^{-2} = c^{-1} = 3.335640952 \cdot 10^{-9} \, (J)$$

The mere existence of this fundamental quantum involves unavoidably compositeness —or structurefulness— in subatomic particles.

INTERVALINO ELECTRIC CHARGE

The *intervalino*, **I**, is the very last fundamental massful particle postulated by the IT, from which are structurefully composed all subatomic particles of Nature. According to IT the huge explosion at the Big Bang was due to the liberation of energy caused by the aggregation of intervalinos to compose material particles. All massive particles of the Standard Model and beyond can be easily explained as different assemblies of intervalinos.

The intervalino is intended to be the singlest primordial particle according to the values of the intervalic quanta. It is a particle which charge is the intervalic quantum of electric charge, $q_I = \sqrt{-(c^{-1}\hbar)} \, (i^{-1/2} L^{1/2})$:

$$q_I = \sqrt{-(c^{-1}\hbar)} \, (i^{-1/2} L^{1/2}) = 1 \, (i^{-1/2} L^{1/2}) = 5.9339900 \cdot 10^{-22} \, (C)$$

Thus, the elementary charge is exactly defined as $e = 270 \, q_I$, which matches exactly with the actual experimental value, and therefore the magnitude of the fine structure constant has an exact theoretical value: $\alpha = 270^2 \cdot 10^{-7} = 7.29 \cdot 10^{-3} = 1/137.17421$.

The sign of the electric charge of an *isolated* intervalino is, according to IT, undetermined, that is to say, positive and negative *at once*: $\{\pm\}$.

Only when an intervalino interacts with another intervalino the electric sign is *realized*: {+ +, + -, - +, - -}. This beautiful feature can be considered as the quintessence of the quantum uncertainty; moreover it does not involve a paradox like those of Einstein-Podolsky-Rosen or Schroedinger's cat, but it defines clearly a physical state. "God does not play dice".

If the electric charge of an *isolated* intervalino is {±}, it would imply that the electromagnetic energy of a single intervalino could not be defined, but it is just the case are we are going to see straightaway. Besides, it is clear that such a quantum uncertainty must be found at the *last* structure level of the intervalic structure in order to avoid the apparition of a *singularity* in the electromagnetic energy as the *radius* of the constituent subparticles becomes smaller.

INTERVALINO ELECTROMAGNETIC ENERGY

It can be remembered that the concept of *electromagnetic potential energy* is not applicable to a *single* particle, according to the well based traditional view, but only to *composite* particles which can disaggregate in fragments. Whereupon there is a serious contradiction inside SM since leptons and quarks are considered to be single particles without structure.

On the contrary, according to IT all subatomic particles are composite particles with structure (with the exception of the unique primordial particle: the *intervalino*, from which are derived all the rest particles). Then, all subatomic particles have both *intervalic* energy and *electromagnetic* potential energy, while intervalino must have only *intervalic* energy because it can't have the electromagnetic one. This deduction fully matches with the uncertainty of the electric charge sign in the *isolated* intervalino, which now makes further sense.

ORIGIN OF INTERVALINO MASS

The intervalic energy of the intervalino at rest is:

$$I(\mathbf{I}) = c^{\pm 2}\hbar\, q_I^{-2} = c^{-1} = 3.335640952 \cdot 10^{-9}\, (J)$$

Since the electric charge of the intervalino, **I**, is by definition the intervalic quantum of charge, the intervalic mass —formerly named "intrinsic" mass— of the intervalino is just the equivalent intervalic energy of the intervalic charge, q_I:

$$m(\mathbf{I}) = c^{\pm 4}\hbar\, Q^{-2} = c^{-3} = 3.711401092 \cdot 10^{-26}\,(kg) = 20{,}819.42423\, (MeV/c^2)$$

Thus, the 100% of the intervalino mass has an *intervalic* origin and it has no *electromagnetic* mass. This is the simplest *creation* of matter starting from the two fundamental constants, c and ℏ, that has ever been proposed in a physical theory.

INTERVALINO SPIN ENERGY

The Intervalic Theory in Particle Physics is a theory of *composite and structureful* particles, not a theory of *single and structureless* particles. Therefore, apart from the traditional physical interpretations of spin, IT adds one more interpretation of that fundamental quantum number. According to IT the energy of the intrinsic angular momentum of a subatomic particle is a *structural energy* which makes a perfect balance with the other two structural energies: the intervalic and the electromagnetic ones. This important law has been named *intervalic principle of energy balance*. It states the following balance for the total energy of subatomic particles:

$I - I^{-1} - E(J) = 0$

where I^{-1} U is the electromagnetic energy and the last term is the spin energy. It is important to note that I and U are *manifested* as *mass energies*, while $E(J)$ is manifested as *kinetic energy*.

LINEAR VELOCITY DUE TO SPIN ON SURFACE

Applied to the intervalino, we have seen that it can't have electromagnetic energy. Therefore, its intervalic structure shows a balance between the intervalic energy in one part, and only the energy of the intrinsic angular momentum on the other part. Being v_I the maximum linear velocity on intervalino's surface and ω_I its angular velocity, we have:

$I(I) - 0 = E(J_I) = 3.3356409 \cdot 10^{-9}$ (J)
$c^{\pm 2} \hbar\, q_I^{-2} - 0 = m_I v_I^2 = m_I r_I^2 \omega_I^2$

As $c^{\pm 2} \hbar\, q_I^{-2} = c^{\pm 2} m_I$, we obtain a very beautiful result which makes that intervalino becomes a *"geometric"* particle:

$v_I = c$

This means that the constituent particles of intervalino —two photons in antisymmetric state under interchange— are turning at a linear velocity of just the speed of light due to spin, as may be expected of any light particle.

And of course we have:

$v_I = r_I\, \omega_I = c$

INTERVALIC SCHWARZSCHILD INTERVALINO RADIUS AND INTERVALINO AS A GEOMETRIC LIMIT TO BE A BLACK HOLE

It can be noted the meaningful parallelism between this equation and the gravitational Schwarzschild radius, $r_{Schw} = 2Gm/c^2$, since the *intervalic* energy plays the same role as the *gravitational* one in the Schwarzschild equation, and the *spin* energy of the subatomic particle is analogous to the *kinetic* energy of the star. The maximum linear velocity on intervalino's surface can be interpreted as the *escape velocity* from intervalino; as it is just the speed of light, $v = c$, intervalino in just the *geometric* limit to be a black hole.

As it is well known, the Schwarzschild radius of a star is closely related with its mass. A star which is a supposed black hole can leave to be a black hole if its radius is augmented. On the contrary, the escape velocity of intervalino does not vary with the magnitude of the intervalino radius because such velocity —the speed of light— is a structural magnitude of intervalino which is, *a fortiori, invariant* since it is a *geometric* feature. Therefore, the intervalic Schwarzschild radius of intervalino may paradoxically take any value and intervalino will continue to be a geometric limit of black hole, with independence of its radius.

According to this, whatsoever particle which leaves intervalino must do it at precisely the speed of light (!). Now, what particles may leave the intervalino? The only particles which may leave the intervalino are just those equally composed, like the intervalino, by closed strings: they are only two: the *photon* and the *graviton*, and they must leave intervalino at just the speed of light (!).

In this way IT shows in a beautifully simple and intuitive manner the interaction between matter and electromagnetic and gravitational waves at the last scale of the intervalic length, $\hbar$. Electromagnetic and gravitational waves interact with matter because the interacting block of matter is the *intervalino*, and all of them are equally composed by closed strings. Now Richard Feynman could know what to answer to his father, written in his autobiographic book **Surely You're Joking, Mr. Feynman**, when he questioned to his son: "how can photons go out from electron?"

I believe that when we do not have an intuitive understanding of a physical feature, this unavoidably means that we do no know the underlying foundations of the feature, because our *intuition* always grow and go closely along with our *knowledge* of Nature. To pretend, as someone does, that intuition does not grow in mankind and that it is a primitive capacity of mind, is really a "primitive" assertion... Any knowledge of Physics which is still comprehended without intuition is really not comprehended, because such knowledge is with certainty a *partial* one yet (and the complexity of maths in modern Physics is not a valid objection because intuition —by definition— sees not under but over that complexity).

INTERVALINO RADIUS

We will begin with some approximations. Starting from previously viewed data we have:

$r_I = c / \omega_I$

There is an inferior limit for the magnitude of the intervalino radius, which is geometrically determined by the intervalic angular velocity ω_I:

$\omega_I = c\, \hbar^{-1} = 2.8399227 \cdot 10^{42}\, (s^{-1})$

Therefore we would have:

$$r_I \geq c / \omega_I = \hbar = 1.0556363 \cdot 10^{-34} \, (m)$$

On the other hand, the maximum value of the intervalino radius can't be greater than the predicted value for the dalino 1, D_1 —which is just composed by one intervalino— in the set G_1 of the 16 allowed dalinos (they are all the possible aggregations of intervalinos, which are those which constituent number of intervalinos is a divisor of 270, the number of the elementary charge: 1, 2, 3, 5, 6, 9, 10, 15, 18, 27, 30, 45, 54, 90, 135, 270). That magnitude is:

$$r_I \leq r(D_1) \approx 6.0102597 \cdot 10^{-25} \, (m)$$

It corresponds to an angular velocity:

$$\omega_I = c / r_I = 4.9880117 \cdot 10^{32} \, (s^{-1})$$

Nevertheless, the intervalino radius postulated by IT is deduced from the epistemological rank of the intervalic dimensions. As the intervalic dimension of spin is *length*, this lead to interpret the spin of a subatomic particle as its resulting *dynamic radius* —from the sum of its constituent particles—. For example, photon has a spin 1 because it is a symmetric assembly of intervalic strings (spin ½), and therefore its resultant spin is:

$$J(\gamma) = J(S) + J(S) = ½ + ½ = 1$$

However, the antisymmetric assembly of intervalic strings yields a spin 0 particle (chi):

$$J(\cent) = J(S) - J(S) = ½ - ½ = 1$$

In a similar way, we have that the spin of intervalino and graviton are respectively:

$J(\mathbf{I}) = J(\gamma) - J(\gamma) = 1 - 1 = 0$

$J(g) = J(\gamma) + J(\gamma) = 1 + 1 = 2$

Herein the intervalino (and graviton) radius is:

$r_\mathbf{I} = 2\hbar = 2.1112726 \cdot 10^{-34}$ (m)

INTERVALINO ANGULAR VELOCITY

The angular velocity of intervalino due to spin will be:

$\omega_J(\mathbf{I}) = c / r_\mathbf{I} = \frac{1}{2}\, c\, \hbar^{-1} = 1.41996092 \cdot 10^{42}$ (s^{-1})

THE EMPTINESS OF SUBATOMIC PARTICLES

The intervalino radius shows that subatomic world is much more empty than it was ever thought. If it is supposed that the intervalino "fills" a quantum sphere of radius $r_\mathbf{I}$, the intervalino volume would be:

$V(\mathbf{I}) = (4/3)\pi\, r_\mathbf{I}^3 = 3.9420447 \cdot 10^{-101}$ (m^3)

On the other hand, supposed a quantum spherical form, the electron volume would be (taking the intervalic electron radius):

$V(e) = (4/3)\pi\, r_e^3 = 1.3650028 \cdot 10^{-43}$ (m^3)

Since $e = 270\ q_I$, the relation between "filled space" and "empty space" inside the electron quantum sphere would be:

$$270 V(\mathbf{I}) / V(e) = 2.8879389 \cdot 10^{-58}$$

This fully demonstrates that subatomic particles are practically hollow, and consistency and other usual properties of matter are due to spin and electromagnetic phenomena.

THE EMPTINESS OF THE INTERVALIC UNIVERSE

THE EMPTINESS OF COSMIC SPACE

The average density of the Universe has been calculated in around one Hydrogen atom per cubic metre of cosmic space. Taking the Bohr radius, $a_0 = \alpha/4\pi R_\infty = 5.291772 \cdot 10^{-11}$ (m) for calculating a value for the volume of an Hydrogen atom, the relation between "filled space" and "empty space" in cosmic space would simply be:

$$R(U) = (4/3)\pi\, a_0^3 / 1^3 = 6.207146 \cdot 10^{-31}$$

THE EMPTINESS OF ATOMS

Being the intervalic proton radius, $r_p \approx 1.2374486 \cdot 10^{-15}$ (m), the relation between "filled and empty" space inside an Hydrogen atom would be:

$$R(H) = (4/3)\pi\, r_p^3 / (4/3)\pi\, a_0^3 = r_p^3 / a_0^3 = 1.278728 \cdot 10^{-14}$$

THE EMPTINESS OF MASSFUL SUBATOMIC PARTICLES

All massful subatomic particles have got an intervalic structure which is ultimately made from an assembly of intervalinos at successive structure levels. The intervalino radius shows that subatomic world is much more empty than it was ever thought. If it is supposed that the intervalino "occupies" a sphere of radius r_I, the *intervalino volume* would be:

$$V(I) = (4/3)\pi\, r_I^3 = 3.9420447 \cdot 10^{-101}\ (m^3)$$

Likewise, supposed a spherical form, the *electron volume* would be — taking the intervalic electron radius, $r_e = 3.194098699 \cdot 10^{-15}\ (m)$—:

$$V(e) = (4/3)\pi\, r_e^3 = 1.3650028 \cdot 10^{-43}\ (m^3)$$

Since $e = 270\ q_I$, the relation between "filled and empty" space inside the sphere "occupied" by the electron would be:

$$R(e) = 270\, V(I)\, /\, V(e) = 2.8879389 \cdot 10^{-58}$$

Following a similar way for nucleons, whose intervalic structure is, namely: $N = M_3 = 3\ L_3 = 9\ G_6 = 54\ D_{45} = 2430\ I = 4860\ \gamma = 9720\ S$, we have got the relation:

$$R(N) = 2430\, V(I)\, /\, V(N) = 1.206862 \cdot 10^{-53}$$

This means that subatomic particles are surprisingly much more empty than the own cosmic space, and fully demonstrates that subatomic particles are practically hollow. Summing the above results we have that the total "filled" space —"filled" with *photons*, which are the constituent particles of intervalino— inside a cubic metre of cosmic space containing one Hydrogen atom would only be:

$$R(U) \cdot R(H) \cdot \tfrac{1}{2}\, [R(e) + R(N)] = 4.789698 \cdot 10^{-98}$$

THE EMPTINESS OF LIGHT

We have seen that matter is absolutely empty. At last it is composed by *intervalinos,* which ultimately are made by two photons in antisymmetric state under interchange. So finally only photons appear to "fill" the space; but do they really fill it?

A *photon* is composed by two intervalic strings assembled in symmetric state under interchange. Although photon is moving at the speed of light forever, it is a particle whose intervalic structure fits entirely inside a *plane*, that is to say, photon has only two space dimensions: it is a bidimensional particle moving inside a tridimensional space.

Hereinafter we can easily calculate the volume "filled" by a photon: as it has got only two dimensions the volume of a photon is *zero*.

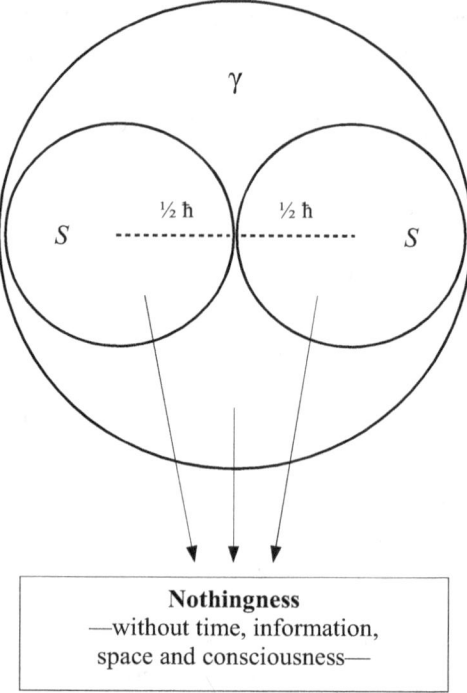

THE EMPTINESS OF THE INTERVALIC STRING

Finally let us try to calculate the surface "filled" by the volumeless *photon*, which will be the surface occupied by its constituent particles: the two intervalic strings. The intervalic structure of the intervalic string fits entirely inside a *line* since it is made only from unidimensional space: actually it is the *intervalic quantum of space*. Therefore the surface "filled" by it is once more *zero*.

Then, what kind of ultimate stuff, if any, fills the circular surface inside the *intervalic string*? The answer is what remains or exists where there is not time, information, space and consciousness: *nothingness*.

EXCHANGE VELOCITY OF ITS CONSTITUENT PARTICLES INSIDE

The maximum exchange velocity of its constituent particles —two photons in antisymmetric state under interchange— inside the intervalino will be:

$$v(\gamma)_I = 2\pi\, r_I\, I(\mathbf{I}) / \hbar = 2\pi\, c^{-1} = 2.095845022 \cdot 10^{-8}\ (m\ s^{-1})$$

Please note that this slow exchange movement is perpendicular to the speed of light direction, that is to say, this movement is in the same plane of the closed string —which is the own photon—. On the contrary, the speed of light is perpendicular to that plane.

INTERVALINO AND DARK MATTER

IT postulates that the huge energy liberated in the primordial aggregation of intervalinos is the origin and the cause of the so named Big Bang. We suppose that all massive particles were created through the aggregation of intervalinos at the Big Bang. As we have explained in other site, the primordial aggregation of intervalinos stopped when a perfect balance was reached among all the structural energies involved in a subatomic particle: the intervalic energy, the electromagnetic energy and the spin energy:

$$I - U - E(J) = 0$$

This balance determines, among others, the magnitude of the so named *elementary charge*, which is now defined in IT as: $e = 270\ q_I$. Since this value is the same for all aggregations of intervalinos, it imposes a powerful symmetry constraint in further aggregation of aggregation, aggregations of aggregations of aggregations, etc., so only the value of the elementary charge and its divisor are allowed as constituent aggregations, and the number of constituent intervalinos of the final subatomic particles that we know only can be 270 or a multiple. Here we probably have the first emergency of spontaneous order in the Universe strongly based on well defined and measurable physical principles.

After this brief introduction to the problem the question that arises is the following: what is the *efficiency* in the primordial aggregation of intervalinos? Did stayed remaining *isolated* intervalinos without aggregating?

As we will explain later in the corresponding chapters, the primordial aggregation of intervalinos —for composing *dalinos*— was made exclusively through gravitational and intervalic interaction, being the own intervalino, which has spin 0, both the source and intermediate particle of the intervalic interaction. Therefore, the intervalinar interaction is an exchange frequorce. Since the intervalic interaction is *short ranged*, the intervalinos which could not assemble with others in primordial times, will not be able to aggregate *after* it. What is the destiny of these intervalinos which remain isolated?

As we have seen, single intervalinos do not have electromagnetic energy and don't interact electromagnetically. The electromagnetic interaction only exist between *aggregations of intervalinos —dalinos—*, but an *isolated* intervalino do not have electromagnetic energy at all. On the other hand, although an intervalino could interact intervalically when shocking with a lepton or other particle, the symmetry constraint imposed by the elementary charge does not allow such aggregation, since it would destroy the electric balance of the particle in an Universe fully ordered according to a strong symmetry derived from the value of the elementary charge.

In resume, we have an isolated intervalino which does not interact electromagnetically, and which does not have a number of neighbour free intervalinos for aggregating with them through *short ranged* intervalic interaction. Therefore, our isolated intervalino only can interact with other

INTERVALIC THEORY:
The Intervalic Structures of Subatomic Particles and the Last Foundations of Physics

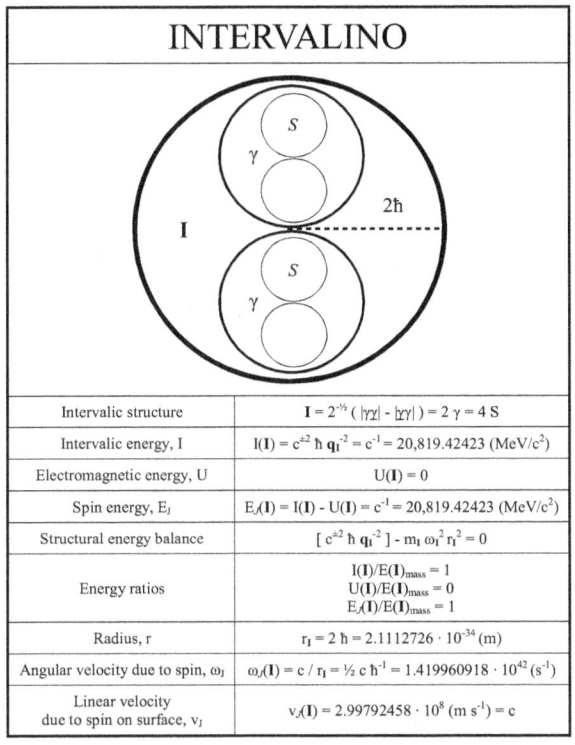

INTERVALINO

Intervalic structure	$I = 2^{-\frac{1}{2}} (|\gamma\gamma| - |\gamma\gamma|) = 2\gamma = 4S$
Intervalic energy, I	$I(I) = c^{\pm 2} \hbar\, q_I^{-2} = c^{-1} = 20,819.42423$ (MeV/c²)
Electromagnetic energy, U	$U(I) = 0$
Spin energy, E_J	$E_J(I) = I(I) - U(I) = c^{-1} = 20,819.42423$ (MeV/c²)
Structural energy balance	$[\, c^{\pm 2} \hbar\, q_I^{-2}\,] - m_I\, \omega_J^2\, r_I^2 = 0$
Energy ratios	$I(I)/E(I)_{mass} = 1$ $U(I)/E(I)_{mass} = 0$ $E_J(I)/E(I)_{mass} = 1$
Radius, r	$r_I = 2\hbar = 2.1112726 \cdot 10^{-34}$ (m)
Angular velocity due to spin, ω_J	$\omega_J(I) = c / r_I = \frac{1}{2}\, c\, \hbar^{-1} = 1.419960918 \cdot 10^{42}$ (s⁻¹)
Linear velocity due to spin on surface, v_J	$v_J(I) = 2.99792458 \cdot 10^8$ (m s⁻¹) = c

The intervalino is the first particle with *mass*, generated by the synthesis of two photons in antisymmetric state under interchange, whose threshold frequency —*coupling frequency of matter*— must be greater than: $\varphi_{cp} = 1 / (4\pi\, c\, \hbar) = 2.51452013 \cdot 10^{24}$ (s⁻¹) for the synthesis may occur, so that intervalinos could eventually be synthesized artificially in the laboratory.

The physical properties of intervalino are extraordinary, because it is the only particle with an electric charge that has no electromagnetic energy — which only particles *with structure* can have, being this a basic definition of the physical laws that has forgotten the standard model—, so it is logically necessary the existence of a fundamental particle, for the electromagnetic interaction, with these extraordinary physical features: an indivisible electric charge and, by this reason, without electromagnetic potential energy.

The electric charge of the intervalino is the *intervalic quantum of electrical charge*: $q_I = \sqrt{-(c^{-1}\hbar)} = 1\, (i^{-1/2} L^{1/2}) = 5.93398995 \cdot 10^{-22}$ (C), which is the fundamental charge of Nature, whose geometric value is exactly 1/270 of the elementary charge, e.

Its mass energy, which comes exclusively from the equivalent energy of the previous intervalic quantum of electric charge, $I\,(\mathbf{I}) = c^{\pm 2}\, \hbar\, q_I^{-2} = c^{-1} = 20,819.42423$ (MeV/c²), is also the *intervalic quantum of mass*: $m_I = 1\, (i) = c^{-1} = 20,819.42423$ (MeV/c²), while its spin energy is equally c^{-1}.

Its radius is twice the photon radius, i.e., twice the *intervalic quantum of length*: $r_I = 2\, l_I = 2\hbar = 2.1112726 \cdot 10^{-34}$ (m).

The antisymmetric state of the two constituents photons of the intervalino can be visualized as two photons traveling in opposite directions which are coupled tangentially, so that opposite ends to the coupling point of each photon —which are situated in the centre of the intervalino— continue moving at the speed of light, c, as it can not be otherwise, since all non massive particle always moves at the speed of light in intervalic space-time, hence the linear velocity on the "surface" of the intervalino is precisely c.

particles via *gravitational* interaction. However, due to the symmetry constraint imposed by the elementary charge, already described, it will never be able to aggregate with other dalinos and will follow its lonely way indefinitely (unless it would meet with a lot of isolated intervalinos to make a dalino). As can be seen, the behaviour and features described for a single intervalino are just those that are intended for the *dark matter*. We will comment this exciting subject when describing the composition of dark matter according to IT, in the chapter on the Intervalic Dark Matter.

INTERVALINAR FUSION

The equations of the intervalic principles of equivalence represent a further step in fission and fusion processes. Whereas the intervalic binding energy liberated in a intervalinar fusion is several orders greater than the binding energy liberated in a nuclear fusion, we would have an almost unconceivable reserve of energy in the *intervalic charge* if its liberation could be controlled. Since the history of science has ever been like a series of impossible task finally reached, there is no reason to think that the intervalinar fusion will not be able to be controlled in the future.

According to the intervalic principle of equivalence $E = c^{\pm 2} h\, Q^{-2}$, the energy liberated in a intervalinar fusion of 270 intervalic charges to conform an elementary one would be:

$$E = (270^3-1)\, 270^{-2} c^{-1} = 9.0062301 \cdot 10^{-7}\, (J) = 5{,}621{,}244.4\, (MeV/c^2)$$

Of course, the problem for practice purposes is that all electric charges known in the actual Universe are already absolutely *degraded* to the state of minimal intervalic energy: the *elementary charge*. If there is really the mentioned ratio of single intervalinos floating across interstellar space, and a spatial navy moving through spacetime could join a number of them by some device up to make elementary charges, there would have an unlimited source of energy at our disposal for interstellar travels.

INTERVALINO AND STRING THEORY

The postulation of the *intervalino* can divide theoretical Physics research on two main fields: physics *before* the intervalino and Physics *after* it. The last can study the features of the successive aggregations of intervalinos to make subatomic particles, since all matter is made only from the assembly of intervalinos; this is the aim of IT in Particle Physics. The first one can study the mystery of the creation of the own intervalino, the last fundamental block of Nature; this will have to be likewise the aim of the String Intervalic Theory. Really it does not need to explain anything more since all particles can be explained starting from the intervalino and as simple assemblies of intervalinos in IT. Intervalino is really the unique primordial particle needed for yielding all subatomic particles in Physics. Henceforth, it might be the main area of research in String Theory. But to make this possible, it is unavoidable that a new Intervalic String Theory start to work from the very beginning, reformulating Physics in intervalic dimensions, with the intervalic quanta and limits, and expressing it in the intervalic system of units.

Finally, I regret to say that some parts of String Theory are lamentably *inconsistent* under intervalic units. This failure is, of course, due to the introduction of the so named "geometrized units" —$c = 1$ (1), $\hbar = 1$ (1)— which are absolutely geometryless and I have serious doubts about if such *non independent* system of measurement can be properly named "units". It is not necessary to point out that starting from clearly inconsistent foundations, it is impossible to obtain any right result, unless it is yielded by chance. Anyone can read in a Physics textbook by Anthony Zee (***Quantum Field Theory in a Nutshell***), that being T some operator to be determined, time reversal verifies:

$$T^{-1}(-i)T = i$$

As Zee comments: "Speaking colloquially, we can say that in quantum physics time goes with an i and so flipping time means flipping i as well". But this just coincides with the intervalic dimension of time: $T = iL$, which is in the foundations of the Intervalic Theory.

I am afraid that String Theory should reorient its research, forgetting the zoo of subatomic particles and focusing exclusively on the intervalino features, since all massive particles can be explained starting only from the intervalino. In other way, it is possible that String Theory does not be the mathematics of XXI century felt on the XX century, as someone have said, but only a fantastic mathematical crossword without any epistemological junction with Physics.

As we will see when describing the intervalic primordial aggregation of the Universe, all massive particles of nature can be fully explained as assemblies of intervalinos, this means that massless particles —photon, graviton— are not composed by intervalinos. This destroys the beauty of a faultless theory because the most simple assumption for such a theory is to find *one* fundamental particle from which are derived all the remaining ones, both massless and massful ones.

Richard Feynman says in his autobiography that his father asked him the following: how can be possible that an electron emits a lot of photons, have the electron a great store of photons inside? Feynman could not gave a satisfactory answer to his father. Our ignorance of a feature often lead us to accept blindly that feature without question it. The case is that the question of Feynman's father is absolutely relevant, and it is similar to asking how can liquid water make vapour of water. Of course, it is only possible because liquid water have a number of molecules of H_2O. Similarly, we say that matter is like "solidified" energy, therefore matter should be composed by a number of "blocks" of energy. If we continue this reasoning up to the end, as ancient Greeks taught us, we will obtain an apparently absurd conclusion: intervalino must be composed by some kind of photons. Even worst, as electromagnetic waves have inertia, photons and gravitons must have a similar composition. In resume, intervalinos, photons and gravitons should have a similar composition.

There are several powerful reasons —which will be understood along this book— to postulate that intervalino is a spin 0 particle, photon a spin 1 particle, and graviton a spin 2. Supposing that photon is a closed string, it is clear that there is only one manner to relate those three particles with full symmetry under interchange: spin 0 intervalino would be an assembly of two closed strings in an antisymmetric state under interchange, and spin 2 graviton would be an assembly of two closed strings in a symmetric state under interchange. We will develop later the details of this picture in a wider context. By the moment, we can say that if this relation was consistent with the primordial aggregation of intervalinos, the fundamental interactions, the symmetry breakings, etc., we would have got a very simple and elegant faultless theory which could explain all fundamental particles of Nature starting from *one* particle: a *closed string* —later named *photon*, from the dalino synthesis and onwards, when acting as carrier of the electromagnetic interaction—. Moreover, the own closed string could still come from an open string —the intervalic string—, some of whose main physical features are intended to be the same ones as of vacuum.

INTERVALINO COUPLING FREQUORCE

We already know that the intervalino mass energy comes completely from the intervalic energy. If we equalize the intervalino intervalic energy with the *transversal energy* of a pair of massless closed strings, in an antisymmetric state under interchange, coupled at a distance of the intervalic length, $\hbar = 1$ (L) $= 1.0556363 \cdot 10^{-34}$ (m), we have:

$$E(\mathbf{I}) = c^{\pm 2} \hbar\, q_I^{-2} = c^{\pm 2} \hbar\, [\sqrt{-(c^{-1}\hbar)}]^{-2} = c^{-1} = 3.335640952 \cdot 10^{-9} (J) =$$
$$= 20{,}819.42423\ (MeV/c^2)$$
$$E(\gamma) + E(\chi) = 2 \cdot 2\pi J \varphi = 4\pi \hbar \varphi = E(\mathbf{I})$$

Therefore, the corresponding frequorce of the constituent photons is:

$$\varphi_m = 1/(4\pi\, c\hbar) = 2.51452013 \cdot 10^{24}\,(s^{-1})$$

This magnitude would be the intervalino coupling frequorce, which can be interpreted as the *coupling frequorce of matter*, since all intervalinos of the Universe are coupled at that frequorce. It is also clear that it will be the minimal allowed energy of the resultant pair photon-antiphoton in the theoretical electromagnetic decay of an intervalino (regardless the threshold energy). Therefore, we can affirm that the frequorce of the primordial isolated closed strings —ulterior photons— was necessarily *greater* that this one, since in other case the primordial aggregation of closed strings was not happened because it was not led to a state of lesser energy.

We can know the related *coupling temperature of matter*, Θ_m, which is the minimal temperature needed for the assembly of intervalinos. By means of the geometrical properties of the intervalic units, taking dimensionless intervalic light-units for the coupling frequorce and for the coupling temperature, being φ_I and Θ_I respectively the *intervalic frequorce* and the *intervalic temperature* (namely, the geometric *intervalic limits* in the Intervalic Space), we have:

$$\varphi_c = \varphi_m / \varphi_I = 1/(4\pi\, c\hbar) / c\, h^{-1} = 1/(4\pi\, c^2)$$
$$\Theta_c = \Theta_m / \Theta_I = \Theta_\gamma\, c^{-1}\, k_B$$
$$\Theta_m = 1/(4\pi\, c\, k_B) = 1.922575127 \cdot 10^{13}\,(K)$$

As we will see opportunely, the *threshold temperature* of annihilation-materialization of intervalinos in the primordial Universe, considered as a gas in thermodynamic equilibrium, is around 10 times greater than the above one.

INTERVALINO DECAY AND LIFETIME

The fact is that some matter decays electromagnetically. This means that we necessarily have some intervalic structure which decays into *photons*. But we know that all particles are composed finally by *intervalinos*. According to the principles of economy and simplicity in logical assumptions, this means that we should be compelled to investigate the possibility that the intervalino should be composed by some assembly of photons. It appears that the most simple possible decay of intervalino would be the separation of its constituent closed strings. In this case the genuine block of matter, namely the intervalino, which is the fundamental boson of the intervalic interaction with spin 0, would be at the same time a particle composed by two closed strings, that we can suppose with spin 1 and radius h. It appears that the most simple assumption is to identify these two closed strings with *photons*:

$$I \to \gamma\gamma$$

Since the lifetime of photon is intended to be infinite, we have no reason to suppose the lifetime of intervalino (and also of graviton) was not likewise infinite. Therefore we may suppose that the decay of intervalino into its two constituent closed strings (photons) only comes when reaching some geometric limit of temperature. That limit should be somehow related with the intervalic temperature, $\Theta_I = c\, k_B^{-1} = 2.1713738 \cdot 10^{31}\,(K)$, a geometric magnitude which points the beginning of the Universe in traditional cosmologic models, or the interval between the Big Crunch and the next Big Bang in an oscillating Universe. By the way, the oscillating Universe is the unique model that satisfies the ancient wisdom about the deep parallelism existing between macro cosmos and micro cosmos. Since elegance and simplicity are indispensable features of a faultless theory of Nature, we believe that the inner geometry of Nature should involve the model of an oscillating Universe, as the two other possibilities —an open and a closed Universes— only can be really postulated by fragmented, inelegant and tasteless minds of bonkers, which surely can not understand the wonderful mind of Nature.

This is the way postulated by IT. However we still have another logical possibility inside the Intervalic Space that we are going to comment briefly. The intervalino lifetime could be related with the intervalic pseudo limit of time: $t_I^{-1} = c\hbar^{-1} = 2.8399227 \cdot 10^{42}$ (s), which is supposed to be the greatest magnitude of time expected in the Universe. We say pseudo limit instead of limit because time is not geometrically *limited* in the Intervalic Geometry, but it is *quantized* by the intervalic time, t_I. However t_I^{-1} is the inverse period of the intervalic frequency, namely, the largest oscillation allowed in the Intervalic Space, which may be supposed to be the oscillation of the own Universe from a Big Bang to a Big Crunch, being that magnitude $8.9991693 \cdot 10^{34}$ years. If this was the case, it is clear that the lifetime of proton postulated by SUSY could not be by no means greater than the intervalino lifetime.

To conclude we will consider the lifetime of intervalino as a *virtual* particle. According to well known physics principles it is:

$$T(I) = c^{-1}\hbar / c\, m_I = c\,\hbar = 3.1647180 \cdot 10^{-26}\,(s^{-1})$$

It can be noted that at the Big Bang, due to the, to say in an old style, intervalic Lorentz-Einstein transformation regarding *temperature*, the dilation of time could increase considerably such magnitude since the initial temperature of the Universe began with the *intervalic temperature*, Θ_I, being k_B the Boltzmann constant and φ_I the intervalic frequorce:

$$\Theta_I = E / k_B = \hbar\, \varphi_I / k_B = \hbar\, c\, \hbar^{-1} / k_B = c\, k_B^{-1} = 2.171374 \cdot 10^{31}\,(K)$$

This temperature is the intervalic geometrical limit which determines the highest one allowed in the Intervalic Space. Such temperature plays a role analogous to the speed of light with respect to the measurement of spacetime and physical quantities in Lorentz-Einstein transformations. The *c-temperature* is defined as:

$$\Theta_c = \Theta/\Theta_I$$

And the gamma factor regarding temperature is:

$$\gamma(\Theta) = 1 / \sqrt{(1 - \Theta_c^2)}$$

Therefore, the intervalic Lorentz-Einstein transformations of space and time regarding temperature are, written in the classic mode of Relativity:

$$L = L_0 \gamma^{-1}(\Theta)$$
$$T = \gamma(\Theta) T_0$$

This means that if we suppose a primordial temperature $\Theta_c \approx 1$, the dilatation of intervalino's lifetime considered as *virtual* particle could be as large as the c-temperature was close to 1. This may have surprising consequences for the genesis of the Universe and clearly makes irrelevant the assumption of inflation which lead to the uneconomic existence of a lot of a lot of a lot of multiple Universes. That is neither the way of Art by no means, nor the way of Nature insofar as the unique art work made by Nature —the Universe— is necessarily still much more intelligent than the artificial art works —the musical compositions— made by human beings.

THE INTERVALIC STRUCTURE OF SPACE

THE TWO OLD CONCEPTIONS OF SPACE

There have been two different conceptions of *space* in the history of Physics:
- *Continuous*, completely full, made from an infinite number of infinitesimal space-time cells. This is the traditional paradigm of space in Physics, hold by both Classic and Quantum Mechanics.
- *Discrete, discontinuous*, granular, completely empty, made from the juxtaposition of a finite number of space blocks. It has been rigorously avoided by traditional Physics.

INTERVALIC STRUCTURE OF SPACE

According to the Intervalic Theory space is not continuous at quantum scale. Academic physics can not accept the concept of a granular space as it seems to involve that space is woven as a fixed net. However things are far to be that way. Tridimensional space is made by the cosmic composition of a finite number of intervalic quanta of length, $\hbar$ (L). As any surface is made by an infinite number of lines, this means that the sum of any *finite* number of intervalic lengths is neither enough to make any surface nor any volume. Therefore *the total volume of space of the Universe is zero*, and *the total surface of the Universe is equally zero*.

The spatial net which makes what we know as tridimensional space is composed exclusively by *lines* ($\hbar$) at quantum scale. This means that the volume of space is absolutely *empty* at quantum scale and contains no other than *nothingness* between a tridimensional net of *lines of space* composed by a number of intervalic quanta of length.

Insofar as such quanta are the latest minimal blocks of *information*, which at once are the foundations of consciousness, *space* itself can be interpreted in some ways as equivalent and identical to *consciousness*. This means that 'space', as it is usually imagined, does not exist, but only a vast *cobweb of space-consciousness*, whose volume and surface is zero, which makes the illusion of a filled space. This involve that 'vacuum' is a misleading concept and that the quantum concept of vacuum is completely wrong.

Moreover, the space which makes that cobweb is just the stuff of which are made all subatomic particles —which are ultimately made from intervalic strings, that is to say, from space (L)—. Therefore we surprisingly find that *it does not exist any space in the Universe apart from the own one-dimensional space of which are made all subatomic particles*. This **one-dimensional physical** space is unfolded or curved in a **three-dimensional mathematical** space, making the great illusion of continuum we usually see with our senses.

If we could see a truthful image of the real stuff of the Universe during an interval of time, we would see but *a mess of straight lines of space crossing in all directions* (the track left by intervalinoless particles) and *a number of big cosmic nests* also made from *lines of space* (the track left by intervalinoful particles, gathered in vast bulks forming stars). By the way, the famous picture of the lead ball sinking on a rubber cloth used to explain the curvature of space in General Relativity has to be substituted by this one.

FINITENESS AND DISCRETENESS OF SPACE

By these reasons the total value in the Universe of everyone of those strongly related terms —space, consciousness and information— is necessarily *finite*.

The opposite is logically contradictory. If *space* was continuous —and therefore infinite— as it is intended by Classic Mechanics and Quantum Mechanics, it involves that *consciousness* and *information* would also be *infinite*. However, it goes against the higher principles of logical economy and in this way it seems to be a logical impossibility that consciousness was infinite. Moreover, since there is a definite geometric equation between *information* and *energy*, this involves that the total energy of the Universe would necessarily be infinite, what is absolutely impossible in an intervalic Universe which has got all its physical quantities geometrically limited by the intervalic quanta and limits, derived straightforwardly from the Intervalic System of Dimensions.

This means that the cosmic space in the Universe, although unlimited, it is however finite, which involves two things: the cosmic space is *finite* in size, and it is also *discrete* (as if it was *continuous* it would be anew *infinite*).

Chapter 7

INTERVALIC DALINO

INTERVALIC COMPOSITENESS

Intervalic compositeness —or structurefulness— is an obliged feature in IT since the fundamental geometry of the Intervalic Space quanta define logically the intervalic quantum of electric charge: $q_I = \sqrt{-(c^{-1}\hbar)} = 1\ (i^{-1/2}L^{1/2}) = 5.9339900 \cdot 10^{-22}$ (C). Therefore, all charges of Nature must unavoidably be multiple of the intervalic charge, q_I. Starting from here, when developing in a systematic way the infinite possibilities of the aggregation of that electric quantum of charge to create the particles of Nature, we arrive to the surprising conclusion that there is only *one* way to make such primordial aggregation of intervalic charges according to the intervalic geometry. We start postulating only the existence of the geometric relations derived from the existence of the two fundamental constants of Nature, c (i^{-1}) and ℏ (L), which intervalic dimensions are, at once, just and precisely the unique two components of the *dimensional basis* of the Intervalic Theory. All those geometric relations are simply the *geometric equivalences* between all the 40 physical quantities existing in the Intervalic Space. Among such *geometric statements*, we find the intervalic principles of equivalence between electric charge, energy and matter:

$$I = c^{\pm 2} \hbar\, Q^{-2}$$
$$E = c^{\pm 2} m$$
$$Q = c^{\pm 2} \sqrt{(\hbar/m)}$$

These important formulas are fundamental equations in classic Physics, but in IT they have the epistemological rank of geometric statements —in a very different manner from the degenerate "geometrized units", which have not any geometric meaning, and hardly can be named independent units—.

Apart form this, we only need to postulate the validity of one theorem in order to develop logically the intervalic primordial aggregation of intervalinos: it is the *spin-statistics theorem*, which can be supposed to be valid before the apparition of the four supposed traditional interactions. Thus *spin* can be considered as the first degree of freedom existing in Nature, and therefore the first interaction between particles should be a spin dependent interaction. But this is just the *intervalic interaction*, which is a spin dependent exchange frequorce. Moreover, it relies and is defined starting from the *intervalic energy*, $I = c^{\pm 2} \hbar\, Q^{-2}$, which can be viewed as a geometric statement in the Intervalic Space.

In a remarkable show of simplicity, the existence of traditional interactions and other classic physical features is not necessary by no means to make such primordial aggregation. Moreover, the four supposed forces of Nature were born as a result of such primordial aggregation, appearing at successive steps and being related with corresponding *intervalic structures* and *symmetry breakings*, as we will opportunely see.

According to these simple and powerful constraints, the systematic aggregations of intervalinos led to the assembly of successive particles, all of them defined univocally by its *intervalic structure*. Thus, the intervalino based compositeness of Nature yields up to six intervalic structure levels, each one determined by an intervalic structure which name correspond to its related particle. They are, by order: *intervalino*, *dalino*, *gaudino*, *lisztino*, *monteverdino* and *palestrino* —these terms are taken from prominent artists of different epochs: Salvador Dalí, Antonio Gaudí, Franz Liszt, Claudio Monteverdi and Giovanni Pierluigi da Palestrina—.

STRUCTURE OF DALINOS

We define the *dalino* as an intervalic structure composed by the aggregation of a determine number of intervalinos. The term 'dalino' is taken in honour of the great Spanish surrealist artist Salvador Dalí (1904-1989).

Due to intervalic principles of symmetry, that number only can take 16 values, which are precisely the divisors of the elementary charge, since the magnitude of the elementary charge is the main state of lowest energy, reached through the accurate balance between the intervalic energy, electromagnetic and spin energies. These values are the 16 multiples of the intervalic charge, $q_I = \sqrt{-(c^{-1}\hbar)}$:

1, 2, 3, 5, 6, 9, 10, 15, 18, 27, 30, 45, 54, 90, 135, 270.

Of course, the aggregations of intervalinos could take any other exotic values at the very beginning of the Big Bang, but none of them survived after the primordial synthesis of subatomic particles because its energy could not reach a balance imposing a different set of symmetrical constraints than those derived from the elementary charge, would could play like a strong and powerful attractor. Therefore, we are compelled to think that any particle which symmetries do not fit with the constraints imposed by the state of lowest energy of the elementary charge could not stay after primordial times.

According to the number of constituent intervalinos, we name the dalinos as, i.e., dalino 18, dalino 27, dalino 30, etc., including likewise this number behind its symbol as a subindex: D_{18}, D_{27}, D_{30}, etc. Need not to say that the dalino 270 (D_{270}) is the *electron*.

INTERVALIC STRUCTURE LEVELS

1st.	2nd.	3rd.	4th.	5th.	6th.
Intervalinar level	*Dalinar level*	*Gaudinar level*	*Lisztinian level*	*Monteverdic level*	*Palestrinian level*
INTER-VALINO	Symmetric aggregation of intervalinos → **DALINO** (ELECTRON)	Symmetric aggregation of dalinos → **GAUDINO** (MUON, TAU, CHARGED MASSIVE BOSONS)	Symmetric aggregation of gaudinos → **LISZTINO** (ZERO CHARGED MASSIVE BOSONS)		
			Antisymmetric aggregation of gaudinos → **FRACTIONAL CHARGED LISZTINO** (QUARKS)	Symmetric aggregation of lisztinos → **MONTEVERDINO** (MESONS)	
				Antisymmetric aggregation of lisztinos → **MONTE-VERDINO** (BARYONS)	Antisymmetric aggregation of monteverdinos → **PALES-TRINO**
					—
Antisymmetric aggregation of intervalinos → DARK MATTER				Antisymmetric aggregation of dalinos → DARK MATTER	

DALINOS INTERVALIC ENERGY

The intervalic energy of dalino constitutes its formerly named "intrinsic mass" of the particle. It has to be remarked that surprisingly the intervalic mass of dalinos do not rely on their radii, so it is a *geometric* magnitude which may have great importance to understand some physical features, as we will see along this book. Here are shown the intervalic masses of all dalinos with the magnitudes expressed in traditional units:

$I(D_{270}) = I(e) = c^{\pm 2} \hbar (270 \, q_I)^{-2} = c^{\pm 2} \hbar [270 \sqrt{-(c^{-1}\hbar)}]^{-2} = 270^{-2} \, c^{-1} =$
$= 4.575639166 \cdot 10^{-14} \, (J) = 0.285588809 \, (MeV/c^2)$

$I(D_{135}) = c^{\pm 2} \hbar (135 \, q_I)^{-2} = c^{\pm 2} \hbar [135 \sqrt{-(c^{-1}\hbar)}]^{-2} = 135^{-2} \, c^{-1} =$
$= 1.8302557 \cdot 10^{-13} \, (J) = 1.1423553 \, (MeV/c^2)$

$I(D_{90}) = c^{\pm 2} \hbar (90 \, q_I)^{-2} = c^{\pm 2} \hbar [90 \sqrt{-(c^{-1}\hbar)}]^{-2} = 90^{-2} \, c^{-1} =$
$= 4.1180752 \cdot 10^{-13} \, (J) = 2.5702993 \, (MeV/c^2)$

$I(D_{54}) = c^{\pm 2} \hbar (54 \, q_I)^{-2} = c^{\pm 2} \hbar [54 \sqrt{-(c^{-1}\hbar)}]^{-2} = 54^{-2} \, c^{-1} =$
$= 1.1439098 \cdot 10^{-12} \, (J) = 7.1397203 \, (MeV/c^2)$

$I(D_{45}) = c^{\pm 2} \hbar (45 \, q_I)^{-2} = c^{\pm 2} \hbar [45 \sqrt{-(c^{-1}\hbar)}]^{-2} = 45^{-2} \, c^{-1} =$
$= 1.6472301 \cdot 10^{-12} \, (J) = 10.281197 \, (MeV/c^2)$

$I(D_{30}) = c^{\pm 2} \hbar (30 \, q_I)^{-2} = c^{\pm 2} \hbar [30 \sqrt{-(c^{-1}\hbar)}]^{-2} = 30^{-2} \, c^{-1} =$
$= 3.7062677 \cdot 10^{-12} \, (J) = 23.132694 \, (MeV/c^2)$

$I(D_{27}) = c^{\pm 2} \hbar (27 \, q_I)^{-2} = c^{\pm 2} \hbar [27 \sqrt{-(c^{-1}\hbar)}]^{-2} = 27^{-2} \, c^{-1} =$
$= 4.5756391 \cdot 10^{-12} \, (J) = 28.558881 \, (MeV/c^2)$

$I(D_{18}) = c^{\pm 2} \hbar (18 \, q_I)^{-2} = c^{\pm 2} \hbar [18 \sqrt{-(c^{-1}\hbar)}]^{-2} = 18^{-2} \, c^{-1} =$
$= 1.0295188 \cdot 10^{-11} \, (J) = 64.257483 \, (MeV/c^2)$

$I(D_{15}) = c^{\pm 2} \hbar (15 \, q_I)^{-2} = c^{\pm 2} \hbar [15 \sqrt{-(c^{-1}\hbar)}]^{-2} = 15^{-2} \, c^{-1} =$
$= 1.4825071 \cdot 10^{-11} \, (J) = 92.530775 \, (MeV/c^2)$

$I(D_{10}) = c^{\pm 2} \hbar (10 \, q_I)^{-2} = c^{\pm 2} \hbar [10 \sqrt{-(c^{-1}\hbar)}]^{-2} = 10^{-2} \, c^{-1} =$
$= 3.3356409 \cdot 10^{-11} \, (J) = 208.19424 \, (MeV/c^2)$

$I(D_9) = c^{\pm 2} \hbar (9 \, q_I)^{-2} = c^{\pm 2} \hbar [9 \sqrt{-(c^{-1}\hbar)}]^{-2} = 9^{-2} \, c^{-1} =$
$= 4.1180752 \cdot 10^{-11} \, (J) = 257.02993 \, (MeV/c^2)$

$I(D_6) = c^{\pm 2} \hbar (6 \, q_I)^{-2} = c^{\pm 2} \hbar [6 \sqrt{-(c^{-1}\hbar)}]^{-2} = 6^{-2} \, c^{-1} =$

$$= 9.2656692 \cdot 10^{-11} \,(J) = 578.31735 \,(MeV/c^2)$$
$$I(D_5) = c^{\pm 2}\, \hbar\, (5\, q_I)^{-2} = c^{\pm 2}\, \hbar\, [5\, \sqrt{-(c^{-1}h)}]^{-2} = 5^{-2}\, c^{-1} =$$
$$= 1.3342564 \cdot 10^{-10} \,(J) = 832.77698 \,(MeV/c^2)$$
$$I(D_3) = c^{\pm 2}\, \hbar\, (3\, q_I)^{-2} = c^{\pm 2}\, \hbar\, [3\, \sqrt{-(c^{-1}h)}]^{-2} = 3^{-2}\, c^{-1} =$$
$$= 3.7062677 \cdot 10^{-10} \,(J) = 2{,}313.2694 \,(MeV/c^2)$$
$$I(D_2) = c^{\pm 2}\, \hbar\, (2\, q_I)^{-2} = c^{\pm 2}\, \hbar\, [2\, \sqrt{-(c^{-1}h)}]^{-2} = 2^{-2}\, c^{-1} =$$
$$= 8.3391023 \cdot 10^{-10} \,(J) = 5{,}204.8561 \,(MeV/c^2)$$
$$I(D_1) = I(I) = c^{\pm 2}\, \hbar\, (1\, q_I)^{-2} = c^{\pm 2}\, \hbar\, [1\, \sqrt{-(c^{-1}h)}]^{-2} = 1^{-2}\, c^{-1} =$$
$$= 3.3356409 \cdot 10^{-9} \,(J) = 20{,}819.424 \,(MeV/c^2)$$

The total sum of the intervalic energy of the set of dalinos is $4.87763005 \cdot 10^{-9} \,(J) = 30{,}443.759 \,(MeV/c^2)$.

DALINOS RADIUS

We already know the radius of some dalinos: among others, those of the electronic dalino (D_{270}) and the muonic dalino (D_{45}). They are respectively:

$$r(D_{270}) = r_e = 3.0257565 \cdot 10^{19} \,(L) = 3.1940984 \cdot 10^{-15} \,(m)$$
$$r(D_{45}) = 2.5850690 \cdot 10^{16} \,(L) = 2.7288927 \cdot 10^{-18} \,(m)$$

which correspond to the following electromagnetic potential energies:

$$U(D_{270}) = U(e) = \tfrac{1}{2}\, (1/4\pi\varepsilon_0)\, (270\, q_I)^2 \,/\, r_e =$$
$$= 3.6114723 \cdot 10^{-14} \,(J) = 0.22541028 \,(MeV/c^2)$$
$$U(D_{45}) = \tfrac{1}{2}\, (1/4\pi\varepsilon_0)\, (45\, q_I)^2 \,/\, r(D_{45}) =$$
$$= 1.1742041 \cdot 10^{-12} \,(J) = 7.3288027 \,(MeV/c^2)$$

Instead to adjust a curve of interpolation, which is always inelegant, and since the radius is obviously proportional to the electric charge raised to the power 4, we will take intuitively as a first approximation the following formula with the electron radius ratio as reference: $r(D_n) \approx n^4 \cdot r(D_{270}) / 270^4$:

$r(D_{270}) = r_e = 3.0257565 \cdot 10^{19}\,(L) = 3.1940984 \cdot 10^{-15}\,(m)$
$r(D_{135}) \approx 1.8910978 \cdot 10^{18}\,(L) = 1.9963115 \cdot 10^{-16}\,(m)$
$r(D_{90}) \approx 3.7355018 \cdot 10^{17}\,(L) = 3.9433313 \cdot 10^{-17}\,(m)$
$r(D_{54}) \approx 4.8412106 \cdot 10^{16}\,(L) = 5.1105576 \cdot 10^{-18}\,(m)$
$r(D_{45}) \approx 2.3346887 \cdot 10^{16}\,(L) = 2.4645821 \cdot 10^{-18}\,(m)$
$r(D_{30}) \approx 4.6117307 \cdot 10^{15}\,(L) = 4.8683104 \cdot 10^{-19}\,(m)$
$r(D_{27}) \approx 3.0257565 \cdot 10^{15}\,(L) = 3.1940984 \cdot 10^{-19}\,(m)$
$r(D_{18}) \approx 5.9768030 \cdot 10^{14}\,(L) = 6.3093302 \cdot 10^{-20}\,(m)$
$r(D_{15}) \approx 2.8823317 \cdot 10^{14}\,(L) = 3.0426940 \cdot 10^{-20}\,(m)$
$r(D_{10}) \approx 5.6934947 \cdot 10^{13}\,(L) = 6.0102597 \cdot 10^{-21}\,(m)$
$r(D_9) \approx 3.7355018 \cdot 10^{13}\,(L) = 3.9433313 \cdot 10^{-21}\,(m)$
$r(D_6) \approx 7.3787689 \cdot 10^{12}\,(L) = 7.7892963 \cdot 10^{-22}\,(m)$
$r(D_5) \approx 3.5584341 \cdot 10^{12}\,(L) = 3.7564122 \cdot 10^{-22}\,(m)$
$r(D_3) \approx 4.6117307 \cdot 10^{11}\,(L) = 4.8683104 \cdot 10^{-23}\,(m)$
$r(D_2) \approx 9.1095914 \cdot 10^{10}\,(L) = 9.6164154 \cdot 10^{-24}\,(m)$
$r(D_1) \approx 5.6934947 \cdot 10^{9}\,(L) = 6.0102597 \cdot 10^{-25}\,(m)$

DALINOS EQUIVALENT ELECTROMAGNETIC ENERGY

Taking the preceding values of the radius, the equivalent electromagnetic energy of dalinos would be:

$U(D_{270}) = U(e) = 3.6114723 \cdot 10^{-14}\,(J) = 0.22541028\,(MeV/c^2)$
$U(D_{135}) = 1.4445889 \cdot 10^{-13}\,(J) = 0.90164111\,(MeV/c^2)$
$U(D_{90}) = 3.2503251 \cdot 10^{-13}\,(J) = 2.0286925\,(MeV/c^2)$
$U(D_{54}) = 9.0286804 \cdot 10^{-13}\,(J) = 5.6352567\,(MeV/c^2)$
$U(D_{45}) = 1.3001300 \cdot 10^{-12}\,(J) = 8.1147699\,(MeV/c^2)$
$U(D_{30}) = 2.9252925 \cdot 10^{-12}\,(J) = 18.258232\,(MeV/c^2)$
$U(D_{27}) = 3.6114723 \cdot 10^{-12}\,(J) = 22.541028\,(MeV/c^2)$

$U(D_{18}) = 8.1258126 \cdot 10^{-12}$ (J) $= 50.717312$ (MeV/c²)
$U(D_{15}) = 1.1701170 \cdot 10^{-11}$ (J) $= 73.032928$ (MeV/c²)
$U(D_{10}) = 2.6327633 \cdot 10^{-11}$ (J) $= 164.32409$ (MeV/c²)
$U(D_9) = 3.2503251 \cdot 10^{-11}$ (J) $= 202.86925$ (MeV/c²)
$U(D_6) = 7.3132316 \cdot 10^{-11}$ (J) $= 456.45582$ (MeV/c²)
$U(D_5) = 1.0531053 \cdot 10^{-10}$ (J) $= 657.29638$ (MeV/c²)
$U(D_3) = 2.9252925 \cdot 10^{-10}$ (J) $= 1,825.8232$ (MeV/c²)
$U(D_2) = 6.5819083 \cdot 10^{-10}$ (J) $= 4,108.1023$ (MeV/c²)
$U(D_1) = U(\mathbf{I}) = 0$

Please note that according to the right (and frequently forgotten) definition of electromagnetic potential energy, only can have potential energy the *composite* particles, that is to say, those which can be separated in fragments. Since the intervalino is a *single* particle —and the unique fundamental block of which are composed all subatomic particles—, it can not have electromagnetic potential energy. In the same way, the dalino 1 which is composed by *one* intervalino does not have either. In other case, its electromagnetic potential energy would be:

$U(D_1) = 5.26552606 \cdot 10^{-9}$ (J) $= 32,864.8144$ (MeV/c²)

We will see with further detail this uncanny feature when describing the intervalic dark matter.

DALINOS TOTAL MASS

Since the total mass of a subatomic particle at rest is the sum of its intervalic and electromagnetic energies, $m(D_n) = c^{\pm 2}[I(D_n) + U(D_n)]$, we have from the below results:

$m(D_{270}) = 8.1871114 \cdot 10^{-14}$ (J) $= 0.51099909$ (MeV/c²)
$m(D_{135}) = 3.2748446 \cdot 10^{-13}$ (J) $= 2.0439964$ (MeV/c²)

$m(D_{90}) = 7.3684003 \cdot 10^{-13}$ (J) $= 4.5989918$ (MeV/c²)
$m(D_{54}) = 2.0467778 \cdot 10^{-12}$ (J) $= 12.774977$ (MeV/c²)
$m(D_{45}) = 2.9473601 \cdot 10^{-12}$ (J) $= 18.395967$ (MeV/c²)
$m(D_{30}) = 6.6315602 \cdot 10^{-12}$ (J) $= 41.390926$ (MeV/c²)
$m(D_{27}) = 8.1871114 \cdot 10^{-12}$ (J) $= 51.099909$ (MeV/c²)
$m(D_{18}) = 1.8421001 \cdot 10^{-11}$ (J) $= 114.97480$ (MeV/c²)
$m(D_{15}) = 2.6526240 \cdot 10^{-11}$ (J) $= 165.56370$ (MeV/c²)
$m(D_{10}) = 5.9684041 \cdot 10^{-11}$ (J) $= 372.51833$ (MeV/c²)
$m(D_9) = 7.3684003 \cdot 10^{-11}$ (J) $= 459.89918$ (MeV/c²)
$m(D_6) = 1.6578901 \cdot 10^{-10}$ (J) $= 1,034.7732$ (MeV/c²)
$m(D_5) = 2.3873618 \cdot 10^{-10}$ (J) $= 1,490.0734$ (MeV/c²)
$m(D_3) = 6.6315602 \cdot 10^{-10}$ (J) $= 4,139.0926$ (MeV/c²)
$m(D_2) = 1.4921011 \cdot 10^{-9}$ (J) $= 9,312.9584$ (MeV/c²)
$m(D_1) = 3.3356409 \cdot 10^{-9}$ (J) $= 20,819.424$ (MeV/c²)

The total mass of the set of dalinos is 38,040.09338 (MeV/c²).

Of course, it has to be remembered that we have taken the ratio of energies of electron structure for all dalinos for yielding these calculations, but we totally know that the energy balance between intervalic, electromagnetic and spin energy in each particle has to be calculated individually, since there are slight deviations from this value in the energy ratios of all subatomic particles due to obvious reasons depending on the own intervalic structure of each particle. Nevertheless, it is a great pleasure to be able to write down a complete set with the approximate allowed values for all possible dalinos existing in Nature.

INTERVALIC PRINCIPLE OF ENERGY BALANCE FOR SUBATOMIC PARTICLES

The Lagrangian formalism usually deals with the difference between kinetic and potential energies, $L = K - V$. It leads to a law of conservation of energy: $E = K + V$, where E is interpreted as the *total energy* of the system. This total energy is not very useful because it is only the sum of other two amounts and does not have a formulation related with a single physical feature.

On the contrary, in IT this "total" energy —which is not total as we are going to see— is substituted by the *intervalic energy*, which is defined by means of the *intervalic principle of equivalence between electric charge and energy*: $I = c^{\pm 2} \hbar Q^{-2}$. The intervalic energy of any system is ever *invariant*, unless there is a change in the *structure* of the system. And it is invariant *a fortiori* because it states a *geometrical relation* among physical quantities in the Intervalic Space, which validity does not depend therefore on the system of reference chosen. This is one great difference between the intervalic energy and all the rest of physical quantities: the intervalic energy involves determine *levels of structure* in the system. It have to be applied in an *intervalic* way to anyone of those structures and can not be merged over different levels, as all the rest of physical quantities do. It can be noted that it is the first physical quantity that incorporates this feature in the last foundations of Physics.

Henceforth, the traditional balance in Lagrangian formalism has to be substituted by the *intervalic principle of energy balance* when calculating the total structural energy of subatomic particles: $I = U + E(J)$, where the last term is the intrinsic kinetic energy, namely, the spin energy. It is important to note that I and U are *manifested* as *mass energy*, while $E(J)$ is manifested as *kinetic energy*.

Moreover, the electromagnetic energy —as all the four supposed "forces" of Nature— is derived from the intervalic energy, being just its *inverse* in intervalic units: $U \quad I^{-1} = c^{\pm 2}(n\hbar)^{-1} Q^2 = c^{\pm 2} Q^2 / r$. As we have seen, in SI units it has to be added to the equation the permeability of vacuum factor, which was conventionally set by definition as $\mu_0/4\pi = 10^{-7}(-1)$, instead of $1\,(-1)$ as in intervalic units. Thus, subatomic particles are the final *balance* reached between two inverse equations, a relation of paramount elegance:

$$I - I^{-1} - E(J) = 0$$

This fundamental intervalic principle give us the basic relations existing among the *structural energies* —intervalic, electromagnetic and spin enegies— involved in any subatomic particle, and will be thoroughly used along this book.

In IT is defined the *energy interval* as the difference of energy between two states. That interval is the difference of energy between two *intervalic structures*, therefore, it is ever and *a fortiori* invariant, because it is like a geometrical interval. From the energy interval can be derived other classic physical quantities —force, field, potential, etc.— which dress the interaction. On the contrary, the Lagrangian way in quantum field theory tries to tell perhaps the same but in an old-fashioned overloaded mode, which is most limited since it is still involved in avoiding the "infinite paths" between those two states, which are irrelevant in IT because we have precisely avoid them in the definition: it has little sense to talk about the infinite paths between two structures since we are dealing with *geometrical relations*.

DALINO SPIN ENERGY

All stable subatomic particles must have a perfect *balance* in its involved structural energies conforming the particle. According to this principle of intervalic balance for subatomic particles, the intervalic energy must be always equal or greater than electromagnetic energy. In the second case, which is the more usual, the remaining energy up to balance the intervalic one is realized by the intrinsic angular momentum of the particle. The spin energy is manifested as *kinetic energy*, while the two other energies, the intervalic and the electromagnetic ones, are manifested as the *mass energy* of the particle:

$$I(D_n) - U(D_n) = E_j(D_n)$$

According to this important intervalic principle, the spin energy of all possible dalinos would be approximately, taken likewise the electron spin energy ratio as reference:

$E_J(D_{270}) = E_J(e) = 9.6416680 \cdot 10^{-15}$ (J) $= 0.060178533$ (MeV/c²)
$E_J(D_{135}) = 3.8566680 \cdot 10^{-14}$ (J) $= 0.24071418$ (MeV/c²)
$E_J(D_{90}) = 8.6775010 \cdot 10^{-14}$ (J) $= 0.54160679$ (MeV/c²)
$E_J(D_{54}) = 2.4104176 \cdot 10^{-13}$ (J) $= 1.5044637$ (MeV/c²)
$E_J(D_{45}) = 3.4710004 \cdot 10^{-13}$ (J) $= 2.1664271$ (MeV/c²)
$E_J(D_{30}) = 7.8097525 \cdot 10^{-13}$ (J) $= 4.874462$ (MeV/c²)
$E_J(D_{27}) = 9.6416677 \cdot 10^{-13}$ (J) $= 6.017853$ (MeV/c²)
$E_J(D_{18}) = 2.1693755 \cdot 10^{-12}$ (J) $= 13.540171$ (MeV/c²)
$E_J(D_{15}) = 3.1239008 \cdot 10^{-12}$ (J) $= 19.497847$ (MeV/c²)
$E_J(D_{10}) = 7.0287760 \cdot 10^{-12}$ (J) $= 43.87015$ (MeV/c²)
$E_J(D_9) = 8.6775014 \cdot 10^{-12}$ (J) $= 54.16068$ (MeV/c²)
$E_J(D_6) = 1.9524378 \cdot 10^{-11}$ (J) $= 121.8615$ (MeV/c²)
$E_J(D_5) = 2.8115104 \cdot 10^{-11}$ (J) $= 175.4806$ (MeV/c²)
$E_J(D_3) = 7.8097525 \cdot 10^{-11}$ (J) $= 487.4462$ (MeV/c²)
$E_J(D_2) = 1.7571941 \cdot 10^{-10}$ (J) $= 1{,}096.7538$ (MeV/c²)
$E_J(D_1) = E_J(\mathbf{I}) = 3.3356409 \cdot 10^{-9}$ (J) $= 20{,}819.424$ (MeV/c²)

The rotational kinetic energy due to the intrinsic angular momentum of dalino is: $K(D_n) = \frac{1}{2} I_D \omega_D^2$, being I_D the dalino's moment of inertia and ω_D the angular velocity due to its spinning. But as a *composite* particle is like a micro galaxy at quantum scale —in this case the intervalinos would be like the stars and the dalino would play the role of the whole galaxy—, we can unexpectedly introduce a very beautiful application of a law from Cosmology, the Virial theorem, by means of which the potential energy due to dalino's spinning is: $U(D_n) = 2 K(D_n)$. Therefore the spin energy of dalino, $E(D_n)_J$, will be:

$$E(D_n)_J = m_D \, r_D^2 \, \omega_D^2$$

From here we will be able to deduce the linear velocity due to spin on dalino's surface, v_D:

$$v_D = \omega_D \, r_D = 1.028801396 \cdot 10^8 \ (m \ s^{-2}) = 0.343171206 \ c$$

And the acceleration on dalino's surface due to spin will be:

$$a_D = v_D^2 \, / \, r_D$$

Once more, needless to say that these are rough values because it has been taken the magnitude of the dalino radius according to the electron's ratio. Accurate values can be deduced when treating individually each particle, although it is expected that the deviation from electron's energy ratio will be very small, as we will see when studying the lepton-massive bosons, which energy ratios are very similar to the electron's one.

THE PRIMORDIAL AGGREGATION OF INTERVALINOS

The definition of the elementary charge in IT as exactly $e = 270 \ q_I$ is based on superbly solid basis: the intervalic system of dimensions, the intervalic quanta and the intervalic limits, and the intervalic system of units. All of them conform the foundations of the Intervalic Theory in Physics and it is hard —or perhaps impossible— to refuse any of them without challenging seriously the very last foundations of Mathematical Logic. Against all expected, the intervalic interaction states that the *electric charge is inversely proportional to energy* at quantum scale. This equivalence explains why the minimal state of energy of the 16 subatomic charges allowed by the Intervalic Theory (the divisors of the elementary charge: 1, 2, 3, 5, 6, 9, 10, 15, 18, 27, 30, 45, 54, 90, 135, 270, values that are very different than those presumed by String Theory) is that of the elementary charge —which is the greater single charge: 270 q_I—, instead of the intervalic quantum of electric charge: q_I —the smaller one—. By the same reason we also can infer that there is no any *single* electric charge *greater* than the electron's, since it would exist, it would be the elementary one, because it would be then the main state of lesser energy.

Primordial intervalic charges will aggregate among them if, and only if, the aggregation lead to a state of lesser energy. This only is possible when the total *intervalic energy* of the aggregation of n intervalic charges is greater than its total *electromagnetic potential energy*, that is to say, when the intervalic energy is greater than its *inverse*, $I > I^{-1}$:

$$I(nq_I) > I^{-1}(nq_I)$$

Since the subatomic particles are dynamical structures, the equality between I and I^{-1} can never be reached because there is a remaining structural energy shared by the intrinsic angular momentum of the particle, which is not manifested as *mass* energy but as *spin* energy. It is also clear that Nature balance for subatomic particles tries to maximize the magnitude of the dalino radius, since the electron radius is around 10^{10} orders greater than the maximum intervalino radius. It can be supposed that the magnitude of the radius can not increase indefinitely, but it has a limit determined, at least, by the short range of the intervalinar interaction: the distance among intervalinos inside the dalino must be comprised inside the effective range of the intervalic interaction. Those distances are different in all dalinos. For example, the intervalic distance between intervalinos inside dalino 45 gives smaller values than the analogous distance inside dalino 270. By the moment and at first sight, we can think that the possible biggest dalino radius could not be much more greater than the actual electron radius because it could be expected that according to a general law Nature usually tries to reach an energy equilibrium maximizing the volume of subatomic particles. (Of course these are still rough appreciations, but they are sufficient for our general introductory purpose at this moment; the intervalic interaction will be more thoroughly described later, in the two chapters devoted to that subject).

Henceforth, given supposedly an approximate magnitude of the biggest dalino radius the inequality $I(n\mathbf{q}_I) > I^{-1}(n\mathbf{q}_I)$ would be valid until $n = 286$ inclusive. For the value $n = 287$ we already have: $I(287\mathbf{q}_I) < U(287\mathbf{q}_I)$, and further aggregations will not happen anyway. Of course, if $n = 286$ there was not a remaining structural energy to be shared by the intrinsic angular momentum of the dalino, which maintains dynamically the quantum structure of the intervalic charges inside the dalino, so n must necessarily be lesser than 286. Moreover, it is hoped that it should be enough lesser, since in other case the intrinsic angular momentum of electron would be ridiculously small (compared with the speed of light). Actually, the linear velocity on electron's surface is $v(e)_J = 1.028801396 \cdot 10^8$ (m s^{-2}) = 0.343171206 c.

On the other hand, due to symmetry principles it is expected that n should be divisible by the first integers because it ensures more rich symmetries, a very important constraint which is still difficult to quantify by human minds, but that Nature clearly knows and applies to. Well, the first number below 286 which is divisible by 2, 3, and 5 is 270. The next number is 240 which is too far from the expected values. And above the closed number would be 300. Therefore the maximum allowed aggregation of intervalinos, which at the same time is the state of minimal energy which tend all the aggregations of intervalinos, would be 240, 270 or 300. As we experimentally know by means of the magnitudes of the intervalic and elementary charges, it is just 270, and such dalino is called *electron*.

Since the state of minimal energy in the number of aggregation of intervalinos was fixed at 270, a lot of aggregations of intervalinos were made reaching just that number in the primordial Universe. The repetition and wide spreading of this physical fact along the Universe involves the emergency of an spontaneous order, or in other words, a new symmetry constraint based on that value, which now is named *elementary charge*, since all intervalic structures assembled after it will have to be exactly balanced with that value. In this way can be clearly understood the uniformity imposed by the elementary charge, which magnitude plays a role as an attractor to the main state of minimal energy in the primordial fabric of subatomic particles.

As we will explain opportunely when describing the intervalic primordial aggregation, the spin 0 dalino is an intervalic structure *defined* as the assembly of intervalinos *in symmetric state under interchange*. Therefore, it only can exist dalinos composed by intervalinos with like charges, because a "dalino" composed by intervalinos with unlike charges would be necessarily in an *antisymmetric* state, that is to say, it would not really be a 'dalino' but another primordial particle, which will be described at due course.

INTERVALIC CHAOS

Really, the complete understanding of the value of the attractor of the elementary charge is one of the most exciting problems in IT. Please note that at this precise point there is a confluence of meeting point of all the, say, main systems of thought in Physics, by chronological order: relativity, quantum mechanics, theory of chaos and intervalic geometry, and all of them must probably be satisfied here, inside IT.

Since there are non linear equations necessarily involved in the making of dalino, we must be aware of the developments of the theory of chaos. It can be proposed a lot of chaotic models for the primordial assembly of intervalinos which can match *ad hoc* with the dalino 270, and they may likely be interesting. However, only one model among all of them should be the unique entirely reliable one.

It is clear that there might be a trouble in the theory of chaos inasmuch as the vast majority of papers on chaos are infected by Bohr's or Heisenberg's philosophy in the sense that there try to prove — erroneously— the supposed "uncertainty" of Nature. On the contrary, the theory of chaos should be interpreted as a wide way of Nature to reach a state of minimal energy. Basically we can distinguish two principal features in any system to appraise its degree of chaotic behaviour: the difference between the energy of the initial and final states, and the number of physical variables involved in the system. When there is a great difference of energy or there are few variables, the involved physical quantities reach their corresponding state of minimal energy in a quick and straight way. We usually call this behaviour as 'determined'. As the number of variables increases or the difference of energy is small, the paths to reach the state of minimal energy augment enormously and this final state is reached slower and the multiple existing paths become more evident than in the first case. In that case we usually hear that the behaviour of the system is 'undetermined'.

However it is clear that there is no a net barrier between both behaviours, but a complete gradation between both behaviours of Nature. We can both say that a "determined" behaviour is like a very strong and narrow attractor; or we could also say that a "undetermined" behaviour is like a very long geodesic which has multiple paths to reach the final state of equilibrium. It is only the unlucky philosophy of the Standard Model which introduces unneeded troubles in the subject. We are accustomed to find the first "determined" behaviour in Physics and the second in Biology, and to make a fracture between both sides. Nevertheless this is not true: theory of chaos and IT have found "undetermined" behaviours in Physics, and the explanation of that query is very simple: the physical laws of Physics and of Biology are just the same and comes from the same source: Nature. Both are inside the Universe, and anything existing in the Universe has been created by Nature. Can it be really simpler?

To conclude, whatsoever it be the details of the process to reach the state of minimal energy represented by the elementary charge, the incontrovertible fact is that the dalino 270 is the first strongest state of minimal energy which was reached in the intervalic primordial aggregation, which imposed a powerful symmetry constraint to the next assemblies of intervalinos and further intervalic structures which made the subatomic particles of the actual Universe.

THE INTERVALIC UNIVERSE BEFORE THE BIG BANG

0	1	2	3	4
POINT —quantum of being— (Share = 1)	**INTERVALIC STRING** —quantum of space— (Share = 1) = Antisymmetric assembly of points An *interval* is, by definition, the distance between two points = Antisymmetric assembly of points Intervalic string state $S = \{\uparrow, \downarrow\}$ Intervalic string radius $r_S = \frac{1}{2} h$ Intervalic string spin $J_S = \frac{1}{2} h$ Intervalic string length $l_S = \pi \mathbf{l}_S = \pi h$ Composition of the finite group of Physical Quantities from (i, L^n). No. of Ph.Q. for $n = 3$: $4 + 12n = 40$	**PHOTON** —quantum of light— (Share = 1/2) = Symmetric assembly of Intervalic Strings Photon state $\gamma = \{\uparrow\uparrow, \downarrow\downarrow\}$ $2^{-\frac{1}{2}}(\|\uparrow\uparrow\rangle + \|\downarrow\downarrow\rangle)$ Photon radius $r_\gamma = h = 1.0556363 \cdot 10^{-34}$ (m) Frequence of primordial photon $\varphi_\gamma = \mathbf{\varphi}_\gamma = c\, h^{-1} = 2.83992 1837 \cdot 10^{42}$ (s^{-1}) Temperature of primordial photon $\Theta_\gamma = \mathbf{\Theta}_\gamma = c\, k_B^{-1} = 2.17138589 \cdot 10^{31}$ (K)	**INTERVALINO** —quantum of matter— (Share = 1/4) = Antisymmetric assembly of photons Intervalino state $\mathbf{I} = 2^{-\frac{1}{2}}(\|\uparrow\gamma\rangle - \|\downarrow\gamma\rangle)$ Intervalino charge $q_\mathbf{I} = \sqrt{(-(c^{-2} h))} = 5.93398995 \cdot 10^{-22}$ (C) Intervalino intervalic energy $I(\mathbf{I}) = c^{-2} h\, q_\mathbf{I}^{-2} = c^{-1} = 20,819.42423$ (MeV/c^2) Electromagnetic energy: $U(\mathbf{I}) = 0$ Intervalino mass: $m(\mathbf{I}) = I(\mathbf{I})$ Spin energy: $E_J(\mathbf{I}) = I(\mathbf{I}) - U(\mathbf{I}) = c^{-1}$ Intervalino radius: $r_\mathbf{I} = c / \omega_\mathbf{I} = 2\, h$	**16 DALINOS** —quanta of electric charge— (Share = 1/8) $D^{d\,\{4/1,2,3,5,6,9,10,15,18,27,30,45,54,90,135,270\}}$ = Symmetric assembly of Intervalinos (Big Bang origin) **Electron** $e = G_I = D_{270} = 270\, \mathbf{I}$ Electron charge $e = 270\, q_\mathbf{I} = 270\, \sqrt{(-(c^{-1} h))}$ **BINTERVALINO** (Share = 1/8) = Antisymmetric assembly of intervalinos Dark matter $BI = 2^{-\frac{1}{2}}(\|\mathbf{I}\downarrow\rangle - \|\mathbf{I}\uparrow\rangle)$
		CHI (Share = 1/2) = Antisymmetric assembly of i.s. Dark energy $\varsigma = 2^{-\frac{1}{2}}(\|\uparrow\downarrow\rangle - \|\downarrow\uparrow\rangle)$	**GRAVITON** (Share = 1/4) = Symmetric assembly of photons Dark energy $g = \|\gamma\gamma\rangle, 2^{-\frac{1}{2}}(\|\uparrow\gamma\rangle + \|\downarrow\gamma\rangle)$	**ELECTRIC CHARGE** → ELECTROMAGNETIC INTERACTION Third symmetry breaking
POINT (Share = 0) = Symmetric assembly of points		**SPIN** → INTERVALIC CHANGELESS INTERACTION (formerly strong interaction) LIGHT UNIVERSE (SPACETIME UNIVERSE). First symmetry breaking	**MASS** → GRAVITATIONAL INTERACTION DARK MATTER UNIVERSE Second symmetry breaking	
NOTHINGNESS	**INFORMATION** → INFORMATIONAL INTERACTION. INFORMATION (TIMELESS) UNIVERSE. Zero symmetry breaking			

SYDNEY D'AGVILO

Chapter 8

INTERVALIC ELECTRON

INTERVALIC ELECTRON

The intervalic structure of electron is:

$G_1 = D_{270} = 270 \, \mathbf{I}$

INTERVALIC ELECTRON MASS

According to IT, the magnitude of the elementary charge, e, is by definition —and univocally determined by the intervalic quanta and units—: $e = 270 \, \mathbf{q_I}$, where the intervalic charge is $\mathbf{q_I} = \sqrt{-(c^{-1}\hbar)} = 5.9339900 \cdot 10^{-22}$ (C). Hence, the *intervalic energy* of the elementary charge is:

$$I(e) = c^{\pm 2} \, \hbar \, e^{-2} = c^{\pm 2} \, \hbar \, (270 \, \mathbf{q_I})^{-2} = c^{\pm 2} \, \hbar \, [270 \, \sqrt{-(c^{-1}\hbar)}]^{-2} = 270^{-2} \, c^{-1} =$$
$$= 4.575639166 \cdot 10^{-14} \, (J) = 0.285588809 \, (MeV/c^2)$$

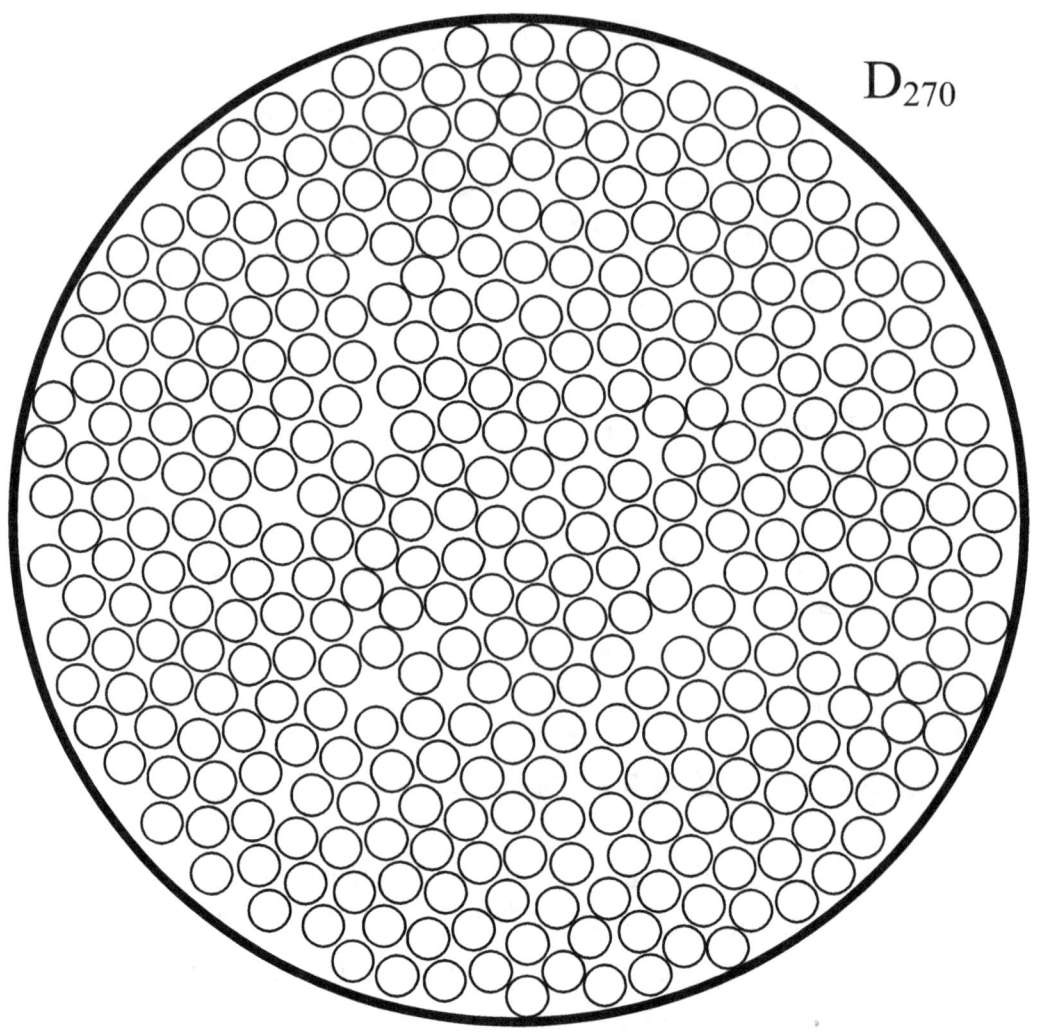

Figured intervalic structure of *electron*:
$L_1 = G_1 = D_{270} = 270\ \mathbf{I}$

Applied to the case of the electron it can be easily deduced the value of the *intervalic mass* —which is obviously the formerly named "*intrinsic mass*" attending to its unexplained origin—. The remaining mass will be named *material mass* and it has an electromagnetic origin. Remembering the *intervalic principle of energy balance* for subatomic particles, we can deduce that the *total mass* of electron must be the sum of its intervalic and material —electromagnetic— masses, since the other structural energy —the spin energy— does not contribute to the mass of the particle:

$I(e) = c^{\pm 2} \, h \, e^{-2} = 4.575639166 \cdot 10^{-14} \, (J) = 0.28558881 \, (MeV/c^2)$

$U(e) = E(e)_{mass} - I(e) = 3.611471925 \cdot 10^{-14} \, (J) = 0.22541025 \, (MeV/c^2)$

It must be pointed out that the intervalic mass of electron is a geometric magnitude which does not rely on any other physical feature but only on the symmetry of its intervalic structure.

As we will see immediately, this material mass has completely an *electromagnetic* origin. The ratios between their involved energies are:

$I(e)/U(e) = 1.266973485 \approx 5/4$

$I(e)/E(e)_{mass} = 0.558883239 \approx 5/9$

$U(e)/E(e)_{mass} = 0.44111676 \approx 4/9$

INTERVALIC ELECTRON RADIUS

IT postulates a right correspondence between the *material* energy of the electron and its *electromagnetic* potential energy —instead of its *total* energy, which is a great mistake, as it is postulated in the definition of the classic electron radius—. Actually, taking the doubtful magnitude of the classical electron radius the electromagnetic potential energy of electron would be slightly greater than its material mass:

$$U(e) = \tfrac{1}{2} (1/4\pi\varepsilon_0) e^2 / r_e(m_e) = 4.0935555 \cdot 10^{-14} \, (J)$$

This result clearly shows that the idea of the classical electron radius is absolutely wrong. The obvious reason is that it can't be merged the *intervalic* and *electromagnetic* energies because these contributions are totally different (although both ones come from the same source: the electric charge). The magnitude of the intervalic electron radius does not depend directly on the contribution to mass of the *intervalic* energy, but only on the magnitude of the *electromagnetic* energy. And not to mention that if electron is postulated to be *structureless*, as SM holds, it can not have got any electromagnetic potential energy because this physical quantity is exclusively well defined as the energy derived from the assembly of two or more electric charges which make a composite particle. Therefore the classical electron radius would be erroneous twice.

In resume, the *intervalic electron radius* is:

$$r_e = \tfrac{1}{2} (1/4\pi\varepsilon_0) e^2 / U(e) = 3.194098699 \cdot 10^{-15} \, (m)$$

which is ~13.35% greater than the traditional magnitude of the classical electron radius, a result which ensures that no other appreciable energies are involved in the electron's energy mass.

ELECTRON MINIMAL RADIUS

The electromagnetic energy of a particle must ever be equal or lesser than its intervalic energy. In other case the particle could not be stable by no means. Therefore, any subatomic particle, X, has a minimal allowed radius determined by:

$$r_{min} = \tfrac{1}{2} (1/4\pi\varepsilon_0) X^2 / I(X) = 2.5210462 \cdot 10^{-15} \, (m)$$

which would be the radius for the limit case: I(X) = U(X). In the case of electron we have:

$$r_e \geq \tfrac{1}{2} (1/4\pi\varepsilon_0) \, e^2 / I(e) = 2.5210462 \cdot 10^{-15} \text{ (m)}$$

With a radius just equal to the minimal one, the electron could not have got any spin energy and therefore it would not have got intrinsic angular momentum. However, since according to IT *spin* is the very first and most fundamental *degree of freedom* existing in Nature —from which is derived the own *electric charge* as we will see a due course—, we can not postulate the existence of a fundamental particle without spin. Moreover, having got the intervalino a great spin energy, it would not be possible that any dalino was spinless. Therefore we can affirm that the intervalic electron radius must be greater (but not only equal) to the minimal one.

INTERVALIC ELECTRON SPIN ENERGY

Contrarily to what is spitefully intended by SM, electron is not a *single* particle but it is composed by 270 *intervalinos* —the primordial block of Nature which electric charge is the intervalic quantum of charge, $q_I = \sqrt{-(c^{-1}\hbar)}$—. As already commented, intervalic and electromagnetic energies of electron can not be equal, as it could perhaps have been expected if no other energies were involved. So, a part of the electron's structural energy will be consumed by its intrinsic angular momentum and will be manifested as kinetic energy. The electron spin energy, $E_J(e)$, is just the difference between intervalic and electromagnetic energies, which is not *manifested* as *mass* but as *spin*. This can be postulated because the electron is obviously a state of minimal energy, and therefore it must be the final result of a *perfect energetic balance* in all its involved structural energies:

$$E_j(e) = I(e) - U(e) = 9.64167241 \cdot 10^{-15} \text{ (J)} = 0.060178559 \text{ (MeV/}c^2\text{)}$$

From here it can be easily deduced a lot of the dynamical features of electron, absolutely unknown up to date, as I have described with detail in other site. Likewise, the spin energy of leptons is the starting point to determine the neutrinos masses.

INTERVALIC ELECTRON STRUCTURAL ENERGY

The total intrinsic energy of electron is not only that due to mass, but it also has to include the energy due to the intrinsic angular momentum. As we already know all of them, the total structural energy of electron, $A(e)$, is:

$$A(e) = E(e)_{mass} + E_j(e) = I(e) + U(e) + E_j(e) = 2\,I(e) =$$
$$= 9.151277835 \cdot 10^{-14} \text{ (J)} = 0.571177588 \text{ (MeV/}c^2\text{)}$$

As can be easily seen, the sum of all the intrinsic energies of electron is just equal to the double of the intervalic energy. Therefore, the total structural energy of electron is a geometric magnitude which does not rely on the share between the electromagnetic and spin energies. Whatever it be, the total structural energy will not be affected by such share of energy.

Thus we have got the total structural energy ratios:

$E_j(e)/E(e)_{mass} = 0.117766476$
$E_j(e)/A(e) = 0.105358748$
$E(e)_{mass}/A(e) = 0.894641251$

Chapter 9

INTERVALIC GAUDINO

INTERVALIC ENERGY OF THE INTERVALIC CHARGE

Let us remember the equation of the *intervalic principle of equivalence between energy and electric charge*: $I = c^{\pm 2} \, \hbar \, Q^{-2}$. Showing its intervalic dimensions between brackets:

$$I \, (i^{-1}) = c^{\pm 2} \, (-1) \, \hbar \, (L) \, Q^{-2} \, (iL^{-1})$$

In not-singular units, as the traditional ones, the half symmetry which is conserved is only the following: $I = c^{-2} \, \hbar \, Q^{-2}$, which is the expression of the same principle written in a non-singular system of units (those with $c \neq 1$).

Now we can know which is the intervalic energy of the intervalic quantum of electric charge, $q_I = \sqrt{-(c^{-1}\hbar)}$:

$$I(q_I) = c^{\pm 2} \, \hbar \, q_I^{-2} = c^{\pm 2} \, \hbar \, [\sqrt{-(c^{-1}\hbar)}]^{-2} = c^{-1} = 3.335640952 \cdot 10^{-9} \, (J) =$$
$$= 20{,}819.42423 \, (MeV/c^2)$$

INTERVALIC ENERGY OF THE ELEMENTARY CHARGE

According to IT, the magnitude of the elementary charge, e, is by definition —and univocally determined by the magnitudes of the intervalic quanta and units—: $e = 270\, \mathbf{q_I}$, where the intervalic charge is $\mathbf{q_I} = \sqrt{(c^{-1}\hbar)} = 5.9339900 \cdot 10^{-22}$ (C). Hence, the intervalic energy of the elementary charge is:

$$I(e) = c^{\pm 2}\, \hbar\, e^{-2} = c^{\pm 2}\, \hbar\, (270\, \mathbf{q_I})^{-2} = c^{\pm 2}\, \hbar\, [270\, \sqrt{-(c^{-1}\hbar)}]^{-2} = 270^{-2}\, c^{-1} =$$
$$= 4.575639166 \cdot 10^{-14}\, (J) = 0.285588809\, (MeV/c^2)$$

ELECTROMAGNETIC POTENTIAL ENERGY OF SUBATOMIC PARTICLES

Although well known, sometimes it is forgotten that there has no physical sense to talk about the electromagnetic potential energy of a *single* structureless particle, but only of a *composite* particle. Now we can define the electromagnetic potential energy of electron because it is not a single particle yet, as in the mischievous SM, but it is composed by the aggregation of 270 intervalic charges.

In the calculation of that energy appears a coefficient that depends on the details of the geometric form of the particle, which is 3/5 in the case of a *rigid* sphere. Nevertheless, we will suppose that leptons are "quantum" spheres which have slightly less energy than a rigid sphere, since they are dynamical structures, as we will see later. In resume, we work on the assumption that the coefficient for a quantum sphere is ½, a simple detail that anyway does not affect the model.

INTERVALIC STRUCTURE OF LEPTON-CHARGED MASSIVE BOSON

A physical theory can be satisfactory only if its structures are composed of elementary foundations.

ALBERT EINSTEIN
To Arnold Sommerfeld, 14.01.1908

The best which can be said about the Lagrangian formalism in SM is that it is so general that it can "explain" almost any particle observed by means of adjusting some rows of parameters, which number is always *greater* than the own number of particles which is trying to "explain" (!). The shoddy repair of Weinberg-Salam electroweak theory of leptons is a good example. We "only" have to adjust by hand more than three constants to the Higgs field for yielding the masses of the lepton family. In this formalism it can be easily predicted *infinite* number of particles by means of adjusting by hand a parameter for each particle observed. (I wonder if it was not more simple to adjust by hand directly the own masses instead of the same —and greater— number of meaningless parameters). It is sure that if, say, a leptonic flying pig was observed inside a collider, SM will be able to "predict" it simply by adjusting some constants and, surely, with an astonishing exactitude of more than nine digits (!). Being Nature constructed according to number, as Pythagoras said, it is not surprising that its symmetries can arise through any enough powerful mathematical formalism, say the unhappy Lagrangian as well as any other. But this is very different to have find by far the correct symmetries and the right underlying geometry of Nature. What we need is just the contrary: a theory that can't predict infinite particles, but only the few ones which really exist now or could exist at the Big Bang, that is to say, the only one *symmetries allowed by Nature*. And a faultless theory will have to reach those deductions without adjusting by hand *anyone* parameter. In other way, it can not be considered faultless, nor to a lesser extent aesthetically elegant.

Besides, the basic physical principles on which lies the fake Lagrangian formalism, although richly dressed with mathematical clothes, are no other than the same old-fashioned and classic —*macroscopic*— principles applied to quantum —*subatomic*— world, plus some naïve quantum numbers. Although that macroscopic principles can run at atomic scale, like in Quantum Mechanics, it is doubtful that they can be similarly applied at subatomic scale. Such a proposal hardly can be considered as fundamental, simple and elegant. Therefore it is almost sure that at least *a fundamental principle of Physics is still missing*, and it is clear that such a principle will do show an outstanding simplicity (as all fundamental physical principles do). In an ideal case, the very last foundations of Physics should rely in the system of dimensions, but till today it was almost unimaginable that a physical principle could be related with or derived from the dimensional system. If when starting from such a dimensional system, it could be logically and unavoidably deduced the magnitudes of subatomic particles, it is sure that this system of dimensions and units must be the *natural* one.

Since the structureful elementary charge, $e = 270\ \mathbf{q_I}$, is a state of lowest energy reached by Nature, the only allowed fractional values for the primordial aggregation of intervalinos would be, as I have explained in other site, those ones which sum could give 270 in a simple way, that is to say, the compositeness will be the set of divisors of the elementary charge:

1, 2, 3, 5, 6, 9, 10, 15, 18, 27, 30, 45, 54, 90, 135, 270

It is clear that once the value of the elementary charge has been reached through the balance between the intervalic energy and its inverse —the electromagnetic energy—, any other different values from the above 16 divisors would lead to an scenario with greater energy and lesser order. And we think that Nature ever choose the way of *lowest energy* and *spontaneous order*. The first is traditionally accepted. To understand the second it should be rightly understood as auto organization, which is the physical state that can be reached with the lowest amount of *information*. Therefore, Nature choose the maximum economy: lowest energy and information, two physical quantities deeply linked. To resume this picture in a word, we call it *symmetry*. The 16 allowed values for the aggregation of intervalic charges determine completely the *intervalic symmetries* of the electric charge.

Henceforth, the only allowed kinds of dalinos are just 16, corresponding to the 16 divisors of the elementary charge. Dalino can be view as the fundamental "cell" in nature since all subatomic particles are equally composed by simple aggregations of dalinos. These aggregations of dalinos are named *gaudinos* in honour of the great Spanish modernist artist Antonio Gaudí (1852-1926). Due to powerful symmetry constraints explained in other site, the principal gaudinos are those which have a number of constituent intervalinos multiple of 270 (the number of the elementary charge). Thus, there are only 16 kinds of such allowed gaudinos, G_n, where the subindex means the number of its constituent dalinos.

Gaudinos can aggregate likewise among them, being named the resultant subatomic particle *lisztino*, L_n, in honour of the great Hungarian musician Franz Liszt (1811-1886).

If it is supposed, in a roughly approximation, that the ratio intervalic energy/electromagnetic energy of dalinos is similar to that of the electron, ~5/4, the masses of the 16 possible gaudinos are as follows (in MeV/c^2):

$L_1 = G_1 = 1\,D_{270} = 270\,I = 0.51099909$
$L_1 = G_2 = 2\,D_{135} = 270\,I = 4.0879928$
$L_1 = G_3 = 3\,D_{90} = 270\,I = 13.796975$
$L_1 = G_5 = 5\,D_{54} = 270\,I = 63.874885$
$L_1 = G_6 = 6\,D_{45} = 270\,I = 110.37580$
$L_1 = G_9 = 9\,D_{30} = 270\,I = 372.51833$
$L_1 = G_{10} = 10\,D_{27} = 270\,I = 510.99909$

$L_1 = G_{15} = 15\, D_{18} = 270\, I = 1{,}724.6220$
$L_1 = G_{18} = 18\, D_{15} = 270\, I = 2{,}980.1466$
$L_1 = G_{27} = 27\, D_{10} = 270\, I = 10{,}057.995$
$L_1 = G_{30} = 30\, D_9 = 270\, I = 13{,}796.975$
$L_1 = G_{45} = 45\, D_6 = 270\, I = 46{,}564.794$
$L_1 = G_{54} = 54\, D_5 = 270\, I = 80{,}463.964$
$L_1 = G_{90} = 90\, D_3 = 270\, I = 372{,}518.33$
$L_1 = G_{135} = 135\, D_2 = 270\, I = 1{,}257{,}249.4$
$L_1 = G_{270} = 270\, D_1 = 270\, I = 5{,}621{,}244.5$

The total mass of the set of gaudinos is 7,407,675.889 (MeV/c²), a value around 194 times greater than the total mass of the set of dalinos, which was, as we have seen, 38,040.09338 (MeV/c²).

It can be pointed out that is wonder to see how the same set composed by just 270 intervalinos is transformed in all existing particles by means of a mere change in the internal structure, being the only difference among them the different intervalic symmetry in which the aggregation of intervalinos has been made, that is to say, its *intervalic structure*. The balance among its involved energies changes drastically according to the adopted intervalic structure, that is to say, according to the inner form of the particle. And all this play is being performed under almost one unique physical law, the intervalic energy: $I = c^{\pm 2}\, \hbar\, Q^{-2}$. The intervalic structures of the light *leptons-elementary charged massive bosons* that appear in the above table are described by the following intervalic symmetries:

$L_1 = G_1 = 1\, D_{270} = 0.51099909 \rightarrow e$, electron
$L_1 = G_6 = 6\, D_{45} = 110.37580 \rightarrow \mu$, muon
$L_1 = G_{15} = 15\, D_{18} = 1{,}724.6220 \rightarrow \tau$, tau

Please remember that these values are yielded taking the structural energy ratio of electron for all of them, then the slight deviation from experimental results.

Need not to say, likewise, that the above particles are *logical and di-*

rectly yielded starting from a mere change in the system of units and dimensions used —really the only reliable *natural* ones—, but not through an overloaded Lagrangian formalism. It is hard to believe that all (and more) results reached by SM through complex and sometimes doubtful mechanisms can be deduced *directly from the dimensional system* in a much more reliable way in IT, without using any formalism and only applying the new intervalic principles and symmetries. All systems of units and dimensions was supposed to be equivalent in Physics, but clearly they are not since they involve very different symmetries of Nature.

It appears that the intervalic symmetries existing in primordial times between G_1, G_6, G_{15} and G_{45} have not lasted up to present times of low energy. On the other hand, the last five gaudinar structures correspond to the allowed heavy *elementary charged massive bosons* of changeful —weak— intervalic interaction, as it has been explained in other site:

$L_1 = G_{45} = 45\ D_6 = 46{,}564.794 \to Z^\pm$ boson
$L_1 = G_{54} = 54\ D_5 = 80{,}463.964 \to W^\pm$ boson
$L_1 = G_{90} = 90\ D_3 = 372{,}518.33 \to Y^\pm$ boson
$L_1 = G_{135} = 135\ D_2 = 1{,}257{,}249.4 \to X^\pm$ boson
$L_1 = G_{270} = 270\ D_1 = 5{,}621{,}244.5 \to I^\pm$ boson

Thus, in a similar mode as SUSY postulates —although within a very different foundations— leptons and elementary charged massive bosons are members of the same intervalic family. It can be pointed that according to IT there are 16 allowed symmetries and it is not given beforehand the mystic number of 3 families of leptons and quarks, pathetically intended by SM in order to avoid some infinites in a poor and precarious formulation.

Finally, according to IT, neutrino must be *intervalinoless*, because such so small mass can not be obtained through any aggregation of intervalinos because the intervalic and electromagnetic energies are inversely proportional. Thus, as the number of intervalinos increases the intervalic energy becomes smaller but the electromagnetic energy becomes greater, and we have the vice versa as the number of intervalinos decreases. Moreover, after the fundamental intervalic balance between the I and I^{-1} energies was reached at primordial times (balance which is now named *elementary charge*) it is not possible to make any stable particle out of the intervalic symmetries. Therefore, the neutrino must be *intervalinoless*, coming its mass from the non linear contribution of the intervalic energy, as we will describe later in the chapter on the Intervalic Neutrino.

INTERVALIC SYMMETRIES OF ELEMENTARY CHARGED GAUDINOS			
DALINAR SYMMETRY	INTERVALIC STRUCTURE	MASS (MeV/c^2)	PARTICLES WHICH STAY BELOW THE THRESHOLD TEMPERATURE: LEPTONS-CHARGED MASSIVE BOSONS
{D270}	G1D270$^{(\pm)}$	0.5	e, electron
{D135}	G2D135$^{(\pm)}$	4	
{D90}	G3D90$^{(\pm)}$	14	
{D54}	G5D54$^{(\pm)}$	64	
{D45}	G6D45$^{(\pm)}$	106	μ, muon
{D30}	G9D30$^{(\pm)}$	373	
{D27}	G10D27$^{(\pm)}$	511	
{D18}	G15D18$^{(\pm)}$	1,777	τ, tau
{D15}	G18D15$^{(\pm)}$	2,980	
{D10}	G27D10$^{(\pm)}$	10,058	
{D9}	G30D9$^{(\pm)}$	13,797	
{D6}	G45D6$^{(\pm)}$	46,565	
{D5}	G54D5$^{(\pm)}$	80,423	W$^\pm$ boson
{D3}	G90D3$^{(\pm)}$	372,518	Y$^\pm$ boson
{D2}	G135D2$^{(\pm)}$	1,257,249	X$^\pm$ boson
{D1}	G270D1$^{(\pm)}$	5,621,245	

Chapter 10

INTERVALIC LEPTON- -CHARGED MASSIVE BOSON

INTERVALIC MUON

The intervalic structure of *muon* is:

$$\mu \rightarrow L_1 = G_6 = 6\,D_{45} = 270\,I$$

INTERVALIC MUON MASS

As any other subatomic particle, the intervalic muon mass is due to the contribution of the intervalic and electromagnetic energies. The first is:

$$I(\mu) = \Sigma(c^{\pm 2}\hbar Q^{-2}) = c^{\pm 2}\hbar\,e^{-2} + 6\,(c^{\pm 2}\hbar(45q_I)^{-2}) = (270^{-2}c^{-1}) + 6\,(45^{-2}c^{-1}) = 9.92913699 \cdot 10^{-12}\,(J) = 61.9727717\,(MeV/c^2)$$

It can be noted that this magnitude is a geometric one which relies exclusively on the symmetry of the intervalic structures involved.

The material mass of muon is, by definition, the difference between the total mass, 105.658389 (MeV/c²), and the intervalic mass, as has likewise an electromagnetic origin:

$$U(\mu) = E(\mu)_{mass} - I(\mu) = 6.999210568 \cdot 10^{-12} (J) = 43.6856173 \ (MeV/c^2)$$

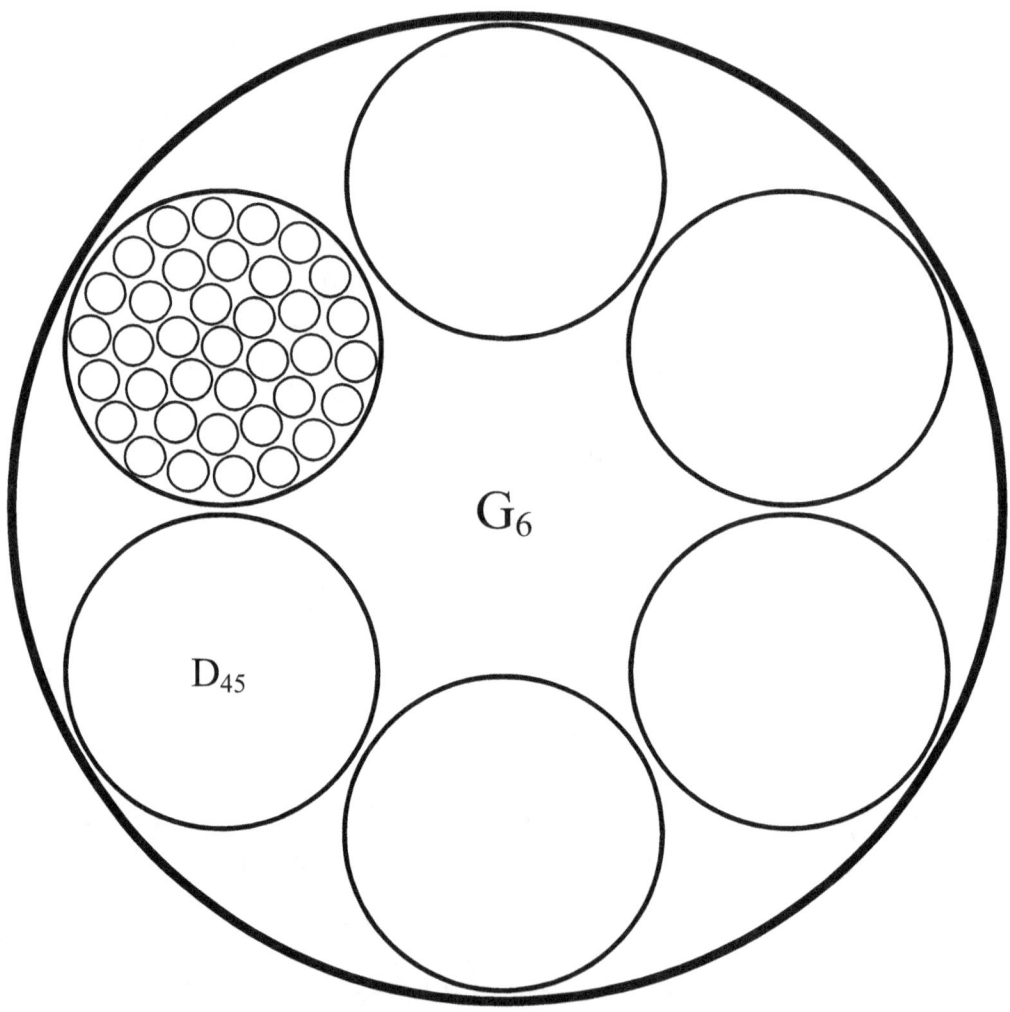

Figured intervalic structure of *muon*:
$L_1 = G_6 = 6\ D_{45} = 270\ \mathbf{I}$

INTERVALIC MUON RADIUS

The magnitude of the *intervalic muon radius* can be deduced through the electromagnetic potential energy:

$$r_\mu = \tfrac{1}{2}\,(1/4\pi\varepsilon_0)\,(270\,\mathbf{q}_I)^2 / U(\mu) = 1.648099834 \cdot 10^{-17}\,(m)$$

In a traditional view, the material energy of muon would be supposed to come from the 15 electromagnetic interactions existing between all dalinos 45. Due to the constituent hexagonal symmetry of muon, the distances between its constituent dalinos must be very similar, and therefore it makes sense to calculate the value of the average distance, $d_{D45}(\mu)$, among its constituent dalinos 45:

$$d_{D45}(\mu) = 15 \cdot (1/4\pi\varepsilon_0)\,(45\,\mathbf{q}_I)^2 / E_q(\mu) = 1.373416477 \cdot 10^{-17}\,(m)$$

MUONIC DALINO 45 RADIUS

Since the value of the electromagnetic energy of intervalic muon is known, we can deduce the magnitude of the *muonic dalino 45 radius*, r_{D45}:

$$U(\mu) = 6 \cdot \tfrac{1}{2}\,(1/4\pi\varepsilon_0)\,(45\,\mathbf{q}_I)^2 / r_{D45} = 7.0452248 \cdot 10^{-12}\,(J)$$
$$r_{D45} = 2.746832954 \cdot 10^{-18}\,(m)$$

Curiously, we find the following perfect relations:

$$r_{D45} / r_\mu = 1/6$$
$$r_{D45} / d_{D45}(\mu) = 1/5$$
$$r_\mu / d_{D45}(\mu) = 6/5$$

INTERVALIC MUON SPIN ENERGY

As in electron, a part of muon's *structural energy* is manifested as intrinsic angular momentum. Thus, muon spin energy, $E_J(\mu)$, is the difference between its intervalic and electromagnetic energies:

$$E_J(\mu) = I(\mu) - U(\mu) = 2.929926422 \cdot 10^{-12} \text{ (J)} = 18.28715441 \text{ (MeV/}c^2\text{)}$$

INTERVALIC MUON STRUCTURAL ENERGY RATIOS

As in the case of intervalic electron, we find the energy ratios:

$I(\mu)/U(\mu) = 1.418608126$
$I(\mu)/E(\mu)_{mass} = 0.586539055$
$U(\mu)/E(\mu)_{mass} = 0.413460944$
$E_J(\mu)/E(\mu)_{mass} = 0.173078111$

INTERVALIC MUON STRUCTURAL ENERGY

The total intrinsic energy of muon is the sum of the energy due to mass plus the energy due to the intrinsic angular momentum. As we already know both, the total structural energy of muon, $A(\mu)$, will be:

$$A(\mu) = E(\mu)_{mass} + E_J(\mu) = I(\mu) + U(\mu) + E_J(\mu) = 2\,I(\mu) =$$
$$= 1.985827398 \cdot 10^{-11} \text{ (J)} = 123.9455434 \text{ (MeV/}c^2\text{)}$$

Once more, it have to be remarked that sum of all the intrinsic energies of muon is just equal to the double of its intervalic energy. Therefore, the total structural energy of muon is a geometric magnitude which does not rely on the share between the electromagnetic and spin energies.

MUON INTERVALIC ARCHITECTURE

In a similar way as in electron we have got:

$$A(\mu) = 2\,I(\mu)$$
$$A(\mu) = c^{\pm 2} m(\mu)_{Schw}$$
$$E(\mu)_{mass} / A(\mu) = m(\mu) / m(\mu)_{Schw} = 0.852458153$$

INTERVALIC TAU

The intervalic structure of *tau* is:

$$\tau \to L_1 = G_{15} = 15\,D_{18} = 270\,I$$

INTERVALIC TAU MASS

The intervalic tau mass is due to the contribution of intervalic and electromagnetic energies. The first is:

$$I(\tau) = \Sigma(c^{\pm 2}\hbar Q^{-2}) = c^{\pm 2}\hbar\,e^{-2} + 15\,(c^{\pm 2}\hbar(18q_I)^{-2}) = (270^{-2}c^{-1}) +$$
$$+ 15\,(18^{-2}c^{-1}) = 1.544735782 \cdot 10^{-10}\,(J) = 964.1478215\,(MeV/c^2)$$

The electromagnetic mass of tau is the difference between the total mass, 1,777 (MeV/c²), and the intervalic mass:

$$U(\tau) = E(\tau)_{mass} - I(\tau) = 1.302333333 \cdot 10^{-10} (J) = 812.8521785 \text{ (MeV/c}^2\text{)}$$

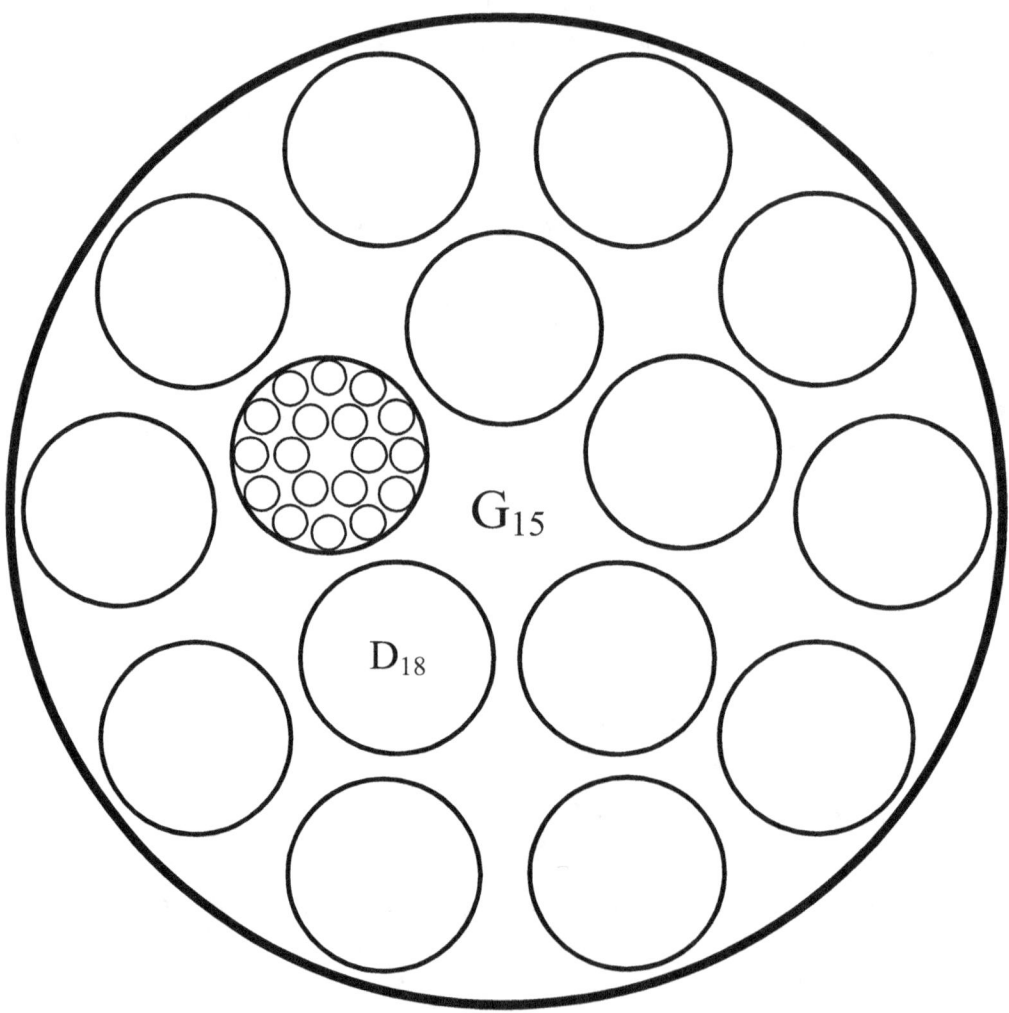

Figured intervalic structure of *tau*:
$L_1 = G_{15} = 15\ D_{18} = 270\ \mathbf{I}$

INTERVALIC TAU RADIUS

The *intervalic tau radius* can be deduced from the electromagnetic potential energy:

$$r_\tau = \tfrac{1}{2} (1/4\pi\varepsilon_0) (270\, q_I)^2 / U(\tau) = 8.857484858 \cdot 10^{-19}\,(m)$$

TAUONIC DALINO 18 RADIUS

Since the value of the electromagnetic energy of intervalic tau is known, we can deduce the tauonic dalino 18 radius, r_{D18}:

$$r_{D18} = 15 \cdot \tfrac{1}{2} (1/4\pi\varepsilon_0) (18\, q_I)^2 / U(\tau) = 5.904989684 \cdot 10^{-20}\,(m)$$

INTERVALIC TAU SPIN ENERGY

As in the other lepton-elementary charged massive bosons, tau spin energy, $E_J(\tau)$, is the difference between the intervalic and electromagnetic energies, which is *manifested* not as *mass* but as *spin*:

$$E_J(\tau) = I(\tau) - U(\tau) = 2.42402449 \cdot 10^{-11}\,(J) = 151.2956428\,(MeV/c^2)$$

INTERVALIC TAU STRUCTURAL ENERGY RATIOS

As in the case of intervalic electron and muon, we find the ratios:

$I(\tau)/U(\tau) = 1.186129344$
$I(\tau)/E(\tau)_{mass} = 0.542570524$
$U(\tau)/E(\tau)_{mass} = 0.457429475$
$E_J(\tau)/E(\tau)_{mass} = 0.085141048$

INTERVALIC TAU STRUCTURAL ENERGY

The total intrinsic energy of tau is the sum of the energy due to mass plus the energy due to the intrinsic angular momentum:

$$A(\tau) = E(\tau)_{mass} + E_J(\tau) = I(\tau) + U(\tau) + E_J(\tau) = 2\,I(\tau) =$$
$$= 3.089471564 \cdot 10^{-10}\,(J) = 1{,}928.295643\,(MeV/c^2)$$

Once more, it have to be noted that the total structural energy of muon is just equal to the double of the intervalic energy. Therefore it is a geometric magnitude which does not rely on the share between the electromagnetic and spin energies. Whatever it be, the total structural energy will not be affected by such share of energy.

TAU INTERVALIC ARCHITECTURE

In a similar way as in electron and muon we have got:

$A(\tau) = 2\,I(\tau)$
$A(\tau) = c^{\pm 2} m(\tau)_{Schw}$
$E(\tau)_{mass} / A(\tau) = m(\tau) / m(\tau)_{Schw} = 0.921539187$

INTERVALIC $W^{\pm}$ BOSON

The intervalic structure of $W^{\pm}$ boson is:

$$W^{\pm} \to L_1 = G_{54} = 54\,D_5 = 270\,I$$

INTERVALIC $W^{\pm}$ BOSON MASS

Remembering the *intervalic principle of energy balance* for subatomic particles, $I - I^{-1} - E(J) = 0$, the intervalic $W^{\pm}$ boson mass will be due to the contribution of intervalic and electromagnetic energies. The first is:

$$I(W^{\pm}) = \Sigma(c^{\pm 2}\hbar Q^{-2}) = c^{\pm 2}\hbar\,e^{-2} + 54\,(c^{\pm 2}\hbar\,(5q_I)^{-2}) = 270^{-2}c^{-1} +$$
$$+ 54\,(5^{-2}c^{-1}) = 7.205030213 \cdot 10^{-9}\,(J) = 44{,}970.24192\,(MeV/c^2)$$

As we know, this magnitude is a geometric one which relies only on the symmetry of the intervalic structure.

The electromagnetic mass of $W^{\pm}$ boson is the difference between the total mass, $80{,}423.002\,(MeV/c^2)$, and the intervalic mass:

$$U(W^{\pm}) = E(W^{\pm})_{tot} - I(W^{\pm}) = 5.6801605 \cdot 10^{-9}\,(J) = 35{,}452.76008\,(MeV/c^2)$$

INTERVALIC $W^{\pm}$ BOSON RADIUS

Since the value of the electromagnetic potential energy of the $W^{\pm}$ boson is known, we can deduce the magnitude of the intervalic $W^{\pm}$ boson radius, r_W:

$$r_W = \frac{1}{2} (1/4\pi\varepsilon_0) (270\, q_I)^2 / U(W^\pm) = 2.0308225 \cdot 10^{-20}\, (m)$$

BOSONIC DALINO 5 RADIUS

We can deduce likewise the magnitude of the constituent bosonic dalino 5 radius, r_{D5}:

$$r_{D5} = 54 \cdot \frac{1}{2} (1/4\pi\varepsilon_0) (5\, q_I)^2 / U(W^\pm) = 3.7607823 \cdot 10^{-22}\, (m)$$

A magnitude which is very close to the first approximation made through the electron's ratio: $r(D_5) \approx 3.7564122 \cdot 10^{-22}$ (m).

INTERVALIC $W^\pm$ BOSON SPIN ENERGY

The spin energy of $W^\pm$ boson is the unique structural energy of any subatomic particle which is not manifested as mass but as intrinsic angular momentum energy. Applying once more the intervalic structural balance for subatomic particles, we have for the $W^\pm$ boson:

$$I(W^\pm) - U(W^\pm) - E_J(W^\pm) = 0$$

Being known the magnitudes of the intervalic and electromagnetic energies of $W^\pm$ boson, its spin energy will be:

$$E_J(W^\pm) = I(W^\pm) - U(W^\pm) = 1.524869713 \cdot 10^{-9}\, (J) = 9{,}517.484017\, (MeV/c^2)$$

INTERVALIC W± BOSON STRUCTURAL ENERGY RATIOS

We find the following ratios between intervalic, electromagnetic and mass energies (ratios which surprisingly are almost identical to those ones of intervalic electron):

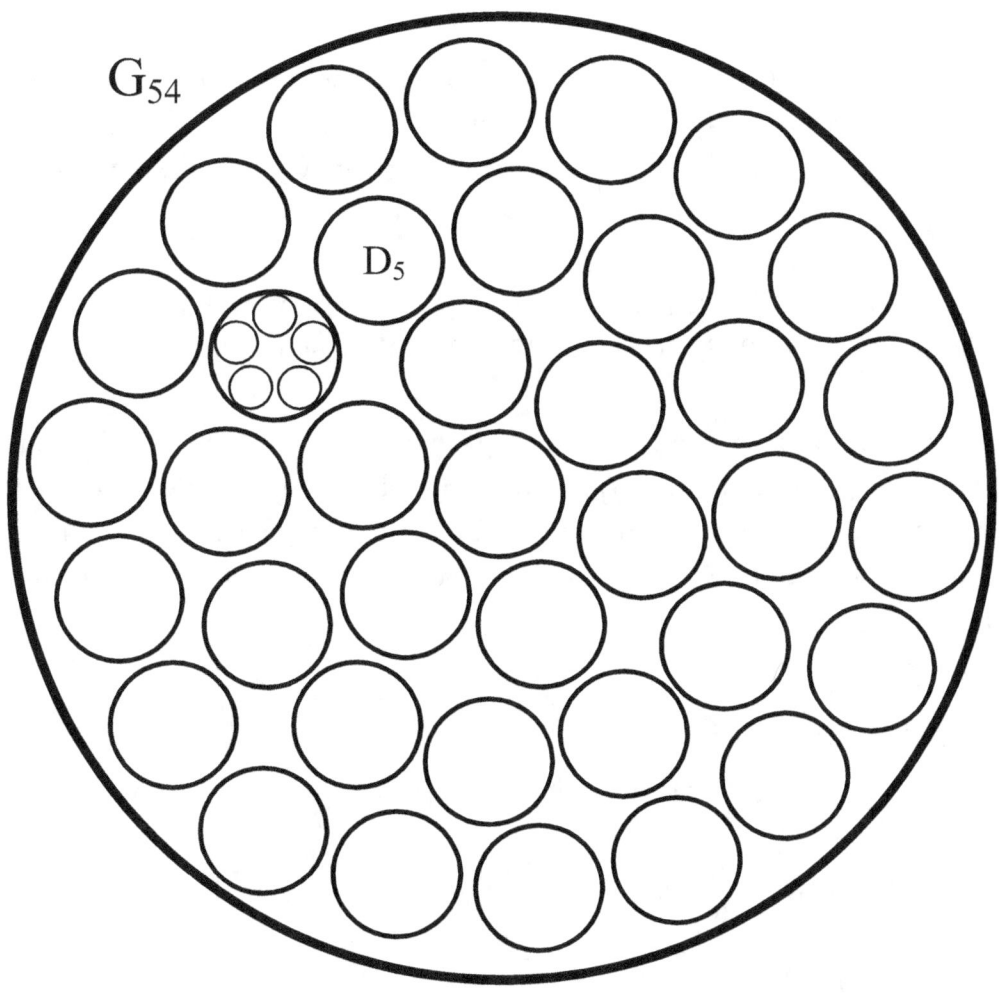

Figured intervalic structure of W± boson :
$L_1 = G_{54} = 54\ D_5 = 270\ \mathbf{I}$

$I(W^\pm)/U(W^\pm) = 1.268455314$
$I(W^\pm)/E(W^\pm)_{tot} = 0.559171391$
$U(W^\pm)/E(W^\pm)_{tot} = 0.440828608$
$E_J(W^\pm)/E(W^\pm)_{tot} = 0.118342809$

INTERVALIC $W^\pm$ BOSON STRUCTURAL ENERGY

The total intrinsic energy of $W^\pm$ boson is the sum of the energy due to mass plus the energy due to the intrinsic angular momentum. As we already know both, the total structural energy of $W^\pm$ boson, $A(W^\pm)$, will be:

$$A(W^\pm) = E(W^\pm)_{mass} + E_J(W^\pm) = I(W^\pm) + U(W^\pm) + E_J(W^\pm) = 2\,I(W^\pm) = 1.441006043 \cdot 10^{-8}\,(J) = 89{,}940.48387\,(MeV/c^2)$$

Once more, it have to be repeated that sum of all the intrinsic energies of $W^\pm$ boson is just equal to the double of its intervalic energy. Therefore, the total structural energy of $W^\pm$ boson is a geometric magnitude which does not rely on the share between the electromagnetic and spin energies.

$W^\pm$ BOSON INTERVALIC ARCHITECTURE

In a similar way we have got:

$A(W^\pm) = 2\,I(W^\pm)$
$A(W^\pm) = c^{\pm 2} m(W^\pm)_{Schw}$
$E(W^\pm)_{mass} / A(W^\pm) = m(W^\pm) / m(W^\pm)_{Schw} = 0.894180223$

INTERVALIC $Y^\pm$ BOSON

The intervalic structure of $Y^\pm$ boson is:

$$Y^\pm \to L_1 = G_{90} = 90\, D_3 = 270\, I$$

The features of $Y^\pm$ boson can be easily deduced in a similar way as in the previous leptons-charged massive bosons. Resuming results, we have obtained:

Intervalic energy:
$I(Y^\pm) = \Sigma(c^{\pm 2}hQ^{-2}) = c^{\pm 2}h\, e^{-2} + 90\, (c^{\pm 2}h\, (3q_I)^{-2}) = 270^{-2}c^{-1} +$
$+ 90\, (3^{-2}c^{-1}) = 3.335645528 \cdot 10^{-8}\, (J) = 208{,}194.5279\, (MeV/c^2)$

Mass energy:
$m(Y^\pm) = 5.968404233 \cdot 10^{-8}\, (J) = 372{,}518.33\, (MeV/c^2)$

Electromagnetic energy:
$U(Y^\pm) = m(Y^\pm) - I(Y^\pm) = 2.632758703 \cdot 10^{-8}\, (J) = 164{,}323.802\, (MeV/c^2)$

Radius:
$r_Y = \tfrac{1}{2}\, (1/4\pi\varepsilon_0)\, (270\, q_I)^2 / U(Y^\pm) = 4.381482156 \cdot 10^{-21}\, (m)$

Spin energy:
$E_j(Y^\pm) = I(Y^\pm) - U(Y^\pm) = 7.02886825 \cdot 10^{-9}\, (J) = 43{,}870.726\, (MeV/c^2)$

INTERVALIC LEPTONS-CHARGED MASSIVE BOSONS

1Dimensional Basis of the Intervalic System of Units: (L, I) dim (ℏ) = L dim (c) = I^{-1}	ELECTRON (0.51099906 MeV/c²)	MUON (105.658389 MeV/c²)	TAU (1,777 MeV/c²)
	$e = D_{270} = 270\,\mathbf{I} = 540\,\gamma = 1080\,S$	$\mu = G_6 = 6\,D_{45} = 270\,\mathbf{I} = 540\,\gamma = 1080\,S$	$\tau = G_{15} = 15\,D_{18} = 270\,\mathbf{I} = 540\,\gamma = 1080\,S$
Intervalic structure	D_{270}	G_6, D_{45}	G_{15}, D_{18}
Intervalic energy, I	$I(e) = c^{-2}\hbar\,e^2 = 4.575639166 \cdot 10^{-14}$ (J) $= 0.285588809$ (MeV/c²)	$I(\mu) = \sum(c^{-2}\hbar\,Q^2) = c^{-2}\hbar\,e^2 +$ $+ 6\,(c^{-2}\hbar\,(45\,\mathbf{q}_0)^2) = 61.9727717$ (MeV/c²)	$I(\tau) = \sum(c^{-2}\hbar\,Q^2) = c^{-2}\hbar\,e^2 +$ $+ 15\,(c^{-2}\hbar\,(18\mathbf{q}_0)^2) = 964.1478215$ (MeV/c²)
Electromagnetic energy, U	$U(e) = c^{-2}m(e) \cdot I(e) = 3.611471925 \cdot 10^{-14}$ (J) $= 0.22541025$ (MeV/c²)	$U(\mu) = c^{-2}m(\mu) \cdot I(\mu) = 6.999210568 \cdot 10^{-12}$ (J) $= 43.6856173$ (MeV/c²)	$U(\tau) = c^{-2}m(\tau) \cdot I(\tau) = 1.302333333 \cdot 10^{-10}$ (J) $= 812.8521785$ (MeV/c²)
Spin energy, E_s	$E_s(e) = I(e) - U(e) = 9.64167241 \cdot 10^{-15}$ (J)	$E_s(\mu) = I(\mu) - U(\mu) = 2.929926422 \cdot 10^{-12}$ (J)	$E_s(\tau) = I(\tau) - U(\tau) = 2.42402449 \cdot 10^{-11}$ (J)
Radius, r	$r_e = \frac{1}{2}(1/4\pi\varepsilon_0)\,e^2 / U(e) = 3.194098699 \cdot 10^{-15}$ (m)	$r_\mu = \frac{1}{2}(1/4\pi\varepsilon_0)\,(270\,\mathbf{q}_0)^2 / U(\mu) = 1.648099834 \cdot 10^{-17}$ (m)	$r_\tau = \frac{1}{2}(1/4\pi\varepsilon_0)\,(270\,\mathbf{q}_0)^2 / U(\tau) = 8.857484858 \cdot 10^{-19}$ (m)
Angular velocity due to spin, ω_s	$\omega_s(e) = (E_s(e)/m_s\,r_e^2)^{\frac{1}{2}} = 3.220944289 \cdot 10^{22}$ (s⁻¹)	$\omega_s(\mu) = (E_s(\mu)/m_s\,r_\mu^2)^{\frac{1}{2}} = 7.567601581 \cdot 10^{24}$ (s⁻¹)	$\omega_s(\tau) = (E_s(\tau)/m_s\,r_\tau^2)^{\frac{1}{2}} = 9.87597042 \cdot 10^{25}$ (s⁻¹)
Linear velocity due to spin on surface, v_s	$v_s(e) = \omega_s(e)\,r_e = 1.028801396 \cdot 10^8$ (m s⁻¹) $= 0.343171206\,c$	$v_s(\mu) = \omega_s(\mu)\,r_\mu = 1.247216291 \cdot 10^8$ (m s⁻¹) $= 0.416026573\,c$	$v_s(\tau) = \omega_s(\tau)\,r_\tau = 8.747625846 \cdot 10^7$ (m s⁻¹) $= 0.29178939\,c$
Intervalic quantum of charge $\mathbf{q}_0 = \sqrt{(c^{-1}\hbar)} = 1\,(i^{1/2}L^{1/2}) = 5.93398995 \cdot 10^{-22}$ (C)	W± MASSIVE BOSON (80,423 MeV/c²)	Y± MASSIVE BOSON (372,518 MeV/c²)	X± MASSIVE BOSON (1,257,249 MeV/c²)
	$W^\pm = G_{54} = 54\,D_5 = 270\,\mathbf{I} = 540\,\gamma = 1080\,S$	$Y^\pm = G_{90} = 90\,D_3 = 270\,\mathbf{I} = 540\,\gamma = 1080\,S$	$X^\pm = G_{135} = 135\,D_2 = 270\,\mathbf{I} = 540\,\gamma = 1080\,S$

	W± MASSIVE BOSON (80,423 MeV/c²)	Y± MASSIVE BOSON (372,518 MeV/c²)	X± MASSIVE BOSON (1,257,249 MeV/c²)
Intervalic quantum of charge $q_I = \sqrt{(c^{-1}\hbar)} \cdot 1 (l^{1/2}L^{1/2}) = 5.93398995 \cdot 10^{-22}$ (C)	$W^\pm = G_{54} = 54\ D_3 = 270\ I = 540\ \gamma = 1080\ S$	$Y^\pm = G_{90} = 90\ D_3 = 270\ I = 540\ \gamma = 1080\ S$	$X^\pm = G_{135} = 135\ D_2 = 270\ I = 540\ \gamma = 1080\ S$
Intervalic structure			
Intervalic energy, I	$I(W^\pm) = \sum(c^{-2}\hbar\ Q^2) = c^{-2}\hbar\ e^2 + 54\ (c^{-2}\hbar\ (5\ q_0)^2) = 44{,}970.24192$ (MeV/c²)	$I(Y^\pm) = \sum(c^{-2}\hbar Q^2) = c^{-2}\hbar\ e^2 + 90\ (c^{-2}\hbar\ (3\ q_0)^2) = 3.335645528 \cdot 10^{-3}$ (J) $= 208{,}194.5279$ (MeV/c²)	$I(X^\pm) = \sum(c^{-2}\hbar Q^2) = c^{-2}\hbar\ e^2 + 135\ (c^{-2}\hbar\ (2q_0)^2) = 1.125779279 \cdot 10^{-2}$ (J) $= 702{,}655.8532$ (MeV/c²)
Electromagnetic energy, U	$U(W^\pm) = c^{-2}m(W^\pm) \cdot I(W^\pm) = 5.6801605 \cdot 10^{-9}$ (J) $= 35{,}452.76$ (MeV/c²)	$U(Y^\pm) = c^{-2}m(Y^\pm) \cdot I(Y^\pm) = 2.632758703 \cdot 10^{-8}$ (J) $= 164{,}323.802$ (MeV/c²)	$U(X^\pm) = m(X^\pm) \cdot I(X^\pm) = 8.88557208 \cdot 10^{-8}$ (J) $= 554{,}593.547$ (MeV/c²)
Spin energy, E_J	$E_J(W^\pm) = I(W^\pm) - U(W^\pm) = 1.524869713 \cdot 10^{-9}$ (J) $= 9{,}517.484017$ (MeV/c²)	$E_J(Y^\pm) = I(Y^\pm) - U(Y^\pm) = 7.02886825 \cdot 10^{-9}$ (J) $= 43{,}870.726$ (MeV/c²)	$E_J(X^\pm) = I(X^\pm) - U(X^\pm) = 2.37222071 \cdot 10^{-8}$ (J) $= 148{,}062.307$ (MeV/c²)
Radius, r	$r_W = \frac{1}{2}(1/4\pi\varepsilon_0)(270\ q_0)^2 / U(W^\pm) = 2.0308225 \cdot 10^{-20}$ (m)	$r_Y = \frac{1}{2}(1/4\pi\varepsilon_0)(270\ q_0)^2 / U(Y^\pm) = 4.381482156 \cdot 10^{-21}$ (m)	$r_X = \frac{1}{2}(1/4\pi\varepsilon_0)(270\ q_0)^2 / U(X^\pm) = 1.298215261 \cdot 10^{-21}$ (m)
Angular velocity due to spin, ω_J	$\omega_J(W^\pm) = (E_J(W^\pm) / m_W\ r_W^2)^{\frac{1}{2}} = 5.0783155 \cdot 10^{27}$ (s⁻¹)	$\omega_J(Y^\pm) = (E_J(Y^\pm) / m_Y\ r_Y^2)^{\frac{1}{2}} = 2.34808191 \cdot 10^{28}$ (s⁻¹)	$\omega_J(X^\pm) = (E_J(X^\pm) / m_X\ r_X^2)^{\frac{1}{2}} = 7.92474925 \cdot 10^{20}$ (s⁻¹)
Linear velocity due to spin on surface, v_J	$v_J(W^\pm) = \omega_J(W^\pm)\ r_W = 1.03131574 \cdot 10^8$ (m s⁻¹) $= 0.34400990\ c$	$v_J(Y^\pm) = \omega_J(Y^\pm)\ r_Y = 1.02880790 \cdot 10^8$ (m s⁻¹) $= 0.343173375\ c$	$v_J(X^\pm) = \omega_J(X^\pm)\ r_X = 1.02880304 \cdot 10^8$ (m s⁻¹) $= 0.343171755\ c$

INTERVALIC BILEPTONS-ZERO CHARGED MASSIVE BOSONS

	Z⁰ MASSIVE BOSON (91,188 MeV/c²)	Y⁰ MASSIVE BOSON (745,037 MeV/c²)	X⁰ MASSIVE BOSON (2,514,499 MeV/c²)
Structural energy balance for subatomic particles $I - I^+ - E_J = 0$ $c^{-2}\hbar Q^2 - [\frac{1}{2}(1/4\pi\varepsilon_0)Q^2/r] - m\ r^2\omega_J^2 = 0$	$Z^0 = L_2 = 2\ G_{45} = 90\ D_6 = 540\ I = 1080\ \gamma = 2160\ S$	$Y^0 = L_2 = 2\ G_{90} = 180\ D_3 = 540\ I = 1080\ \gamma = 2160\ S$	$X^0 = L_2 = 2\ G_{135} = 270\ D_2 = 540\ I = 1080\ \gamma = 2160\ S$
Intervalic structure	$G_{45}^{(\pm)}$ $L_2^{(0)}$	$G_{90}^{(\pm)}$ $L_2^{(0)}$	$G_{135}^{(\pm)}$ $L_2^{(0)}$

INTERVALIC $X^{\pm}$ BOSON

The intervalic structure of $X^{\pm}$ boson is:

$$X^{\pm} \to L_1 = G_{135} = 135\ D_2 = 270\ I$$

We have got likewise the following results:

Intervalic energy:
$$I(X^{\pm}) = \Sigma(c^{\pm 2}\hbar Q^{-2}) = c^{\pm 2}\hbar\ e^{-2} + 135\ (c^{\pm 2}\hbar\ (2q_I)^{-2}) = 270^{-2}c^{-1} +$$
$$+ 135\ (2^{-2}c^{-1}) = 1.125779279 \cdot 10^{-7}\ (J) = 702{,}655.8532\ (MeV/c^2)$$

Mass energy:
$$m(X^{\pm}) = 2.014336487 \cdot 10^{-7}\ (J) = 1{,}257{,}249.4\ (MeV/c^2)$$

Electromagnetic energy:
$$U(X^{\pm}) = m(X^{\pm}) - I(X^{\pm}) = 8.88557208 \cdot 10^{-8}\ (J) = 554{,}593.547\ (MeV/c^2)$$

Radius:
$$r_X = \tfrac{1}{2}\ (1/4\pi\varepsilon_0)\ (270\ q_I)^2 / U(X^{\pm}) = 1.298215261 \cdot 10^{-21}\ (m)$$

Spin energy:
$$E_j(X^{\pm}) = I(X^{\pm}) - U(X^{\pm}) = 2.37222071 \cdot 10^{-8}\ (J) = 148{,}062.307\ (MeV/c^2)$$

Chapter 11

INTERVALIC DYNAMICS OF LEPTON-CHARGED MASSIVE BOSON

INTERVALIC ELECTRON

The intervalic structure of *electron* is:

$e \to G_1 = 1\ D_{270} = 270\ \mathbf{I}$

ELECTRON INTERVALIC AND ELECTROMAGNETIC ENERGY

The intervalic energy of electron is:

$I(e) = c^{\pm 2}\ \hbar\ e^{-2} = c^{\pm 2}\ \hbar\ (270\ \mathbf{q_I})^{-2} = c^{\pm 2}\ \hbar\ [270\ \sqrt{-(c^{-1}\hbar)}]^{-2} = 270^{-2}\ c^{-1} =$
$= 4.575639166 \cdot 10^{-14}\ (J) = 0.285588809\ (MeV/c^2)$

And the electromagnetic energy is:

$$U(e) = E(e)_{tot} - I(e) = 3.611471925 \cdot 10^{-14} \, (J) = 0.22541025 \, (MeV/c^2)$$

Thus, the *intervalic electron radius* is:

$$r_e = \tfrac{1}{2} \, (1/4\pi\varepsilon_0) \, e^2 / U(e) = 3.194098699 \cdot 10^{-15} \, (m)$$

ELECTRON INTERVALIC EXCHANGE FREQUORCE

The overall exchange frequorce due to intervalic interaction among intervalinos is proportional to the intervalic energy:

$$\varphi(I)_e = I(e) / \hbar = 4.334484487 \cdot 10^{20} \, (s^{-1})$$

The related maximum linear velocity of intervalinos inside electron will be:

$$v(I)_e = 2\pi \, r_e \, \varphi(I)_e = 8.698926337 \cdot 10^6 \, (m \, s^{-1}) = 0.029016494 \, c$$

ELECTRON INTERVALIC POTENTIAL WELL

The relation between the intervalic energy and the intervalic structure make an astonishing simple formulation of the potential well corresponding to the intervalic interaction. The difference between the intervalic energy of electron and the intervalic energy of its *isolated* constituent intervalinos yields the magnitude of the intervalic interaction among intervalinos inside electron. This concept is similar to the definition of the electromagnetic potential energy of a composite particle, although in a new way. Lets us go to see it. The electromagnetic potential of electron is due to the difference between the electromagnetic energy of electron and the electromagnetic energy of its *isolated* constituent intervalinos. In our paper on the intervalino I have explained in detail a beautiful feature of IT, that the electromagnetic energy of an *isolated* intervalino is just zero (being its charge sign *not realized*, that is to say, neither + nor -, but ±). According to this we have:

$$\Delta U(e) = U(e) - 270\, U(\mathbf{I}) = U(270 q_I) - 270\, U(q_I) = U(270 q_I) - 0 =$$
$$= c^{\pm 2} m_e - I(e) = 3.611471925 \cdot 10^{-14}\, (J) = 0.22541025\, (MeV/c^2)$$

And the electromagnetic potential is:

$$V(e) = \Delta U(e) / m_e = 3.964559695 \cdot 10^{16}\, (m^2\, s^{-2})$$

In the case of electromagnetic energy (and *long ranged* interactions) 'isolated' means separate by an infinite length. However, in the case of intervalic interaction (and *short ranged* interactions) 'isolated' means *out of* the intervalic structure, an event which does not relies on the macroscopic distance of the separation as in the case of a long ranged interaction. Therefore, the electron intervalic potential is determined by the difference on the intervalic energy between electron and its constituent disassembled intervalinos:

$$\Delta I(e) = I(e) - 270\, I(\mathbf{I}) = I(270 q_I) - 270\, I(q_I) =$$
$$= [c^{\pm 2} \hbar\, (270\, q_I)^{-2}] - [270\, c^{\pm 2} \hbar\, q_I^{-2}] = 270^{-2} c^{-1} - 270 c^{-1} =$$
$$= -9.006230113 \cdot 10^{-7}\, (J) = -5{,}621{,}244.256\, (MeV/c^2)$$

And the intervalic potential well is:

$$\Phi(e) = \Delta I(e) / m_e = -9.886754667 \cdot 10^{23}\, (m^2\, s^{-2})$$

ELECTRON DYNAMICAL FREQUORCE BALANCE

The magnitude of the corresponding electromagnetic frequorce on surface will be, supposing an overall frequorce:

$$\varphi(e)_{em} = U(e) / r_e = 11.30670109 \text{ (N)}$$

As the intervalic frequorce is short ranged it does not relies on the distance, unlike the long ranged one does. The intervalic interaction exists within its range of interaction, which is determined by the energy interval in each case, which is closely related with the masses of the intermediate bosons of the changeless intervalic interaction. In this case, the range of interaction between electron and an isolated intervalino will be between the following ranges:

$$d_e = c^{-1}\hbar / m_e = 3.865488017 \cdot 10^{-13} \text{ (m)}$$
$$d_I = c^{-1}\hbar / m_I = 9.487585915 \cdot 10^{-18} \text{ (m)}$$

Out of these ranges the magnitude of the intervalic interaction of electron is precisely *zero* since it is short ranged.

The corresponding intervalic frequorce at electron's surface is determined by the intervalic energy of electron:

$$\varphi(e)_{in} = I(e) / r_e = 14.32529047 \text{ (N)}$$

Which is in the same order of the electromagnetic one, but there is a difference between the intervalic —attractive— and the electromagnetic —repulsive— frequorces just at the electron's surface of:

$$\varphi(e)_{in} - \varphi(e)_{em} = 3.01858938 \text{ (N)}$$

This frequorce should be exactly compensated by another structural centrifugal frequorce in order to the stability of electron. There must be a perfect balance among all its involved structural energies. And the only one remaining structural energy of electron is the spin energy. Therefore, the corresponding dynamical frequorce due to spin, $\varphi(e)_J$, have to be ~3.01858938 (N):

$$\varphi(e)_J = m_e \omega_J^2 r_e = 3.01858938 \text{ (N)}$$

The intervalic energy of electron marks its *geodesic line of structural energy*. Inside this geodesic line in energy electron does not "feel" the intervalic field, as like as a planet does not "feel" acceleration due to gravity when following its geodesic line in spacetime. If electron deviates from this energy geodesic through a disassembly of its constituent intervalinos, it will be leaving the intervalic potential well, and the intervalic frequorce due to short ranged intervalic interaction will be in first approximation:

$$\varphi(e)_{in} = \Delta I(e) / r_e = 2.81964678 \cdot 10^8 \text{ (N)}$$

Thus, to make a metaphor, the geodesic of gravitational or electromagnetic interaction is like an infinite soft valley, whilst the geodesic of intervalic interaction is like a crack on the plane soil.

ELECTRON INTERVALIC KINEMATICS

From the dynamical frequorce due to spin, the angular velocity will be:

$$\omega_J = (\varphi(e)_J / m_e r_e)^{1/2} = 3.220944289 \cdot 10^{22} \text{ (s}^{-1}\text{)}$$

The linear velocity due to spin on electron's surface is:

$$v_J = \omega_J r_e = 1.028801396 \cdot 10^8 \text{ (m s}^{-1}\text{)} = 0.343171206 \, c$$

And the acceleration on electron's surface will be:

$$a_J = v_J^2 / r_e = 3.313711982 \cdot 10^{30} \ (m \ s^{-2})$$

ELECTRON STRUCTURAL ENERGY BALANCE

Intervalic and electromagnetic energies of electron are not equal as could have been expected if no other energies were involved. Of course, this is due because electron is not a *single* particle but it is composed by 270 *intervalinos* —the primordial single particle which electric charge is the intervalic quantum of charge, $q_I = \sqrt{-(c^{-1}\hbar)}$—. Therefore, a part of the electron's structural energy will be consumed by its intrinsic angular momentum and will be manifested as kinetic energy. The electron spin energy, $E(J_e)$, is just the difference between intervalic and electromagnetic energies, which is *manifested* not as *mass* but as *spin*. This can be postulated because the electron is obviously a state of minimal energy, and therefore it must be the final result of a *perfect energetic balance* in all its involved structural energies.

The rotational kinetic energy due to the intrinsic angular momentum of electron will be, according to traditional Mechanics: $K(J_e) = \frac{1}{2} I_e \omega_J^2$, being I_e the electron's moment of inertia. By the Virial theorem — since from now on a *composite* particle can be considered like a micro galaxy at quantum scale— the potential energy due to spin is: $U(J_e) = 2 K(J_e)$. Therefore we have (depicting the geometrical coefficient of the moment of inertia):

$$U(J_e) = m_e r_e^2 \omega_J^2 = 9.641672411 \cdot 10^{-15} \ (J)$$

The application of a physical law usually related with star systems or galaxies, as the Virial theorem, at microscopic scale, in a way that a subatomic particle can be compared with a complex star system, is a beautiful example of the legendary correspondence between macrocosmos and microcosmos, which now is materialized from the vagueness of Philosophy to the precision of Physics.

On the other hand the magnitude of the spin energy is under an immovable constraint based on the powerful *intervalic principle of energy balance* for subatomic particles, which establishes the fundamental equation:

$$I(e) - U(e) - E(J_e) = 0$$

And there should be verified the equality: $E(J_e) = U(J_e)$, that is to say:

$$[\, c^{\pm 2}\, h\, (270\, q_I)^{-2} \,] - [\, \tfrac{1}{2}\, (1/4\pi\varepsilon_0)\, e^2 / r_e \,] - m_e\, \omega_J^2\, r_e^2 = 0$$

As may be expected, the value of the left member of the equation is, in stunning agreement with $U(J_e)$ up to the last digit:

$$E(J_e) = m_e\, r_e^2\, \omega_J^2 = 9.64167241 \cdot 10^{-15}\ (J)$$
$$\varphi(e)_J = E(J_e) / r_e = 3.01858938\ (N)$$

This is the delicate balance reached by all subatomic particles at every structure level. In the case of electron, as there is only a level —the dalinar one—, the balance is established between the number of intervalinos (which determine totally the intervalic energy and partially electromagnetic energy) and the magnitude of radius (which is the other variable of the electromagnetic energy) and the angular velocity (which involves the spin energy).

Finally, we can say in advance that the spin energy of leptons is the starting point to determine the neutrinos masses, as it has been explained in the chapter on the intervalic neutrino.

INTERVALIC SCHWARZSCHILD ELECTRON MASS AND RADIUS

Using the traditional concepts introduced in the formulation of the *gravitational* Schwarzschild radius, $r_{Schw} = 2Gm/c^2$, we can transform them substituting the gravitational energy by the intervalic energy — because the intervalic energy is the only attractive one inside electron (as the gravitational one is despicable) and therefore the intervalic energy plays just the role of the attractive energy inside electron—, so we will obtain an *intervalic* Schwarzschild radius for electron. The maximum linear velocity on electron's surface due to spin would now play the role as the escape velocity of the star —which now is converted into the electron or any other subatomic particle—, our electron's spin energy would play as the kinetic energy of the star, and the intervalic energy of electron, $I(e)$, would be analogous to the potential energy of the star. In this way, electron could be considered to be a black hole if its mass, $m_{Schw}(e)$, was slightly greater than the actual mass (substituting v_J by c):

$I(e) = K_{escape}(e)$
$c^{\pm 2} \hbar\, e^{-2} = \frac{1}{2}\, m_{Schw}(e)\, v_J^2 = \frac{1}{2}\, c^{\pm 2}\, m_{Schw}(e)$
$m_{Schw}(e) = 2\, c^{\pm 4} \hbar\, e^{-2} = 1.01821698 \cdot 10^{-30} (kg) = 0.571177588\, (MeV/c^2)$

Please note that the only constants involved in the calculus are just the two fundamental ones, c and $\hbar$, plus the elementary charge, e. Therefore the magnitude of the intervalic Schwarzschild mass is a *geometric* result. Thus the ratio between the electron actual mass and this one might be meaningful:

$m_e / m_{Schw}(e) = 0.894641299$

Alternatively, electron could equally be an imaginary black hole if its radius was slightly smaller than the actual one:

$$r_{Schw}(e) = [\, m_e / m_{Schw}(e) \,]\, r_e = 2.857572613 \cdot 10^{-15} \; (m)$$

This magnitude can be considered as an intervalic Schwarzschild electron radius, which is slightly smaller than the actual intervalic electron radius. Therefore, we can see that according to the old-fashioned value of the absolutely misleading Bohr's classical electron radius, $\alpha^2 a_0 = 2.81794092 \cdot 10^{-15}$ (m), the electron would be a black hole (!), a curious result which proves once more the gross mistake of Bohr's calculus.

Thus we reach a result which may help us to understand better the inner architecture of electron and the balance between the share of structural energies and the dynamical physical quantities of electron: the intervalic Schwarzschild mass (or radius) to actual electron mass (or radius) ratio is equal to the ratio between electron's mass energy —$I(e) + U(e) = c^{\pm 2} m_e$— and the total structural energy —$I(e) + U(e) + E(J_e) = 2\, I(e)$—:

$m_e / m_{Schw}(e) \;\; r_{Schw}(e) / r_e = 0.894641299$
$c^{\pm 2} m_e / 2\, I(e) = 0.894641299$

In other words, we could say that the actual mass (or radius) of electron is as greater than the Schwarzschild mass (or radius) as greater becomes the share of the spin energy with respect to the mass energy of electron, being this proportion roughly ~1/9. As the reader will be able to see quietly along this book, this share becomes greater as it is increased the number of constituent intervalic structure levels of subatomic particles.

This may imply that the mass (or radius) of a subatomic particle that would have zero spin energy would be the Schwarzschild's one. Since all subatomic particles without exception have got spin, we can expect that there are not black holes to be single subatomic particles.

ELECTRON INTERVALIC ARCHITECTURE

From here and in resume we have got the following meaningful results:

Electron mass to Schwarzschild mass ratio
$m(e) / m(e)_{Schw} = 0.894641299$

Electron mass energy to structural energy ratio
$E(e)_{mass} / A(e) = 0.894641299$

Electron Structural Energy
$A(e) = I(e) + U(e) + E_J(e) = E(e)_{mass} + E_J(e) = 2\,I(e) =$
$= 0.571177588\ (MeV/c^2)$

Electron Schwarzschild Mass
$c^{\pm 2}\,\hbar\,e^{-2} = \tfrac{1}{2}\,m(e)_{Schw}\,v_{esc}^{\pm 2}$
$m(e)_{Schw} = 0.571177588\ (MeV/c^2)$

Electron Intervalic Architecture
$A(e) = 2\,I(e)$
$A(e) = c^{\pm 2} m(e)_{Schw}$
$E(e)_{mass} / A(e) = m(e) / m(e)_{Schw} = 0.894641299$

In plain English, the *structural energy* of electron coincides exactly with the *Schwarzschild mass* of electron. And both values are just equal to the double of the *intervalic energy* of electron. Here is the beautiful underlying *geometry* of the electron's structure. It must be noted that most balances of electron constituent energies have a *geometric* origin.

(The calculus is similar to all leptons-charged massive bosons. However it is a little more complex when the particle contains subparticles with both like and unlike charges, because in this case there have to be added to the attractive intervalic energy the electromagnetic energy between unlike subparticles for calculating the Schwarzschild mass of the particle).

MINIMAL ELECTRON RADII: GEOMETRIC RADIUS, SCHWARSCHILD RADIUS, BOSE-EINSTEIN RADIUS

To conclude this study on electron intervalic dynamics, we can make a miscellaneous comment on the minimal electron radius.

The electromagnetic energy of a particle must ever be equal or lesser than its intervalic energy. In other case the particle could not be stable by no means. Therefore, the electron radius has a low limit magnitude, the *minimal geometric radius*:

$$r_e > \tfrac{1}{2} (1/4\pi\varepsilon_0) \, e^2 / I(e) = 2.5210462 \cdot 10^{-15} \,(m)$$

Besides, we can take the magnitude of the electron *Schwarschild radius*, which is another limit minimal radius:

$$r_{b.h.}(e) > [\, m_e / m_{b.h.}(e) \,] \, r_e = 2.85756954 \cdot 10^{-15} \,(m)$$

And we still can calculate another minimal value which is greater than the two previous ones. Please note that the Bose-Einstein condensed state among electrons only could be possible —and it is— if the difference of the intervalic energy of a paired electrons was bigger than its corresponding electromagnetic energy:

$$I(e^-+e^-) - I(2e^-) > U(2e^-)$$

From here it can be deduced another minimal magnitude allowed to the electron radius, r_e, which we will name *Bose-Einstein radius*. Depicting geometrical details:

$$(7/4)\, c^{\pm 2}\, \hbar\, e^{-2} > \tfrac{1}{2}\, (1/4\pi\varepsilon_0)\, (2e)^2 / 2r_e$$
$$r_e > (4/7)\, (1/4\pi\varepsilon_0)\, c^2 \hbar^{-1} e^4 = 2.8811969 \cdot 10^{-15}\ (\mathrm{m})$$

The last two minimal magnitudes exceed clearly the misleading value of the bizarre classical electron radius, which is an obsolete approximation which should already be definitively discarded in Particle Physics.

INTERVALIC MUON

The intervalic structure of *muon* is:

$$\mu \rightarrow L_1 = G_6 = 6\, D_{45} = 270\, \mathbf{I}$$

From here and onwards we will find not only compositeness but *structurefulness*, that is to say, successive levels of compositeness.

MUON INTERVALIC AND ELECTROMAGNETIC ENERGY

As any other subatomic particle, the intervalic muon mass is due to the contribution of the intervalic and electromagnetic energies. The first is:

$$I(\mu) = \Sigma(c^{\pm 2}\hbar Q^{-2}) = c^{\pm 2}\hbar\ e^{-2} + 6\ (c^{\pm 2}\hbar(45q_I)^{-2}) = (270^{-2}c^{-1}) + 6\ (45^{-2}c^{-1}) = 9.92913699 \cdot 10^{-12}\ (J) = 61.9727717\ (MeV/c^2)$$

The material mass of muon is, by definition, the difference between the total mass and the intervalic mass, as has likewise an electromagnetic origin:

$$U(\mu) = E(\mu)_{tot} - I(\mu) = 6.999210568 \cdot 10^{-12}\ (J) = 43.6856173\ (MeV/c^2)$$

The magnitude of the *intervalic muon radius* can be deduced through the electromagnetic energy:

$$r_\mu = \tfrac{1}{2}\ (1/4\pi\varepsilon_0)\ (270\ q_I)^2\ /\ U(\mu) = 1.648099834 \cdot 10^{-17}\ (m)$$

MUON INTERVALIC EXCHANGE FREQUORCE

The overall exchange frequorce due to intervalic interaction among intervalinos inside dalino 45 will be:

$$\varphi(I)_{D45} = I(D_{45})\ /\ \hbar = 1.560414415 \cdot 10^{22}\ (s^{-1})$$

The related maximum linear velocity of intervalinos inside dalino 45 will be, known the muonic dalino 45 radius, $r_{D45} = 2.746832954 \cdot 10^{-18}\ (m)$:

$$v(I)_{D45} = 2\pi r_{D45}\ \varphi(I)_{D45} = 2.693097464 \cdot 10^{5}\ (m\ s^{-1}) = 0.000898320619\ c$$

The overall exchange frequorce of dalinos 45 inside muon will be:

$$\varphi(D_{45})_\mu = I(e)\ /\ \hbar = 4.334484487 \cdot 10^{20}\ (s^{-1})$$

And the corresponding maximum linear velocity of dalinos 45 inside muon is, known the muon radius:

$$v(D_{45})_\mu = 2\pi\, r_\mu\, \varphi(D_{45})_\mu = 44{,}884.95943 \ (m\ s^{-1})$$

MUON INTERVALIC POTENTIAL WELL

The electromagnetic potential energy of muon is due to the difference between the electromagnetic energy of muon and the electromagnetic energy of its *isolated* constituent dalinos 45, plus the difference between the electromagnetic energy of the dalino 45 and the electromagnetic energy of its *isolated* constituent intervalinos. According to this we have:

$$\Delta U(\mu) = [U(\mu) - 6\, U(D_{45})] + [6\, U(D_{45}) - 6\, (45\, U(\mathbf{I}))] = U(\mu) =$$
$$= c^{\pm 2} m_\mu - I(\mu) = 6.999210568 \cdot 10^{-12}\ (J) = 43.6856173\ (MeV/c^2)$$

And the electromagnetic potential is:

$$V(\mu) = \Delta U(\mu)\, /\, m_\mu = 3.716001622 \cdot 10^{16}\ (m^2\ s^{-2})$$

In the case of electromagnetic energy (and *long ranged* interactions) 'isolated' means separate by an infinite length. However, in the case of intervalic interaction (and *short ranged* interactions) 'isolated' means *out* of the intervalic structure, an event which does not relies on the macroscopic distance of the separation as in the case of a long ranged interaction. Therefore, the muon intervalic potential is determined by the difference on the intervalic energy between muon and its constituent disaggregated intervalinos:

$$\Delta I(\mu) = [I(\mu) - 6\,I(D_{45})] + [6\,I(D_{45}) - 6\,(45\,I(I))] =$$
$$= [c^{\pm 2}\hbar\,(270\,q_I)^{-2}] - \{6\,[c^{\pm 2}\hbar\,(45q_I^{-2})]\} + \{6\,[c^{\pm 2}\hbar\,(45q_I^{-2})]\} -$$
$$- \{6\,[45\,(c^{\pm 2}\hbar\,q_I^{-2})]\} = 270^{-2}c^{-1} - 270c^{-1} = -9.006230113 \cdot 10^{-7}\,(J) =$$
$$= -5{,}621{,}244.256\,(MeV/c^2)$$

And the intervalic potential well is:

$$\Phi(\mu) = \Delta I(\mu) / m_\mu = -4.781562918 \cdot 10^{21}\,(m^2\,s^{-2})$$

MUON DYNAMICAL FREQUORCE BALANCE

The magnitudes of the intervalic and electromagnetic overall frequorces on surface will be:

$$\varphi(\mu)_{in} = I(\mu) / r_\mu = 602{,}459.6802\,(N)$$

$$\varphi(\mu)_{em} = U(\mu) / r_\mu = 424{,}683.6523\,(N)$$

As in electron, the dynamical frequorce due to spin, $\varphi(\mu)_J$, is:

$$\varphi(\mu)_{in} - \varphi(\mu)_{em} = 177{,}776.0279\,(N)$$

MUON INTERVALIC KINEMATICS

From the dynamical frequorce due to spin, the angular velocity will be:

$$\omega_J = (\,\varphi(\mu)_J / m_\mu\,r_\mu\,)^{½} = 7.567601581 \cdot 10^{24}\,(s^{-1})$$

The linear velocity of spin on muon's surface is:

$$v_J = \omega_J r_\mu = 1.247216291 \cdot 10^8 \text{ (m s}^{-1}) = 0.416026573 \text{ c}$$

And the acceleration on muon's surface due to spin will be:

$$a_J = v_J^2 / r_\mu = 9.438435974 \cdot 10^{32} \text{ (m s}^{-2})$$

MUON STRUCTURAL ENERGY BALANCE

The rotational kinetic energy due to the intrinsic angular momentum of muon is: $K(J_\mu) = \frac{1}{2} I_\mu \omega_J^2$, being I_μ the muon's moment of inertia. By the Virial theorem, the potential energy due to spin is: $U(J_\mu) = 2 K(J_\mu)$. As in electron we have:

$$U(J_\mu) = m_\mu r_\mu^2 \omega_J^2 = 2.929926422 \cdot 10^{-12} \text{ (J)}$$

By fundamental intervalic principle of energy balance for subatomic particles, the muon spin energy is:

$$I(\mu) - U(\mu) - E(J_\mu) = 0$$
$$E(J_\mu) = I(\mu) - U(\mu) = 2.929926422 \cdot 10^{-12} \text{ (J)}$$

Therefore is verified the equality $E(J_\mu) = U(J_\mu)$, that is to say:

$$[\, c^{\pm 2}\hbar\, (270\mathbf{q_I})^{-2} + 6\, (c^{\pm 2}\hbar\, (45\mathbf{q_I})^{-2}) \,] - [\, \tfrac{1}{2}\, (1/4\pi\varepsilon_0)\, e^2 / r_\mu \,] - m_\mu \omega_J^2 r_\mu^2 = 0$$

INTERVALIC TAU

The intervalic structure of *tau* is:

$$\tau \to L_1 = G_{15} = 15\, D_{18} = 270\, I$$

TAU INTERVALIC AND ELECTROMAGNETIC ENERGY

In close analogy to electron and muon, the corresponding magnitudes of the intervalic and electromagnetic energies of tau are:

$$I(\tau) = \Sigma(c^{\pm 2}\hbar Q^{-2}) = c^{\pm 2}\hbar\, e^{-2} + 15\,(c^{\pm 2}\hbar(18q_I)^{-2}) =$$
$$= (270^{-2} c^{-1}) + 15\,(18^{-2} c^{-1}) = 1.544735782 \cdot 10^{-10}\,(J) =$$
$$= 964.1478215\,(MeV/c^2)$$

$$U(\tau) = E(\tau)_{tot} - I(\tau) = 1.302333333 \cdot 10^{-10}\,(J) = 812.8521785\,(MeV/c^2)$$

Intervalic tau radius:

$$r_\tau = \tfrac{1}{2}\,(1/4\pi\varepsilon_0)\,(270\, q_I)^2 / U(\tau) = 8.857484858 \cdot 10^{-19}\,(m)$$

TAU INTERVALIC EXCHANGE FREQUORCE

Overall exchange frequorce due to intervalic interaction among intervalinos inside dalino 18:

$$\varphi(I)_{D18} = I(D_{18}) / \hbar = 9.752590095 \cdot 10^{22}\,(s^{-1})$$

Related maximum linear velocity of intervalinos inside dalino 18, known the tauonic dalino 18 radius, $r_{D18} = 5.904989684 \cdot 10^{-20}$ (m):

$$v(I)_{D18} = 2\pi \, r_{D18} \, \varphi(I)_{D18} = 36{,}184.20062 \text{ (m s}^{-1}\text{)} =$$
$$= 0.0001206975014 \, c$$

Overall exchange frequorce of dalinos 18 inside tau:

$$\varphi(D_{18})_\tau = I(e) \, / \, \hbar = 4.334484487 \cdot 10^{20} \text{ (s}^{-1}\text{)}$$

Maximum linear velocity of dalinos 18 inside tau:

$$v(D_{18})_\tau = 2\pi \, r_\tau \, \varphi(D_{18})_\tau = 2{,}412.280132 \text{ (m s}^{-1}\text{)}$$

TAU INTERVALIC POTENTIAL WELL

Tau electromagnetic potential energy:

$$\Delta U(\tau) = U(\tau) = 1.302333333 \cdot 10^{-10} \text{ (J)} = 812.8521785 \text{ (MeV/c}^2\text{)}$$

Tau electromagnetic potential:

$$V(\tau) = \Delta U(\tau) \, / \, m_\tau = 4.111171103 \cdot 10^{16} \text{ (m}^2 \text{ s}^{-2}\text{)}$$

Tau intervalic potential energy:

$$\Delta I(\tau) = -9.006230113 \cdot 10^{-7} \text{ (J)} = -5{,}621{,}244.256 \text{ (MeV/c}^2\text{)}$$

Intervalic potential well:

$$\Phi(\tau) = \Delta I(\tau) \, / \, m_\tau = -2.843062682 \cdot 10^{20} \text{ (m}^2 \text{ s}^{-2}\text{)}$$

TAU DYNAMICAL FREQUORCE BALANCE

Intervalic and electromagnetic overall frequorces on tau's surface:

$$\varphi(\tau)_{in} = I(\tau) / r_\tau = 1.743989187 \cdot 10^8 \, (N)$$

$$\varphi(\tau)_{em} = U(\tau) / r_\tau = 1.47031957 \cdot 10^8 \, (N)$$

Dynamical frequorce due to spin, $\varphi(\tau)_J$:

$$\varphi(\tau)_J = \varphi(\tau)_{in} - \varphi(\tau)_{em} = 2.736696167 \cdot 10^7 \, (N)$$

TAU INTERVALIC KINEMATICS

Tau angular velocity:

$$\omega_J = (\varphi(\tau)_J / m_\tau r_\tau)^{\frac{1}{2}} = 9.87597042 \cdot 10^{25} \, (s^{-1})$$

Linear velocity due to spin on tau's surface:

$$v_J = \omega_J r_\tau = 8.747625846 \cdot 10^7 \, (m \, s^{-1}) = 0.29178939 \, c$$

Acceleration on muon's surface due to spin:

$$a_J = v_J^2 / r_\tau = 8.63912941 \cdot 10^{33} \, (m \, s^{-2})$$

TAU STRUCTURAL ENERGY BALANCE

Tau rotational kinetic energy due to the intrinsic angular momentum:

$$K(J_\tau) = \tfrac{1}{2} I_\tau \omega_J^2$$
$$U(J_\tau) = 2 K(J_\tau) = m_\tau r_\tau^2 \omega_J^2 = 2.42402449 \cdot 10^{-11} \, (J)$$

Tau spin energy:

$$E(J_\tau) = I(\tau) - U(\tau) = 2.42402449 \cdot 10^{-11} \, (J)$$
$$E(J_\tau) = U(J_\tau)$$

Intervalic principle of energy balance for tau:

$$[\, c^{\pm 2}\hbar (270 q_I)^{-2} + 15 (c^{\pm 2}\hbar(18 q_I)^{-2}) \,] - [\, \tfrac{1}{2} (1/4\pi\varepsilon_0) e^2 / r_\tau \,] - m_\tau \omega_J^2 r_\tau^2 = 0$$

INTERVALIC W± BOSON

W± BOSON INTERVALIC EXCHANGE FREQUORCE

The overall exchange frequorce due to intervalic interaction among intervalinos inside dalino 5 will be:

$$\varphi(\mathbf{I})_{D5} = I(D_5) / \hbar = 1.263935676 \cdot 10^{24} \, (s^{-1})$$

The related maximum linear velocity of intervalinos inside dalino 5 will be, known the bosonic dalino 5 radius:

$$v(\mathbf{I})_{D5} = 2\pi r_{D5} \, \varphi(\mathbf{I})_{D5} = 2{,}986.6411 \, (m\,s^{-1})$$

The overall exchange frequorce of dalinos 5 inside $W^\pm$ boson will be:

$$\varphi(D_5)_W = I(e) / \hbar = 4.334484487 \cdot 10^{20} \, (s^{-1})$$

And the corresponding maximum linear velocity of dalinos 5 inside $W^\pm$ boson is, known the $W^\pm$ boson radius:

$$v(D_5)_W = 2\pi \, r_W \, \varphi(D_5)_W = 55.3081698 \, (m \, s^{-1})$$

$W^\pm$ BOSON INTERVALIC POTENTIAL WELL

The electromagnetic potential energy of $W^\pm$ boson is due to the difference between the electromagnetic energy of $W^\pm$ boson and the electromagnetic energy of its *isolated* constituent dalinos 5, plus the difference between the electromagnetic energy of the dalino 5 and the electromagnetic energy of its *isolated* constituent intervalinos. According to this we have:

$$\Delta U(W^\pm) = [U(W^\pm) - 54 \, U(D_5)] + [54 \, U(D_5) - 54 \, (5 \, U(I))] = U(W^\pm) - 0 = c^{\pm 2} m_W - I(W^\pm) = 5.6801605 \cdot 10^{-9} \, (J) = 35{,}452.76 \, (MeV/c^2)$$

And the electromagnetic potential is:

$$V(W^\pm) = \Delta U(W^\pm) / m_W = 3.9619698 \cdot 10^{16} \, (m^2 \, s^{-2})$$

In the case of electromagnetic energy (and *long ranged* interactions) 'isolated' means separate by an infinite length. However, in the case of intervalic interaction (and *short ranged* interactions) 'isolated' means *out of the intervalic structure*, an event which does not relies on the macroscopic distance of the separation as in the case of a long ranged interaction. Therefore, the $W^\pm$ boson intervalic potential is determined by the difference on the intervalic energy between $W^\pm$ boson and its constituent disaggregated intervalinos:

$$\Delta I(W^\pm) = [I(W^\pm) - 54\,I(D_5)] + [54\,I(D_5) - 54\,(5\,I(\mathbf{I}))] =$$
$$= [c^{\pm 2}\hbar\,(270\,\mathbf{q_I})^{-2}] - [54 \cdot 5\,(c^{\pm 2}\hbar\,\mathbf{q_I}^{-2})] = 270^{-2}c^{-1} - 270c^{-1} =$$
$$= -9.006230113 \cdot 10^{-7}\,(J) = -5{,}621{,}244.256\,(MeV/c^2)$$

And the intervalic potential well is:

$$\Phi(W^\pm) = \Delta I(W^\pm)\,/\,m_W = -6.28193724 \cdot 10^{18}\,(m^2\,s^{-2})$$

$W^\pm$ BOSON DYNAMICAL FREQUORCE BALANCE

The magnitudes of the intervalic and electromagnetic overall frequorces on surface will be:

$$\varphi(W^\pm)_{in} = I(W^\pm)\,/\,r_W = 3.54783848 \cdot 10^{11}\,(N)$$

$$\varphi(W^\pm)_{em} = U(W^\pm)\,/\,r_W = 2.79697536 \cdot 10^{11}\,(N)$$

As in other leptons, the dynamical frequorce due to spin, $\varphi(W^\pm)_J$, is:

$$\varphi(W^\pm)_{in} - \varphi(W^\pm)_{em} = 7.50863117 \cdot 10^{10}\,(N)$$

$W^\pm$ BOSON INTERVALIC KINEMATICS

From the dynamical frequorce due to spin, the angular velocity will be:

$\omega_J = (\varphi(W^\pm)_J / m_W r_W)^{\frac{1}{2}} = 5.0783155 \cdot 10^{27}$ (s^{-1})

The linear velocity of spin on W$^\pm$ boson's surface is:

$v_J = \omega_J r_W = 1.03131574 \cdot 10^8$ (m s^{-1}) $= 0.34400990$ c

And the acceleration on W$^\pm$ boson's surface due to spin will be:

$a_J = v_J^2 / r_W = 5.23734672 \cdot 10^{35}$ (m s^{-2})

W$^\pm$ BOSON STRUCTURAL ENERGY BALANCE

The rotational kinetic energy due to the intrinsic angular momentum of W$^\pm$ boson is: $K(J_W) = \frac{1}{2} I_W \omega_J^2$, being I_W the W$^\pm$ boson's moment of inertia. By the Virial theorem —since a *composite* particle is like a micro galaxy at quantum scale—, the potential energy due to spin is: $U(J_W) = 2 K(J_W)$. As in other leptons we have:

$U(J_W) = m_W r_W^2 \omega_J^2 = 1.524869705 \cdot 10^{-9}$ (J)

By fundamental intervalic principle of energy balance for subatomic particles, the W$^\pm$ boson spin energy is:

$I(W^\pm) - U(W^\pm) - E(J_W) = 0$
$E(J_W) = I(W^\pm) - U(W^\pm) = 1.524869713 \cdot 10^{-9}$ (J)

Therefore is verified the equality $E(J_W) = U(J_W)$, that is to say:

$[c^{\pm 2}\hbar (270q_I)^{-2} + 54 (c^{\pm 2}\hbar (5q_I)^{-2})] - [\frac{1}{2} (1/4\pi\varepsilon_0) e^2 / r_W] - m_W r_W^2 \omega_J^2 = 0$

ELEMENTARY DALINOS AND GAUDINOS: LEPTONS-CHARGED MASSIVE BOSONS

	Electron	Muon	Tau	W± Massive Boson
Intervalic structure	$e = G_1 = D_{270} = 270\,\mathbf{I} = 540\,\gamma$	$\mu = L_1 = G_6 = 6\,D_{45} = 270\,\mathbf{I} = 540\,\gamma$	$\tau = L_1 = G_{15} = 15\,D_{18} = 270\,\mathbf{I} = 540\,\gamma$	$W^\pm = L_1 = G_{54} = 54\,D_5 = 270\,\mathbf{I} = 540\,\gamma$
Intervalic energy, I	$I(e) = c^{\pm 2}\hbar\,e^{-2} = 4.575639166 \cdot 10^{-14}$ (J) $= 0.285588809$ (MeV/c²)	$I(\mu) = \sum(c^{\pm 2}\hbar Q_i^2) = c^{\pm 2}\hbar\,e^{-2} + 6\,(c^{\pm 2}\hbar (45\mathbf{q_i})^{-2}) = 9.92913699 \cdot 10^{-12}$ (J) $= 61.9727717$ (MeV/c²)	$I(\tau) = \sum(c^{\pm 2}\hbar Q_i^2) = c^{\pm 2}\hbar\,e^{-2} + 15\,(c^{\pm 2}\hbar (18\mathbf{q_i})^{-2}) = 1.544735782 \cdot 10^{-10}$ (J) $= 964.1478215$ (MeV/c²)	$I(W^\pm) = \sum(c^{\pm 2}\hbar Q_i^2) = c^{\pm 2}\hbar\,e^{-2} + 54\,(c^{\pm 2}\hbar (5\mathbf{q_i})^{-2}) = 7.205030213 \cdot 10^{-9}$ (J) $= 44,970.24192$ (MeV/c²)
Electromagnetic energy, U	$U(e) = E(e)_{mass} - I(e) = 3.611471925 \cdot 10^{-14}$ (J) $= 0.22541025$ (MeV/c²)	$U(\mu) = E(\mu)_{mass} - I(\mu) = 6.999210568 \cdot 10^{-12}$ (J) $= 43.6856173$ (MeV/c²)	$U(\tau) = E(\tau)_{mass} - I(\tau) = 1.302333333 \cdot 10^{-10}$ (J) $= 812.8521785$ (MeV/c²)	$U(W^\pm) = E(W^\pm)_{mass} - I(W^\pm) = 5.6801605 \cdot 10^{-9}$ (J) $= 35,452.76$ (MeV/c²)
Spin energy E_J	$E_J(e) = I(e) - U(e) = 9.64167241 \cdot 10^{-15}$ (J) $= 0.060178559$ (MeV/c²)	$E_J(\mu) = I(\mu) - U(\mu) = 2.929926422 \cdot 10^{-12}$ (J) $= 43.6856173$ (MeV/c²)	$E_J(\tau) = I(\tau) - U(\tau) = 2.42402449 \cdot 10^{-11}$ (J)	$E_J(W^\pm) = I(W^\pm) - U(W^\pm) = 1.524869713 \cdot 10^{-9}$ (J)
Energy ratios	$I(e)/U(e) = 1.266973485 \approx 5/4$ $I(e)/E(e)_{mass} = 0.558883239 \approx 5/9$ $U(e)/E(e)_{mass} = 0.44111676 \approx 4/9$	$I(\mu)/U(\mu) = 1.418608126 \approx 7/5$ $I(\mu)/E(\mu)_{mass} = 0.586539055 \approx 7/12$ $U(\mu)/E(\mu)_{mass} = 0.413460944 \approx 5/12$	$I(\tau)/U(\tau) = 1.186129344 \approx 6/5$ $I(\tau)/E(\tau)_{mass} = 0.542570524 \approx 6/11$ $U(\tau)/E(\tau)_{mass} = 0.457429475 \approx 5/11$	$I(W^\pm)/U(W^\pm) = 1.268 \approx 5/4$ $I(W^\pm)/E(W^\pm)_{mass} = 0.559 \approx 5/9$ $U(W^\pm)/E(W^\pm)_{mass} = 0.441 \approx 4/9$
Radius r	$r_e = \frac{1}{2}\,(1/4\pi\varepsilon_0)\,e^2/U(e) = 3.194098699 \cdot 10^{-15}$ (m)	$r_\mu = \frac{1}{2}\,(1/4\pi\varepsilon_0)\,(270\,\mathbf{q_i})^2/U(\mu) = 1.648099834 \cdot 10^{-17}$ (m)	$r_\tau = \frac{1}{2}\,(1/4\pi\varepsilon_0)\,(270\,\mathbf{q_i})^2/U(\tau) = 8.857484858 \cdot 10^{-19}$ (m)	$r_W = \frac{1}{2}\,(1/4\pi\varepsilon_0)\,(270\,\mathbf{q_i})^2/U(W^\pm) = 2.0308225 \cdot 10^{-20}$ (m)
Angular velocity due to spin, ω_J	$\omega_J = (E_J(e)\,/\,m_e\,r_e^2)^{1/2} = 3.220944289 \cdot 10^{22}$ (s⁻¹)	$\omega_J = (E_J(\mu)\,/\,m_\mu\,r_\mu^2)^{1/2} = 7.567601581 \cdot 10^{24}$ (s⁻¹)	$\omega_J = (E_J(\tau)\,/\,m_\tau\,r_\tau^2)^{1/2} = 9.87597042 \cdot 10^{25}$ (s⁻¹)	$\omega_J = (E_J(W^\pm)\,/\,m_W\,r_W^2)^{1/2} = 5.0783155 \cdot 10^{27}$ (s⁻¹)
Linear velocity due to spin on surface, v_J	$v_J = \omega_J\,r_e = 1.028801396 \cdot 10^8$ (m s⁻¹) $= 0.343171206\,c$	$v_J = \omega_J\,r_\mu = 1.247216291 \cdot 10^8$ (m s⁻¹) $= 0.416026573$ (m s⁻¹)	$v_J = \omega_J\,r_\tau = 8.747625846 \cdot 10^7$ (m s⁻¹) $= 0.29178939$ (m s⁻²)	$v_J = \omega_J\,r_W = 1.03131574 \cdot 10^8$ (m s⁻¹) $= 0.34400990$ (m s⁻²)
Intervalic potential well, Φ	$\Phi(e) = \Delta I(e)\,/\,m_e = -9.886754667 \cdot 10^{23}$ (m² s⁻²)	$\Phi(\mu) = \Delta I(\mu)\,/\,m_\mu = -4.781562918 \cdot 10^{21}$ (m² s⁻²)	$\Phi(\tau) = \Delta I(\tau)\,/\,m_\tau = -2.843062682 \cdot 10^{20}$ (m² s⁻²)	$\Phi(W^\pm) = \Delta I(W^\pm)\,/\,m_W = -6.28193724 \cdot 10^{18}$ (m² s⁻²)
Overall exchange frequence of its constituent particles inside, φ	$\varphi(\mathbf{I})_e = I(e)\,/\,\hbar = 4.334484487 \cdot 10^{20}$ (s⁻¹)	$\varphi(D_{45})_\mu = I(e)\,/\,\hbar = 4.334484487 \cdot 10^{20}$ (s⁻¹)	$\varphi(D_{18})_\tau = I(e)\,/\,\hbar = 4.334484487 \cdot 10^{20}$ (s⁻¹)	$\varphi(D_5)_W = I(e)\,/\,\hbar = 4.334484487 \cdot 10^{20}$ (s⁻¹)
Maximum linear velocity of its constituent particles inside, v	$v(\mathbf{I})_e = 2\pi\,r_e\,\varphi(\mathbf{I})_e = 8.698926337 \cdot 10^6$ (m s⁻¹) $= 0.029016494\,c$	$v(D_{45})_\mu = 2\pi\,r_\mu\,\varphi(D_{45})_\mu = 44,884.95943$ (m s⁻¹)	$v(D_{18})_\tau = 2\pi\,r_\tau\,\varphi(D_{18})_\tau = 2,412.280132$ (m s⁻¹)	$v(D_{45})_W = 2\pi\,r_W\,\varphi(D_5)_W = 55.3081698$ (m s⁻¹)
Structural energy balance	$[c^{\pm 2}\hbar\,(270\,\mathbf{q_i})^{-2}]\,-\,[\,\frac{1}{2}\,(1/4\pi\varepsilon_0)\,e^2/r_e\,]\,-\,m_e\,\omega_J^2\,r_e^2 = 0$	$[c^{\pm 2}\hbar\,(270\,\mathbf{q_i})^{-2}]\,+\,6\,(c^{\pm 2}\hbar\,(45\mathbf{q_i})^{-2})\,]\,-\,[\,\frac{1}{2}\,(1/4\pi\varepsilon_0)\,e^2/r_\mu\,]\,-\,m_\mu\,\omega_J^2\,r_\mu^2 = 0$	$[c^{\pm 2}\hbar\,(270\,\mathbf{q_i})^{-2}]\,+\,15\,(c^{\pm 2}\hbar\,(18\mathbf{q_i})^{-2})\,]\,-\,[\,\frac{1}{2}\,(1/4\pi\varepsilon_0)\,e^2/r_\tau\,]\,-\,m_\tau\,\omega_J^2\,r_\tau^2 = 0$	$[c^{\pm 2}\hbar\,(270\,\mathbf{q_i})^{-2}]\,+\,54\,(c^{\pm 2}\hbar\,(5\mathbf{q_i})^{-2})\,]\,-\,[\,\frac{1}{2}\,(1/4\pi\varepsilon_0)\,e^2/r_W\,]\,-\,m_W\,\omega_J^2\,r_W^2 = 0$

ARCHITECTURE OF THE STRUCTURAL ENERGY OF SUBATOMIC PARTICLES

STRUCTURAL ENERGY OF SUBATOMIC PARTICLES: $A = I + I^{-1} + E_J = c^{\pm 2} m_{Schwarzschild}$
Intervalic Principle of Energy Balance for Subatomic Particles: $I - I^{-1} - E_J = 0$

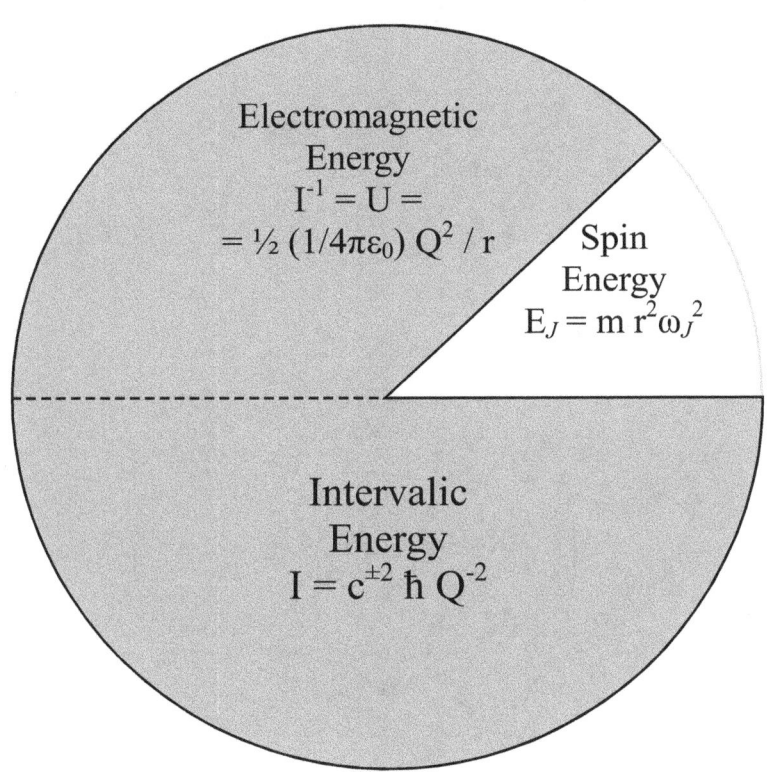

Electron Structural Energy Ratios

$I(e)/U(e) = 1.266973485$
$I(e)/E(e)_{mass} = 0.558883239$
$U(e)/E(e)_{mass} = 0.44111676$
$E_J(e)/E(e)_{mass} = 0.117766479$

Tau Structural Energy Ratios

$I(\tau)/U(\tau) = 1.186129344$
$I(\tau)/E(\tau)_{mass} = 0.542570524$
$U(\tau)/E(\tau)_{mass} = 0.457429475$
$E_J(\tau)/E(\tau)_{mass} = 0.085141048$

Muon Structural Energy Ratios

$I(\mu)/U(\mu) = 1.418608126$
$I(\mu)/E(\mu)_{mass} = 0.586539055$
$U(\mu)/E(\mu)_{mass} = 0.413460944$
$E_J(\mu)/E(\mu)_{mass} = 0.173078111$

$W^{\pm}$ boson Structural Energy Ratios

$I(W^{\pm})/U(W^{\pm}) = 1.268455314$
$I(W^{\pm})/E(W^{\pm})_{mass} = 0.559171391$
$U(W^{\pm})/E(W^{\pm})_{mass} = 0.440828608$
$E_J(W^{\pm})/E(W^{\pm})_{mass} = 0.118342809$

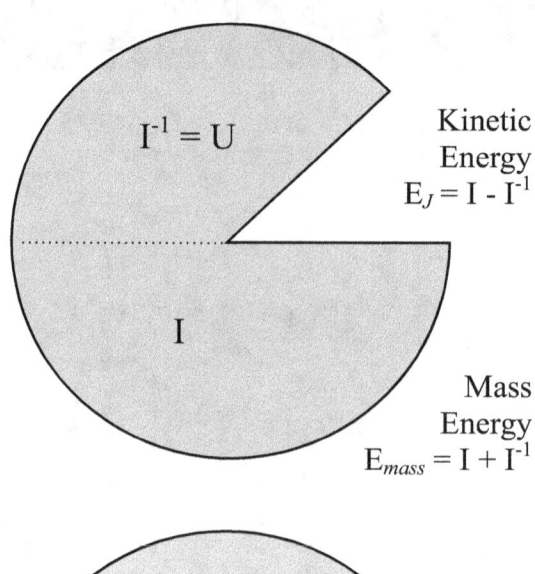

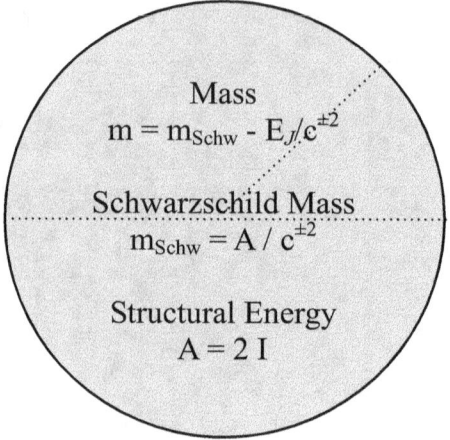

Intervalic Principle of Energy Balance for Lepton-CMB

$$I(l) - U(l) - E_J(l) = 0$$

Lepton-CMB Schwarzschild Mass

$$c^{\pm 2}\, \hbar\, e^{-2} = \tfrac{1}{2}\, m(l)_{Schw}\, v_{esc}^{\pm 2}$$

Lepton-CMB Structural Energy

$$A(l) = E(l)_{mass} + E_J(l) = I(l) + U(l) + E_J(l)$$

Lepton-CMB Intervalic Architecture

$$A(l) = 2\, I(l)$$
$$A(l) = c^{\pm 2} m(l)_{Schw}$$
$$R(l) = E(l)_{mass} / A(l)$$
$$R(l) = m(l) / m(l)_{Schw}$$
$$R(e) = 0.894641299$$
$$R(\mu) = 0.852458153$$
$$R(\tau) = 0.921539187$$
$$R(W^{\pm}) = 0.894180223$$

Chapter 12

INTERVALIC FRACTIONAL GAUDINO

STRUCTURE OF DALINOS

Recapitulating: we define the *dalino* as an intervalic particle composed by the assembly of a determine number of intervalinos. Due to intervalic principles of symmetry, such compositeness only can take 16 values, which are the divisors of the elementary charge, because the elementary charge is defined as the final state of the lowest energy, reached through the balance between intervalic energy, electromagnetic energy and spin energy. These values are the following multiples of the intervalic charge, $q_I = \sqrt{-(c^{-1}\hbar)}$:

1, 2, 3, 5, 6, 9, 10, 15, 18, 27, 30, 45, 54, 90, 135, 270.

Of course, the aggregations of intervalinos could take any other exotic values at the Big Bang, but any of them survived after the primordial synthesis of subatomic particles because its energy could not reach a balance imposing a different set of symmetrical constraints than those of the elementary charge. Therefore, we are compelled to think that any particle which symmetries do not fit with those imposed by the state of lowest energy of the elementary charge could stay after primordial times.

According to the number of constituent intervalinos, we name the dalinos as, i.e., dalino 18, dalino 27, dalino 30, etc., including likewise this number behind its symbol as a subindex: D_{18}, D_{27}, D_{30}, etc. Need not to say that the dalino 1 (D_1) is no other than the *intervalino*, and the dalino 270 (D_{270}) is the *electron*.

It has to be noted that the intervalic interaction has spin 0, as I have shown in other site. Therefore, like charges attract between them, and unlike charges repel, just the contrary to electromagnetic interaction. This fact fully explains why there are not dalinos with intervalic charges of different signs, since intervalic energy is ever much more greater than electromagnetic interaction at scale of subatomic particles.

STRUCTURE OF GAUDINO

We define the *gaudino*, G_n, as an intervalic particle composed by the assembly of a determine number of dalinos. The subindex n means the number of constituent dalinos. Insofar as dalinos are already composite particles, gaudinos involve compositeness of compositeness. Where there are successive levels of compositeness we may talk about *structurefulness*.

With the preceding data about dalinos the question that arises is how can dalinos aggregate in order to compose the subatomic particles that conform the actual Universe. It is clear that there must exist powerful constraints that avoid a great number of possible combinations. We have commented in other site that the magnitude of the elementary charge acts as an attractor of spontaneous order, since it is the state of minimal energy of the intervalic charge. This fact introduces a powerful symmetry constraint in the primordial aggregations of dalinos. Actually, the 16 allowed symmetries of dalinar aggregation can be viewed as a consequence of that constraint, valid for all primordial aggregations of intervalinos.

There are two basic kinds of gaudinos: the 16 *elementary* ones, each of them composed by 270 intervalinos, and the *fractional* ones, which are composed by a number of intervalinos which is a fraction of 270, and which number is postulated to be 12+12 = 24 as we are going to see immediately. In total there would be 16+24 = 40 kinds of gaudinos, regardless gaudinos with poor symmetries which can be discarded for practical purposes.

ELEMENTARY GAUDINOS: L_1

The most simple gaudinar structure is the aggregation of dalinos for constituting an elementary charge with 270 constituent intervalinos. Due to the symmetry constraints of the dalinar structure there are only 16 of such possible particles, which are composed by a determine number of dalinos. This set of 16 symmetries will be named L_1, where the subindex 1 means the number of constituent intervalinos of the elementary charge (270). The masses of all these elementary gaudinos are, taking the electron's ratio between intervalic and electromagnetic energy as reference, the following (in MeV/c^2):

$L_1 = G_1 = 1\ D_{270} = 270\ I = 0.51099909$
$L_1 = G_2 = 2\ D_{135} = 270\ I = 4.0879928$
$L_1 = G_3 = 3\ D_{90} = 270\ I = 13.796975$
$L_1 = G_5 = 5\ D_{54} = 270\ I = 63.874885$
$L_1 = G_6 = 6\ D_{45} = 270\ I = 110.37580$
$L_1 = G_9 = 9\ D_{30} = 270\ I = 372.51833$
$L_1 = G_{10} = 10\ D_{27} = 270\ I = 510.99909$
$L_1 = G_{15} = 15\ D_{18} = 270\ I = 1{,}724.6220$
$L_1 = G_{18} = 18\ D_{15} = 270\ I = 2{,}980.1466$
$L_1 = G_{27} = 27\ D_{10} = 270\ I = 10{,}057.995$
$L_1 = G_{30} = 30\ D_9 = 270\ I = 13{,}796.975$
$L_1 = G_{45} = 45\ D_6 = 270\ I = 46{,}564.794$

$L_1 = G_{54} = 54\,D_5 = 270\,\mathbf{I} = 80{,}463.964$
$L_1 = G_{90} = 90\,D_3 = 270\,\mathbf{I} = 372{,}518.33$
$L_1 = G_{135} = 135\,D_2 = 270\,\mathbf{I} = 1{,}257{,}249.4$
$L_1 = G_{270} = 270\,D_1 = 270\,\mathbf{I} = 5{,}621{,}244.5$

It can be pointed out that all these possible particles are equally composed by 270 intervalinos, being the only difference among them the different *symmetry* or *form* in which the aggregation of intervalinos has been made, that is to say, its *structure*. "God does not play dice". Here we have a clear example of the new importance of structure and form in Physics, importance that increases proportionally as it comes down more and more towards subatomic world.

Thus, all leptons are gaudinos focused on the symmetries 1 D_{270}, 6 D_{45} and 15 D_{18}:

$G_1 = 1\,D_{270} = 0.51099909 \to e$, electron
$G_6 = 6\,D_{45} = 110.37580 \to \mu$, muon
$G_{15} = 15\,D_{18} = 1{,}724.6220 \to \tau$, tauon

Meaningfully, the intermediate boson $W^{\pm}$ is a gaudino, while Z^0 is an aggregation of two gaudinos as we have described in detail in other sites:

$G_{54} = 54\,D_5 = 80{,}463.964 \to W^{\pm}$ intermediate boson

FRACTIONAL GAUDINOS: $L_{2/3}$ and $L_{1/3}$

As in all our papers, we will follow a systematic approach, applying rigorously logical processes up to its latest consequences, without regarding the meaning of the obtained results until theoretical deduction is finalized.

For obvious symmetry reasons, we have begun with the possible combinations of dalinos composed just by the number of intervalinos of the elementary charge. However, it could be proposed other kinds of gaudinos composed, for example, by fractional charge values: $180q_I$, $135q_I$ and $90q_I$ —or in other words ⅔e, ½e, ⅓e—. The states that Nature appears to chose are ever those that shows the *greatest symmetry* and the *lowest energy*. Precisely, the deep understanding and the right measurement about the Nature's preferences on symmetry in relation with energy, or vice versa, is an important pending chapter in our knowledge of Physics. Please note that the values ¼ e and ¾ e are not allowed anyway because ¼ e = 67.5 q_I and ¾ e = 202.5 q_I, values that would imply a *partition* of the elementary charge and, *a fortiori*, a partition of the intervalino.

We can suppose that the fractional gaudinos chosen by Nature are those with the most rich symmetries. As we have seen, the elementary gaudino has 16 symmetries. It can be easily checked that the structures with the greatest number of symmetries among all possible fractional gaudinos are those which have ⅔ 270 = 180 and ⅓ 270 = 90 intervalinos. They are named respectively lisztinos $L_⅔$ and $L_⅓$. Both of them have 12 symmetries, while the next structure is ½ 270 = 135 intervalinos, which has only 8 symmetries, as can be seen right now (masses expressed in MeV/c²). Therefore we will consider only the two first ones. Here is the whole set of lisztinos $L_⅓$:

$L_⅓ = ⅓ G_1 = (1/3) D_{270} = 90\ I$, not allowed
$L_⅓ = ⅓ G_2 = (2/3) D_{135} = 90\ I$, not allowed
$L_⅓ = ⅓ G_3 = 1\ D_{90} = 90\ I = 4.5989918$
$L_⅓ = ⅓ G_5 = (5/3) D_{54} = 90\ I$, not allowed
$L_⅓ = ⅓ G_6 = 2\ D_{45} = 90\ I = 36.791934$
$L_⅓ = ⅓ G_9 = 3\ D_{30} = 90\ I = 124.17278$
$L_⅓ = ⅓ G_{10} = (10/3) D_{27} = 90\ I$, not allowed
$L_⅓ = ⅓ G_{15} = 5\ D_{18} = 90\ I = 574.87400$
$L_⅓ = ⅓ G_{18} = 6\ D_{15} = 90\ I = 993.38220$
$L_⅓ = ⅓ G_{27} = 9\ D_{10} = 90\ I = 3,352.6650$
$L_⅓ = ⅓ G_{30} = 10\ D_9 = 90\ I = 4,598.9918$
$L_⅓ = ⅓ G_{45} = 15\ D_6 = 90\ I = 15,521.598$

$L_{1/3} = \frac{1}{3} G_{54} = 18 D_5 = 90 I = 26,821.321$
$L_{1/3} = \frac{1}{3} G_{90} = 30 D_3 = 90 I = 124,172.78$
$L_{1/3} = \frac{1}{3} G_{135} = 45 D_2 = 90 I = 419,083.13$
$L_{1/3} = \frac{1}{3} G_{270} = 90 D_1 = 90 I = 1,873,748.2$

The entire set of lisztinos $L_{2/3}$ is:

$L_{2/3} = \frac{2}{3} G_1 = (2/3) D_{270} = 180 I$, not allowed
$L_{2/3} = \frac{2}{3} G_2 = (4/3) D_{135} = 180 I$, not allowed
$L_{2/3} = \frac{2}{3} G_3 = 2 D_{90} = 180 I = 9.1979836$
$L_{2/3} = \frac{2}{3} G_5 = (10/3) D_{54} = 180 I$, not allowed
$L_{2/3} = \frac{2}{3} G_6 = 4 D_{45} = 180 I = 73.583868$
$L_{2/3} = \frac{2}{3} G_9 = 6 D_{30} = 180 I = 248.34556$
$L_{2/3} = \frac{2}{3} G_{10} = (20/3) D_{27} = 180 I$, not allowed
$L_{2/3} = \frac{2}{3} G_{15} = 10 D_{18} = 180 I = 1,149.7480$
$L_{2/3} = \frac{2}{3} G_{18} = 12 D_{15} = 180 I = 1,986.7644$
$L_{2/3} = \frac{2}{3} G_{27} = 18 D_{10} = 180 I = 6,705.3299$
$L_{2/3} = \frac{2}{3} G_{30} = 20 D_9 = 180 I = 9,197.9836$
$L_{2/3} = \frac{2}{3} G_{45} = 30 D_6 = 180 I = 31,043.196$
$L_{2/3} = \frac{2}{3} G_{54} = 36 D_5 = 180 I = 53,642.642$
$L_{2/3} = \frac{2}{3} G_{90} = 60 D_3 = 180 I = 248,345.56$
$L_{2/3} = \frac{2}{3} G_{135} = 90 D_2 = 180 I = 838,166.26$
$L_{2/3} = \frac{2}{3} G_{270} = 180 D_1 = 180 I = 3,747,496.3$

And only for curiosity in order to check its lesser symmetries with respect to the preceding sets, the whole set of lisztinos $L_{1/2}$ is:

$L_{1/2} = \frac{1}{2} G_1 = D_{270} = 135 I$, not allowed
$L_{1/2} = \frac{1}{2} G_2 = 1 D_{135} = 135 I = 2.0439964$
$L_{1/2} = \frac{1}{2} G_3 = (3/2) D_{90} = 135 I$, not allowed
$L_{1/2} = \frac{1}{2} G_5 = (5/2) D_{54} = 135 I$, not allowed
$L_{1/2} = \frac{1}{2} G_6 = 3 D_{45} = 135 I = 55.187901$
$L_{1/2} = \frac{1}{2} G_9 = (9/2) D_{30} = 135 I$, not allowed
$L_{1/2} = \frac{1}{2} G_{10} = 5 D_{27} = 135 I = 255.49955$

$L_{½} = ½\, G_{15} = (15/2)\, D_{18} = 135\, I$, not allowed
$L_{½} = ½\, G_{18} = 9\, D_{15} = 135\, I = 1{,}490.0733$
$L_{½} = ½\, G_{27} = (27/2)\, D_{10} = 135\, I$, not allowed
$L_{½} = ½\, G_{30} = 15\, D_{9} = 135\, I = 6{,}898.4877$
$L_{½} = ½\, G_{45} = (45/2)\, D_{6} = 135\, I$, not allowed
$L_{½} = ½\, G_{54} = 27\, D_{5} = 135\, I = 40{,}231.982$
$L_{½} = ½\, G_{90} = 45\, D_{3} = 135\, I = 186{,}259.17$
$L_{½} = ½\, G_{135} = (135/2)\, D_{2} = 135\, I$, not allowed
$L_{½} = ½\, G_{270} = 135\, D_{1} = 135\, I = 2{,}810{,}622.2$

Since the elementary charge plays as an attractor for the aggregations of subatomic particles, these fractional gaudinos will not be able to exist in isolated state but only as constituents of other particles which total charge is the elementary one.

As we will see later with detail, among these symmetries we find the lightest intervalic structures of the allowed quark families:

$L_{⅓} = ⅓\, G_{6} = 2\, D_{45} = 90\, I = 36.791934$
$L_{⅔} = ⅔\, G_{6} = 4\, D_{45} = 180\, I = 73.583868$
$L_{⅓} = ⅓\, G_{9} = 3\, D_{30} = 90\, I = 124.17278$
$L_{⅔} = ⅔\, G_{9} = 6\, D_{30} = 180\, I = 248.34556$
$L_{⅓} = ⅓\, G_{15} = 5\, D_{18} = 90\, I = 574.87400$
$L_{⅔} = ⅔\, G_{15} = 10\, D_{18} = 180\, I = 1{,}149.7480$
$L_{⅓} = ⅓\, G_{45} = 15\, D_{6} = 90\, I = 15{,}521.598$
$L_{⅔} = ⅔\, G_{45} = 30\, D_{6} = 180\, I = 31{,}043.196$
$L_{⅓} = ⅓\, G_{54} = 18\, D_{5} = 90\, I = 26{,}821.321$
$L_{⅔} = ⅔\, G_{54} = 36\, D_{5} = 180\, I = 53{,}642.642$
$L_{⅓} = ⅓\, G_{90} = 30\, D_{3} = 90\, I = 124{,}172.78$
$L_{⅔} = ⅔\, G_{90} = 60\, D_{3} = 180\, I = 248{,}345.56$
$L_{⅓} = ⅓\, G_{135} = 45\, D_{2} = 90\, I = 419{,}083.13$
$L_{⅔} = ⅔\, G_{135} = 90\, D_{2} = 180\, I = 838{,}166.26$

NUCLEONIC GAUDINO 6 INTERVALIC STRUCTURE

It have to be noted that the principal structural difference between leptons and quarks is that the constituent dalinos of leptons have always like electric charge signs, while the nucleonic gaudinos 6 have constituent dalinos with unlike electric charge signs. This fact produces a delicate balance between the intervalic, electromagnetic and spin energies in the nucleonic gaudino 6. It can be also remembered that the electromagnetic energy is just the *inverse* of the intervalic energy, so subatomic particles reach a very elegant balance between a magnitude and its inverse. Due to its hexagonal symmetry, the dynamical electromagnetic interaction inside the gaudino 6 can be calculated as if it was composed by the sum of 15 equal dalino45-to-dalino45 interaction. Then, using the traditional notation of '-' for attractive interaction and '+' for repulsive interaction, and writing down the full set of symmetries, the following result is obtained:

$$U[6(D_{+45}) + 0(D_{-45})] = +6\,U(D_{45})$$
$$U[5(D_{+45}) + 1(D_{-45})] = [(10/15)-(5/15)]\,6\,U(D_{45}) = +(1/3)\,6\,U(D_{45})$$
$$U[4(D_{+45}) + 2(D_{-45})] = [(7/15)-(8/15)]\,6\,U(D_{45}) = -(1/15)\,6\,U(D_{45})$$
$$U[3(D_{+45}) + 3(D_{-45})] = [(6/15)-(9/15)]\,6\,U(D_{45}) = -(1/5)\,6\,U(D_{45})$$
$$U[2(D_{+45}) + 4(D_{-45})] = [(7/15)-(8/15)]\,6\,U(D_{45}) = -(1/15)\,6\,U(D_{45})$$
$$U[1(D_{+45}) + 5(D_{-45})] = [(10/15)-(5/15)]\,6\,U(D_{45}) = +(1/3)\,6\,U(D_{45})$$
$$U[0(D_{+45}) + 6(D_{-45})] = +6\,U(D_{45})$$

The primordial aggregation of intervalinos to make dalinos, and of dalinos to make gaudinos was symmetric under interchange and therefore follows Bose-Einstein statistics, as I have explained in other site. Intervalinos and dalinos are *bosons* with spin 0. On the contrary, the primordial aggregation of gaudinos made symmetric and antisymmetric states under interchange: the first is found in zero charged massive bosons, and the second is found in lisztinos with fractional charge, that is to say, in quarks.

Regarding stable quarks constituent of nucleons, due to powerful symmetry constraints there were only seven modes to make those aggregation of dalinos 45 to make gaudinos 6 at primordial times:

$6 (D_{+45}) + 0 (D_{-45}) = G_6^+$
$5 (D_{+45}) + 1 (D_{-45}) = G_6^{+\frac{2}{3}}$
$4 (D_{+45}) + 2 (D_{-45}) = G_6^{+\frac{1}{3}}$
$3 (D_{+45}) + 3 (D_{-45}) = G_6^0$
$2 (D_{+45}) + 4 (D_{-45}) = G_6^{-\frac{1}{3}}$
$1 (D_{+45}) + 5 (D_{-45}) = G_6^{-\frac{2}{3}}$
$0 (D_{+45}) + 6 (D_{-45}) = G_6^-$

All these gaudinos 6 have the same dalinar mass because the intervalic structure and absolute value of the charge of the constituent dalinos are identical. However the gaudinar masses are different according to their isocharge value, as can be seen in the next table on the isogaudino 6: structural balance between its intervalic and electromagnetic energies.

The last gaudino, $6 (D_{-45})$, is no other than the *muon*, being $6 (D_{+45})$ the *antimuon*. The intervalic structures of $G_6^\pm$ and G_6^0 are unique. On the contrary, fractional charged gaudinos can have different intervalic structures relying on how the aggregation of dalinos is made. Writing only the charge sign of the constituent dalinos 45, we have that the gaudino $G_6^{\pm\frac{2}{3}}$ have only one intervalic structure (writing only, for example, $G_6^{+\frac{2}{3}}$):

$(+++++-) = (++++-+) = (+++-++) = (++-+++) = (+-++++) = (-+++++)$

but the gaudino $G_6^{\pm\frac{1}{3}}$ has got three different structures (writing only $G_6^{+\frac{1}{3}}$):

$(++++--) = (+++--+) = (++--++) = (+---+++) = (--++++) = (-++++-) \neq$
$(+++-+-) = (++-+-+) = (+-+-++) = (-+-+++) = (+-+++-) = (-+++-+) \neq$
$(++-++-) = (+-++-+) = (-+++++) = (++-++-) = (+-++-+) = (-++-++)$

GAUDINAR STRUCTURE OF NUCLEONIC QUARKS

As quarks have fractional charges, the only possible assemblies of isogaudinos (G_6) to compose a quark are the 15 following ones:

$$q(+\tfrac{2}{3}) = G_6^+ + G_6^- + G_6^{+2/3}$$
$$q(+\tfrac{2}{3}) = G_6^+ + G_6^{-2/3} + G_6^{+1/3}$$
$$q(+\tfrac{2}{3}) = G_6^{+2/3} + G_6^{-2/3} + G_6^{+2/3}$$
$$q(+\tfrac{2}{3}) = G_6^+ + G_6^{-1/3} + G_6^0$$
$$q(+\tfrac{2}{3}) = G_6^{+2/3} + G_6^{-1/3} + G_6^{+1/3}$$
$$q(+\tfrac{2}{3}) = G_6^0 + G_6^{+2/3} + G_6^0$$
$$q(+\tfrac{2}{3}) = G_6^{+1/3} + G_6^0 + G_6^{+1/3}$$

$$q(-\tfrac{1}{3}) = G_6^+ + G_6^- + G_6^{-1/3}$$
$$q(-\tfrac{1}{3}) = G_6^+ + G_6^{-2/3} + G_6^{-2/3}$$
$$q(-\tfrac{1}{3}) = G_6^{+2/3} + G_6^- + G_6^0$$
$$q(-\tfrac{1}{3}) = G_6^{+1/3} + G_6^- + G_6^{+1/3}$$
$$q(-\tfrac{1}{3}) = G_6^{+2/3} + G_6^{-2/3} + G_6^{-1/3}$$
$$q(-\tfrac{1}{3}) = G_6^{+1/3} + G_6^{-2/3} + G_6^0$$
$$q(-\tfrac{1}{3}) = G_6^{-1/3} + G_6^{+1/3} + G_6^{-1/3}$$
$$q(-\tfrac{1}{3}) = G_6^0 + G_6^{-1/3} + G_6^0$$

Now we could make some assumptions on the possible physical constraints to select the most stable states among this set. For example:

1) It can be supposed that the constituent charges of these isogaudinos would not be neither the elementary charge not zero charge but fractional charges. If this assumption was held the only remaining combinations would be these four ones:

$$q(+\tfrac{2}{3}) = G_6^{+2/3} + G_6^{-2/3} + G_6^{+2/3}$$
$$q(+\tfrac{2}{3}) = G_6^{+2/3} + G_6^{-1/3} + G_6^{+1/3}$$
$$q(-\tfrac{1}{3}) = G_6^{+2/3} + G_6^{-2/3} + G_6^{-1/3}$$
$$q(-\tfrac{1}{3}) = G_6^{-1/3} + G_6^{+1/3} + G_6^{-1/3}$$

ISOGAUDINO 6:
STRUCTURAL BALANCE BETWEEN INTERVALIC AND ELECTROMAGNETIC ENERGIES

Isogaudino	G_6^+	$G_6^{+2/3}$	$G_6^{+1/3}$	G_6^0	$G_6^{-1/3}$	$G_6^{-2/3}$	G_6^-
Intervalic energy	$I = -(1/36)\ I(D_{45})$	$I = -(1/16)\ I(D_{45})$	$I = -(1/4)\ I(D_{45})$	$I = 0$	$I = -(1/4)\ I(D_{45})$	$I = -(1/16)\ I(D_{45})$	$I = -(1/36)\ I(D_{45})$
Electro-magnetic energy	$U = +6\ U(D_{45})$	$U = +2\ U(D_{45})$	$U = -(2/5)\ U(D_{45})$	$U = -(6/5)\ U(D_{45})$	$U = -(2/5)\ U(D_{45})$	$U = +2\ U(D_{45})$	$U = +6\ U(D_{45})$
Particle	Antimuon μ^+	Constituent gaudino of quarks	Constituent gaudino of quarks	Zero charge: it can't be a constituent of another particle	Constituent gaudino of quarks	Constituent gaudino of quarks	Muon μ^-

2) Since the intervalic energy decreases at each consecutive level of the intervalic structure in all known levels, from the first interior level (intervalinar) to the last exterior one (monteverdic), then it seems that the intervalic energy at the gaudinar structure level should be greater than the intervalic energy at the lisztinian level. This only can be possible if the electric charges of the constituent subparticles are lesser in the gaudinar level than in the lisztinian level. This constraint eliminates the first two intervalic structures of both quarks. Therefore the last two intervalic structures would be the basic states of the nucleonic quarks.

3) We have the partial balance —attraction-repulsion— of the dynamic electromagnetic interaction between the constituent dalinos due to their different charge signs. It is clear that the most stable isogaudinos will be those with the minimal energy. These are by order:

$$G_6^{\pm 1/3} = U[4(D_{\pm 45}) + 2(D_{\pm 45})] = [(7/15)-(8/15)] \; 6 \; U(D_{45}) = -(6/15) \; U(D_{45})$$

$$G_6^0 = U[3(D_{+45}) + 3(D_{-45})] = [(6/15)-(9/15)] \; 6 \; U(D_{45}) = -(6/5) \; U(D_{45})$$

$$G_6^{\pm 2/3} = U[1(D_{\pm 45}) + 5(D_{\pm 45})] = [(10/15)-(5/15)] \; 6 \; U(D_{45}) = +2 \; U(D_{45})$$

$$G_6^{\pm} = U[6(D_{\pm 45}) + 0(D_{\pm 45})] = +6 \; U(D_{45})$$

This constraint clearly eliminates the presence of isogaudinos with elementary charge in the intervalic structure of quarks. However the zero charged gaudino is the second state with lesser energy.

4) We have likewise the sum of this balance of the dynamic electromagnetic interaction between the constituent dalinos at the lisztinian level. The isoquarks with the minimal energy are by order:

$$q(+2/3) = G_6^0 + G_6^{+2/3} + G_6^0 = -(2/5) \; U(D_{45})$$
$$q(-1/3) = G_6^{+1/3} + G_6^{-2/3} + G_6^0 = +(2/5) \; U(D_{45})$$

$$q(+\tfrac{2}{3}) = G_6^{+\tfrac{2}{3}} + G_6^{-\tfrac{1}{3}} + G_6^{+\tfrac{1}{3}} = -(6/5)\,U(D_{45})$$
$$q(-\tfrac{1}{3}) = G_6^{-\tfrac{1}{3}} + G_6^{+\tfrac{1}{3}} + G_6^{-\tfrac{1}{3}} = +(6/5)\,U(D_{45})$$
$$q(+\tfrac{2}{3}) = G_6^{+\tfrac{1}{3}} + G_6^{0} + G_6^{+\tfrac{1}{3}} = -2\,U(D_{45})$$
$$q(-\tfrac{1}{3}) = G_6^{0} + G_6^{-\tfrac{1}{3}} + G_6^{0} = -(14/5)\,U(D_{45})$$

5) Adding the intervalic energy of isogaudinos to the above dynamic electromagnetic interaction between dalinos will give us a wider and stronger constraint. The sum of those energies corresponding to the full set of nucleonic isoquarks can be seen in the next two tables. Its is clear that only the states written in the three last rows of each table can have got any chance to be chosen by Nature to compose nucleonic quarks because the remaining ones have much greater energies and can be discarded at first sight.

6) Since the structural ratio between the intervalic to the electromagnetic energy in nucleonic quarks is roughly: $I/U = \tfrac{2}{3}$, we can weigh up the value of the electromagnetic energy in that way. Then the value of the sum of $(I + \tfrac{2}{3}U)$ for each isoquark will by roughly, with the magnitudes ordered by minimal energy:

$$q(-\tfrac{1}{3}) = G_6^{+\tfrac{1}{3}} + G_6^{-\tfrac{2}{3}} + G_6^{0} = -0.04583$$
$$q(+\tfrac{2}{3}) = G_6^{+\tfrac{2}{3}} + G_6^{-\tfrac{1}{3}} + G_6^{+\tfrac{1}{3}} = +0.2375$$
$$q(+\tfrac{2}{3}) = G_6^{0} + G_6^{+\tfrac{2}{3}} + G_6^{0} = -0.3292$$
$$q(-\tfrac{1}{3}) = G_6^{-\tfrac{1}{3}} + G_6^{+\tfrac{1}{3}} + G_6^{-\tfrac{1}{3}} = -1.5500$$
$$q(+\tfrac{2}{3}) = G_6^{+\tfrac{1}{3}} + G_6^{0} + G_6^{+\tfrac{1}{3}} = -1.8333$$
$$q(-\tfrac{1}{3}) = G_6^{0} + G_6^{-\tfrac{1}{3}} + G_6^{0} = -2.1167$$

7) It can be supposed that states containing a gaudino 6 with zero charge would make unstable structures because they do not have intervalic nor electromagnetic energy at the last structure level, and therefore these states would not be allowed. This powerful constraint eliminates four possible combinations from the above list and the two unique remaining ones would be:

NUCLEONIC ISOQUARK L3D45$^{(+2/3)}$: STRUCTURAL BALANCE BETWEEN INTERVALIC AND ELECTROMAGNETIC ENERGIES

Constituent gaudinos / Attr./repuls. energy	G_6^+	$G_6^{+2/3}$	$G_6^{+1/3}$	G_6^0	$G_6^{-1/3}$	$G_6^{-2/3}$	G_6^-
$-(17/144)\,I(D_{45}) +14\,U(D_{45})$	$-(1/36)\,I(D_{45}) +6\,U(D_{45})$	$-(1/16)\,I(D_{45}) +2\,U(D_{45})$					$-(1/36)\,I(D_{45}) +6\,U(D_{45})$
$-(49/144)\,I(D_{45}) +(38/5)\,U(D_{45})$	$-(1/36)\,I(D_{45}) +6\,U(D_{45})$		$-(1/4)\,I(D_{45}) -(2/5)\,U(D_{45})$			$-(1/16)\,I(D_{45}) +2\,U(D_{45})$	
$-(3/16)\,I(D_{45}) +6\,U(D_{45})$		$-(1/8)\,I(D_{45}) +4\,U(D_{45})$				$-(1/16)\,I(D_{45}) +2\,U(D_{45})$	
$-(5/18)\,I(D_{45}) +(22/5)\,U(D_{45})$	$-(1/36)\,I(D_{45}) +6\,U(D_{45})$			0 $-(6/5)\,U(D_{45})$	$-(1/4)\,I(D_{45}) -(2/5)\,U(D_{45})$		
$-(9/16)\,I(D_{45}) +(6/5)\,U(D_{45})$		$-(1/16)\,I(D_{45}) +2\,U(D_{45})$	$-(1/4)\,I(D_{45}) -(2/5)\,U(D_{45})$		$-(1/4)\,I(D_{45}) -(2/5)\,U(D_{45})$		
$-(1/16)\,I(D_{45}) -(2/5)\,U(D_{45})$		$-(1/16)\,I(D_{45}) +2\,U(D_{45})$		$0 + 0$ $-(12/5)\,U(D_{45})$			
$-(1/2)\,I(D_{45}) -2\,U(D_{45})$			$-(1/2)\,I(D_{45}) -(4/5)\,U(D_{45})$	0 $-(6/5)\,U(D_{45})$			

INTERVALIC THEORY:
The Intervalic Structures of Subatomic Particles and the Last Foundations of Physics

NUCLEONIC ISOQUARK L3D45$^{(-1/3)}$: STRUCTURAL BALANCE BETWEEN INTERVALIC AND ELECTROMAGNETIC ENERGIES

Constituent gaudinos \\ Attr./repuls. energy	G_6^+	$G_6^{+2/3}$	$G_6^{+1/3}$	G_6^0	$G_6^{-1/3}$	$G_6^{-2/3}$	G_6^-
$-(11/36) I(D_{45}) +(58/5) U(D_{45})$	$-(1/36) I(D_{45}) +6 U(D_{45})$				$-(1/4) I(D_{45}) -(2/5) U(D_{45})$		$-(1/36) I(D_{45}) +6 U(D_{45})$
$-(11/72) I(D_{45}) +10 U(D_{45})$	$-(1/36) I(D_{45}) +6 U(D_{45})$						
$-(13/144) I(D_{45}) +(34/5) U(D_{45})$		$-(1/16) I(D_{45}) +2 U(D_{45})$				$-(1/8) I(D_{45}) +4 U(D_{45})$	
$-(19/36) I(D_{45}) +(26/5) U(D_{45})$			$-(1/2) I(D_{45}) -(4/5) U(D_{45})$	0 $-(6/5) U(D_{45})$			$-(1/36) I(D_{45}) +6 U(D_{45})$
$-(3/8) I(D_{45}) +(18/5) U(D_{45})$		$-(1/16) I(D_{45}) +2 U(D_{45})$			$-(1/4) I(D_{45}) -(2/5) U(D_{45})$	$-(1/16) I(D_{45}) +2 U(D_{45})$	$-(1/36) I(D_{45}) +6 U(D_{45})$
$-(5/16) I(D_{45}) +(2/5) U(D_{45})$			$-(1/4) I(D_{45}) -(2/5) U(D_{45})$	0 $-(6/5) U(D_{45})$		$-(1/16) I(D_{45}) +2 U(D_{45})$	
$-(3/4) I(D_{45}) -(6/5) U(D_{45})$			$-(1/4) I(D_{45}) -(2/5) U(D_{45})$		$-(1/2) I(D_{45}) -(4/5) U(D_{45})$		
$-(1/4) I(D_{45}) -(14/5) U(D_{45})$				$0 + 0$ $-(12/5) U(D_{45})$	$-(1/4) I(D_{45}) -(2/5) U(D_{45})$		

$$q(+\tfrac{2}{3}) = G_6^{+\tfrac{2}{3}} + G_6^{-\tfrac{1}{3}} + G_6^{+\tfrac{1}{3}}$$
$$q(-\tfrac{1}{3}) = G_6^{-\tfrac{1}{3}} + G_6^{+\tfrac{1}{3}} + G_6^{-\tfrac{1}{3}}$$

These states can already be postulated to be those that compose the gaudinar structure of nucleonic quarks.

8) The preceding gaudinar structure has the remarkable feature that there is an equality in the values of the dynamical electromagnetic interaction between gaudinos inside the two nucleonic quarks:

$$E_q(u)_G = E_q(G_6^{+\tfrac{2}{3}}G_6^{+\tfrac{1}{3}}G_6^{-\tfrac{1}{3}}) = [(\tfrac{2}{3} \cdot \tfrac{1}{3}) - (\tfrac{2}{3} \cdot \tfrac{1}{3}) - (\tfrac{1}{3})^2]\,(1/4\pi\varepsilon_0)\,e^2/d_{G6}$$
$$= -(1/9)\,(1/4\pi\varepsilon_0)\,e^2/d_{G6} = -1.544278296 \cdot 10^{-13}\,(J) = -0.96386228\,(MeV/c^2)$$

$$E_q(d)_G = E_q(G_6^{-\tfrac{1}{3}}G_6^{+\tfrac{1}{3}}G_6^{-\tfrac{1}{3}}) = [-(\tfrac{1}{3})^2 + (\tfrac{1}{3})^2 - (\tfrac{1}{3})^2]\,(1/4\pi\varepsilon_0)\,e^2/d_{G6} =$$
$$= -(1/9)\,(1/4\pi\varepsilon_0)\,e^2/d_{G6} = -1.544278296 \cdot 10^{-13}\,(J) = -0.96386228\,(MeV/c^2)$$

The minus sign indicates, according to conventional use, that there is an attractive frequorce between the three gaudinos due to the electromagnetic interaction inside quarks up and down. From this chain of compelling results, we can postulate that the *intervalic structure of nucleonic quarks* is:

$$q \to L_3 = 3\,G_6 = 18\,D_{45} = 810\,I$$

$$u^{+\tfrac{2}{3}} \to L_3 = (G_6^{+\tfrac{2}{3}}, G_6^{-\tfrac{1}{3}}, G_6^{+\tfrac{1}{3}}) = 18\,D_{45} = 810\,I$$
$$d^{-\tfrac{1}{3}} \to L_3 = (G_6^{-\tfrac{1}{3}}, G_6^{+\tfrac{1}{3}}, G_6^{-\tfrac{1}{3}}) = 18\,D_{45} = 810\,I$$

This result will be reconfirmed later when describing the intervalic structure of nucleons and their constituent quarks.

Chapter 13

INTERVALIC LISZTINO

LISZTINO INTERVALIC STRUCTURE

Lisztino is defined as the intervalic structure corresponding to the 4th. intervalic level of compositeness or structurefulness, which is made from an assembly of gaudinos.

Lisztinos can be divided among two main sets with different symmetry under interchange of its constituents gaudinos: the symmetric state yield elementary lisztinos, and the antisymmetric state yield fractional charged lisztinos. The phenomenology of both states is represented principally by zero charged massive bosons and quarks respectively.

Recapitulating: we have defined the *gaudino,* G_n, as an intervalic particle composed by the aggregation of a determine number of dalinos. The subindex n means the number of constituent dalinos. As has been described in other site, there are two basic kinds of gaudinos: the *elementary* ones composed by 270 intervalinos, and the *fractional* ones, which are composed by a number of intervalinos which is a fraction of 270. We define likewise the *lisztino* as an assembly of gaudinos. The term lisztino has been taken in honour of the great Hungarian romantic musician Franz Liszt (1811-1886). Lisztinos can be divided among several kinds according to the number of constituent intervalinos:

- *Lisztinos ⅓*: $L_{⅓}$, composed by ⅓ gaudinos (90 intervalinos)
- *Lisztinos ⅔*: $L_{⅔}$, composed by ⅔ gaudinos (180 intervalinos)
- *Lisztinos 1*: L_1, composed by 1 gaudino (270 intervalinos)
- *Lisztinos 2*: L_2, composed by 2 gaudinos (540 intervalinos)
- *Lisztinos 3*: L_3, composed by 3 gaudinos (810 intervalinos)
- *Lisztinos 4*: L_4, composed by 4 gaudinos (1080 intervalinos)
- *Lisztinos 5*: L_5, composed by 5 gaudinos (1350 intervalinos)
- *Lisztinos n*: L_n, composed by n gaudinos

Lisztinos with resultant charge ⅔ or ⅓ of the elementary one are named *fractional lisztinos*, while lisztinos with charge 1 are named *elementary lisztinos*. The constituent gaudinos of fractional lisztinos are in an antisymmetric state under interchange and make the *quarks*, whilst the constituent gaudinos of elementary lisztinos are in a symmetric state under interchange and make the *zero charged massive bosons*.

The detailed intervalic structures of quarks will be described in the chapter devoted to the intervalic quark. Now we are going to give a general view on the lisztinian structures allowed by the intervalic symmetries of compositeness, as the lisztinian structure is perhaps the most complex level of the intervalic structure of subatomic particles. Thus, we are going to list in a systematic way the intervalic structures and masses yielded for all these sets of lisztinos without further commentary, taking the electron's ratio between the intervalic and electromagnetic constituent energies ($I/U \approx 5/4$). Among lisztinos ⅓ and ⅔ there are some intervalic structures which are not allowed because they do not match with the number of constituent intervalinos of the corresponding dalinar structure, or because the gaudinar and dalinar structures are not commeasurable.

LISZTINOS ⅓

$L_{1/3} = \frac{1}{3} G_1 = (1/3) D_{270}$, not allowed
$L_{1/3} = \frac{1}{3} G_2 = (2/3) D_{135}$, not allowed
$L_{1/3} = \frac{1}{3} G_3 = 1 D_{90} = 4.5989918 \ (MeV/c^2)$
$L_{1/3} = \frac{1}{3} G_5 = (5/3) D_{54}$, not allowed
$L_{1/3} = \frac{1}{3} G_6 = 2 D_{45} = 36.791934 \ (MeV/c^2)$
$L_{1/3} = \frac{1}{3} G_9 = 3 D_{30} = 124.17278 \ (MeV/c^2)$
$L_{1/3} = \frac{1}{3} G_{10} = (10/3) D_{27}$, not allowed
$L_{1/3} = \frac{1}{3} G_{15} = 5 D_{18} = 574.87400 \ (MeV/c^2)$
$L_{1/3} = \frac{1}{3} G_{18} = 6 D_{15} = 993.38220 \ (MeV/c^2)$
$L_{1/3} = \frac{1}{3} G_{27} = 9 D_{10} = 3,352.6650 \ (MeV/c^2)$
$L_{1/3} = \frac{1}{3} G_{30} = 10 D_9 = 4,598.9918 \ (MeV/c^2)$
$L_{1/3} = \frac{1}{3} G_{45} = 15 D_6 = 15,521.598 \ (MeV/c^2)$
$L_{1/3} = \frac{1}{3} G_{54} = 18 D_5 = 26,821.321 \ (MeV/c^2)$
$L_{1/3} = \frac{1}{3} G_{90} = 30 D_3 = 124,172.78 \ (MeV/c^2)$
$L_{1/3} = \frac{1}{3} G_{135} = 45 D_2 = 419,083.13 \ (MeV/c^2)$
$L_{1/3} = \frac{1}{3} G_{270} = 90 D_1 = 1,873,748.2 \ (MeV/c^2)$

LISZTINOS ⅔

$L_{2/3} = \frac{2}{3} G_1 = (2/3) D_{270}$, not allowed
$L_{2/3} = \frac{2}{3} G_2 = (4/3) D_{135}$, not allowed
$L_{2/3} = \frac{2}{3} G_3 = 2 D_{90} = 9.1979836 \ (MeV/c^2)$
$L_{2/3} = \frac{2}{3} G_5 = (10/3) D_{54}$, not allowed
$L_{2/3} = \frac{2}{3} G_6 = 4 D_{45} = 73.583868 \ (MeV/c^2)$
$L_{2/3} = \frac{2}{3} G_9 = 6 D_{30} = 248.34556 \ (MeV/c^2)$
$L_{2/3} = \frac{2}{3} G_{10} = (20/3) D_{27}$, not allowed
$L_{2/3} = \frac{2}{3} G_{15} = 10 D_{18} = 1,149.7480 \ (MeV/c^2)$

$L_{2/3} = 2/3 \, G_{18} = 12 \, D_{15} = 1,986.7644 \, (MeV/c^2)$

$L_{2/3} = 2/3 \, G_{27} = 18 \, D_{10} = 6,705.3299 \, (MeV/c^2)$

$L_{2/3} = 2/3 \, G_{30} = 20 \, D_9 = 9,197.9836 \, (MeV/c^2)$

$L_{2/3} = 2/3 \, G_{45} = 30 \, D_6 = 31,043.196 \, (MeV/c^2)$

$L_{2/3} = 2/3 \, G_{54} = 36 \, D_5 = 53,642.642 \, (MeV/c^2)$

$L_{2/3} = 2/3 \, G_{90} = 60 \, D_3 = 248,345.56 \, (MeV/c^2)$

$L_{2/3} = 2/3 \, G_{135} = 90 \, D_2 = 838,166.26 \, (MeV/c^2)$

$L_{2/3} = 2/3 \, G_{270} = 180 \, D_1 = 3,747,496.3 \, (MeV/c^2)$

LISZTINOS 1

$L_1 = 1 \, G_1 = 1 \, D_{270} = 0.51099909 \, (MeV/c^2)$

$L_1 = 1 \, G_2 = 2 \, D_{135} = 4.0879928 \, (MeV/c^2)$

$L_1 = 1 \, G_3 = 3 \, D_{90} = 13.796975 \, (MeV/c^2)$

$L_1 = 1 \, G_5 = 5 \, D_{54} = 63.874885 \, (MeV/c^2)$

$L_1 = 1 \, G_6 = 6 \, D_{45} = 110.37580 \, (MeV/c^2)$

$L_1 = 1 \, G_9 = 9 \, D_{30} = 372.51833 \, (MeV/c^2)$

$L_1 = 1 \, G_{10} = 10 \, D_{27} = 510.99909 \, (MeV/c^2)$

$L_1 = 1 \, G_{15} = 15 \, D_{18} = 1,724.6220 \, (MeV/c^2)$

$L_1 = 1 \, G_{18} = 18 \, D_{15} = 2,980.1466 \, (MeV/c^2)$

$L_1 = 1 \, G_{27} = 27 \, D_{10} = 10,057.995 \, (MeV/c^2)$

$L_1 = 1 \, G_{30} = 30 \, D_9 = 13,796.975 \, (MeV/c^2)$

$L_1 = 1 \, G_{45} = 45 \, D_6 = 46,564.794 \, (MeV/c^2)$

$L_1 = 1 \, G_{54} = 54 \, D_5 = 80,463.964 \, (MeV/c^2)$

$L_1 = 1 \, G_{90} = 90 \, D_3 = 372,518.33 \, (MeV/c^2)$

$L_1 = 1 \, G_{135} = 135 \, D_2 = 1,257,249.4 \, (MeV/c^2)$

$L_1 = 1 \, G_{270} = 270 \, D_1 = 5,621,244.5 \, (MeV/c^2)$

INTERVALIC THEORY:
The Intervalic Structures of Subatomic Particles and the Last Foundations of Physics

LISZTINOS 2

$L_2 = 2\,G_1 = 2\,D_{270} = 1.0219982$ (MeV/c^2)
$L_2 = 2\,G_2 = 4\,D_{135} = 8.1759856$ (MeV/c^2)
$L_2 = 2\,G_3 = 6\,D_{90} = 27.593950$ (MeV/c^2)
$L_2 = 2\,G_5 = 10\,D_{54} = 127.74977$ (MeV/c^2)
$L_2 = 2\,G_6 = 12\,D_{45} = 220.75160$ (MeV/c^2)
$L_2 = 2\,G_9 = 18\,D_{30} = 745.03666$ (MeV/c^2)
$L_2 = 2\,G_{10} = 20\,D_{27} = 1{,}021.9982$ (MeV/c^2)
$L_2 = 2\,G_{15} = 30\,D_{18} = 3{,}449.2440$ (MeV/c^2)
$L_2 = 2\,G_{18} = 36\,D_{15} = 5{,}960.2932$ (MeV/c^2)
$L_2 = 2\,G_{27} = 54\,D_{10} = 20{,}115.990$ (MeV/c^2)
$L_2 = 2\,G_{30} = 60\,D_9 = 27{,}593.950$ (MeV/c^2)
$L_2 = 2\,G_{45} = 90\,D_6 = 93{,}129.588$ (MeV/c^2)
$L_2 = 2\,G_{54} = 108\,D_5 = 160{,}927.93$ (MeV/c^2)
$L_2 = 2\,G_{90} = 180\,D_3 = 745{,}036.66$ (MeV/c^2)
$L_2 = 2\,G_{135} = 270\,D_2 = 2{,}514{,}498.8$ (MeV/c^2)
$L_2 = 2\,G_{270} = 540\,D_1 = 11{,}242{,}489.0$ (MeV/c^2)

LISZTINOS 3

$L_3 = 3\,G_1 = 3\,D_{270} = 1.5329973$ (MeV/c^2)
$L_3 = 3\,G_2 = 6\,D_{135} = 12.263978$ (MeV/c^2)
$L_3 = 3\,G_3 = 9\,D_{90} = 41.390925$ (MeV/c^2)
$L_3 = 3\,G_5 = 15\,D_{54} = 191.62466$ (MeV/c^2)
$L_3 = 3\,G_6 = 18\,D_{45} = 331.12740$ (MeV/c^2)
$L_3 = 3\,G_9 = 27\,D_{30} = 1{,}117.5550$ (MeV/c^2)
$L_3 = 3\,G_{10} = 30\,D_{27} = 1{,}532.9973$ (MeV/c^2)
$L_3 = 3\,G_{15} = 45\,D_{18} = 5{,}173.8660$ (MeV/c^2)
$L_3 = 3\,G_{18} = 54\,D_{15} = 8{,}940.4398$ (MeV/c^2)
$L_3 = 3\,G_{27} = 81\,D_{10} = 30{,}173.985$ (MeV/c^2)
$L_3 = 3\,G_{30} = 90\,D_9 = 41{,}390.925$ (MeV/c^2)

$L_3 = 3\ G_{45} = 135\ D_6 = 139{,}694.38\ (MeV/c^2)$
$L_3 = 3\ G_{54} = 162\ D_5 = 241{,}391.89\ (MeV/c^2)$
$L_3 = 3\ G_{90} = 270\ D_3 = 1{,}117{,}555.0\ (MeV/c^2)$
$L_3 = 3\ G_{135} = 405\ D_2 = 3{,}771{,}748.2\ (MeV/c^2)$
$L_3 = 3\ G_{270} = 810\ D_1 = 16{,}863.734\ (GeV/c^2)$

LISZTINOS 4

$L_4 = 4\ G_1 = 4\ D_{270} = 2.0439964\ (MeV/c^2)$
$L_4 = 4\ G_2 = 8\ D_{135} = 16.3519712\ (MeV/c^2)$
$L_4 = 4\ G_3 = 12\ D_{90} = 55.187900\ (MeV/c^2)$
$L_4 = 4\ G_5 = 20\ D_{54} = 255.49954\ (MeV/c^2)$
$L_4 = 4\ G_6 = 24\ D_{45} = 441.50320\ (MeV/c^2)$
$L_4 = 4\ G_9 = 36\ D_{30} = 1{,}490.0733\ (MeV/c^2)$
$L_4 = 4\ G_{10} = 40\ D_{27} = 2{,}043.9964\ (MeV/c^2)$
$L_4 = 4\ G_{15} = 60\ D_{18} = 6{,}898.4880\ (MeV/c^2)$
$L_4 = 4\ G_{18} = 72\ D_{15} = 11{,}920.5864\ (MeV/c^2)$
$L_4 = 4\ G_{27} = 108\ D_{10} = 40{,}231.980\ (MeV/c^2)$
$L_4 = 4\ G_{30} = 120\ D_9 = 55{,}187.90\ (MeV/c^2)$
$L_4 = 4\ G_{45} = 180\ D_6 = 186{,}259.17\ (MeV/c^2)$
$L_4 = 4\ G_{54} = 216\ D_5 = 321{,}855.85\ (MeV/c^2)$
$L_4 = 4\ G_{90} = 360\ D_3 = 1{,}490{,}073.32\ (MeV/c^2)$
$L_4 = 4\ G_{135} = 540\ D_2 = 5{,}028{,}997.6\ (MeV/c^2)$
$L_4 = 4\ G_{270} = 1080\ D_1 = 22{,}484.978\ (GeV/c^2)$

LISZTINOS 5

$L_5 = 5\ G_1 = 5\ D_{270} = 2.5549955\ (MeV/c^2)$
$L_5 = 5\ G_2 = 10\ D_{135} = 20.439964\ (MeV/c^2)$
$L_5 = 5\ G_3 = 15\ D_{90} = 68.984875\ (MeV/c^2)$

$L_5 = 5\ G_5 = 25\ D_{54} = 319.37442\ (MeV/c^2)$
$L_5 = 5\ G_6 = 30\ D_{45} = 551.87900\ (MeV/c^2)$
$L_5 = 5\ G_9 = 45\ D_{30} = 1,862.5916\ (MeV/c^2)$
$L_5 = 5\ G_{10} = 50\ D_{27} = 2,554.9954\ (MeV/c^2)$
$L_5 = 5\ G_{15} = 75\ D_{18} = 8623.1100\ (MeV/c^2)$
$L_5 = 5\ G_{18} = 90\ D_{15} = 14,900.733\ (MeV/c^2)$
$L_5 = 5\ G_{27} = 135\ D_{10} = 50,289.974\ (MeV/c^2)$
$L_5 = 5\ G_{30} = 150\ D_9 = 68,984.877\ (MeV/c^2)$
$L_5 = 5\ G_{45} = 225\ D_6 = 232,823.97\ (MeV/c^2)$
$L_5 = 5\ G_{54} = 270\ D_5 = 402,319.82\ (MeV/c^2)$
$L_5 = 5\ G_{90} = 450\ D_3 = 1,862,591.7\ (MeV/c^2)$
$L_5 = 5\ G_{135} = 675\ D_2 = 6,286,247\ (MeV/c^2)$
$L_5 = 5\ G_{270} = 1350\ D_1 = 28,106.222\ (GeV/c^2)$

LISZTINOS 6

$L_6 = 6\ G_1 = 6\ D_{270} = 3.0659946\ (MeV/c^2)$
$L_6 = 6\ G_2 = 12\ D_{135} = 24.527956\ (MeV/c^2)$
$L_6 = 6\ G_3 = 18\ D_{90} = 82.781850\ (MeV/c^2)$
$L_6 = 6\ G_5 = 30\ D_{54} = 383.24932\ (MeV/c^2)$
$L_6 = 6\ G_6 = 36\ D_{45} = 662.2548\ (MeV/c^2)$
$L_6 = 6\ G_9 = 54\ D_{30} = 2,235.1099\ (MeV/c^2)$
$L_6 = 6\ G_{10} = 60\ D_{27} = 3,065.9946\ (MeV/c^2)$
$L_6 = 6\ G_{15} = 90\ D_{18} = 10,347.732\ (MeV/c^2)$
$L_6 = 6\ G_{18} = 108\ D_{15} = 17,880.8796\ (MeV/c^2)$
$L_6 = 6\ G_{27} = 162\ D_{10} = 60,347.97\ (MeV/c^2)$
$L_6 = 6\ G_{30} = 180\ D_9 = 82,781.85\ (MeV/c^2)$
$L_6 = 6\ G_{45} = 270\ D_6 = 279,388.76\ (MeV/c^2)$
$L_6 = 6\ G_{54} = 324\ D_5 = 482,783.78\ (MeV/c^2)$
$L_6 = 6\ G_{90} = 540\ D_3 = 2,235,110.0\ (MeV/c^2)$
$L_6 = 6\ G_{135} = 810\ D_2 = 7,543,496.4\ (MeV/c^2)$
$L_6 = 6\ G_{270} = 1620\ D_1 = 33,727.468\ (GeV/c^2)$

CLASSIFICATION OF SUBATOMIC PARTICLES ACCORDING TO ITS LISZTINIAN STRUCTURE

	Elementary charge: e	Fractional charge: ⅔ e, ⅓ e	Zero charge
Lisztinos ⅓ (90 intervalinos)		Quarks	-
Lisztinos ⅔ (180 intervalinos)		Quarks	-
Lisztinos 1 (270 intervalinos)	Leptons-charged massive bosons	Quarks	-
Lisztinos 2 (540 intervalinos)	Mesons	Quarks	Mesons, Zero charged massive bosons
Lisztinos 3 (810 intervalinos)	-	Quarks (principal sequence)	-
Lisztinos 4 (1080 intervalinos)	-	Quarks (principal sequence)	-
Lisztinos 5 (1350 intervalinos)	-	Quarks	-

SYMMETRIES OF NATURE

The vast majority of particles of Nature can be find among these set of intervalic symmetries. And only those ones showing the highest symmetries appear to have been chosen by Nature to make subatomic particles. Indeed, all stable particles pertains only to two families of symmetries: {D270} or {D45}. Quarks are mainly focused on the same families of symmetries as leptons-intermediate massive bosons. Therefore there is a noticeable parallelism among those families, as it has been *supposed or postulated by hand* in SM, but in a very different way since they are involved the intervalic structures of particles.

Till now Nature appears to chose only the symmetries derived from a select number chosen among the 16 divisors of the elementary charge, $270q_I$. It is remarkable that this is a fact realized among an *a priori* infinite number of possibilities. Moreover, among those 16 allowed symmetries there have lasted only a few of them at low energies: in first place $\{D_{270}\}$ and $\{D_{45}\}$; and in second place $\{D_{30}\}$, $\{D_{18}\}$, $\{D_6\}$ and $\{D_5\}$.

As can be seen in the table near this lines, all quarks are lisztinos with fractional charge. Leptons are lisztinos 1. Intermediate massive bosons of weak intervalic interaction are lisztinos 1 (those with elementary charge) or lisztinos 2 (those with zero charge). And finally, the vast majority of mesons have a paired lisztinian structure.

The last level of the intervalic structure is the aggregation of lisztinos which is named *monteverdino*, term chosen in honour of the great Italian musician Claudio Monteverdi (1567-1643).

INTERVALIC STRUCTURE OF ZERO CHARGED MASSIVE BOSONS

While the intervalic structure of all leptons-elementary charged massive bosons is the one of *lisztino 1* —that is to say, they are *gaudinos*—, all zero charged massive bosons have an intervalic structure of *lisztino*

2 —they are composed by a gaudino and its corresponding antigaudino—. The full set of intervalic structures of lisztinos 2 allowed by the intervalic symmetries are just the 16 ones —{L2}— described above in this chapter. Please remember that those masses are yielded taking the electron intervalic to electromagnetic energy ratio for all of them, then the slight deviation from experimental results. On the contrary, it is surprising that this ratio does not vary considerably, since there are lot of very different intervalic structures involved in all leptons and massive bosons. In this set {L2} we find the zero charged massive bosons, which masses in MeV/c² are (taking the Z^0 boson's structural energy ratio):

$L_2 = 2\,G_{45} = 2 \cdot (45\,D_6) = 91{,}187.6 \rightarrow Z^0$ boson
$L_2 = 2\,G_{54} = 2 \cdot (54\,D_5) = 157{,}572.17 \rightarrow W^0$ boson
$L_2 = 2\,G_{90} = 2 \cdot (90\,D_3) = 729{,}500.76 \rightarrow Y^0$ boson
$L_2 = 2\,G_{135} = 2 \cdot (135\,D_2) = 2{,}462{,}065.12 \rightarrow X^0$ boson

Every zero charged massive bosons is composed by an elementary charged massive boson and its corresponding antiboson. Thus, theoretically we could also find here the possible particles composed by a lepton and its corresponding antilepton, which following the usual convention should be named *electronium, muonium* and *tauonium*:

$L_2 = 2\,G_1 = 2 \cdot (1\,D_{270}) = 1.0219982 \rightarrow e_2^0$ electronium
$L_2 = 2\,G_6 = 2 \cdot (6\,D_{45}) = 220.75160 \rightarrow \mu_2^0$ muonium
$L_2 = 2\,G_{15} = 2 \cdot (15\,D_{18}) = 3{,}449.2440 \rightarrow \tau_2^0$ tauonium

However, it seems that Nature does not allow the existence of a dalinar symmetry in both lisztino 1 and lisztino 2 phenomenology. So the leptons-charged massive bosons will not have got their corresponding lisztinos 2 as zero charged massive bosons, and in a similar way, the zero charged massive bosons will not have got their corresponding lisztinos 1 as leptons-charged massive bosons. Up to present we have not found experimentally any exception to this rule. Therefore, in this case the following allowed leptons-massive bosons will not be made at low energies:

$$L_1 = 1\ G_{45} = 45\ D_6 = 46{,}564.794\ (MeV/c^2) \rightarrow Z^{\pm}\ \text{boson}$$
$$L_2 = 2\ G_{54} = 2 \cdot (54\ D_5) = 157{,}572.17\ (MeV/c^2) \rightarrow W^0\ \text{boson}$$

And the heaviest ones will be made only as lisztino 1 or as lisztino 2, but not both states:

$$L_1 = G_{90} = 90\ D_3 = 372{,}518.33 \rightarrow Y^{\pm}\ \text{boson or}$$
$$L_2 = 2\ G_{90} = 2 \cdot (90\ D_3) = 729{,}500.76 \rightarrow Y^0\ \text{boson, and}$$

$$L_1 = G_{135} = 135\ D_2 = 1{,}257{,}249.4 \rightarrow X^{\pm}\ \text{boson or}$$
$$L_2 = 2\ G_{135} = 2 \cdot (135\ D_2) = 2{,}462{,}065.12 \rightarrow X^0\ \text{boson}$$

We will go on these heaviest massive bosons later when explaining the intervalic changeful —weak— interaction.

As can be seen, IT explains easily why Z^0 boson shows a composite lisztinian structure while $W^{\pm}$ boson have a single lisztinian structure. This result has been empirically deduced in SM from the experimental decay of Z^0 but in that mischievous model there is no a satisfactory theoretical explanation of that feature.

Moreover, SM does not have any basic link between leptons and massive bosons —unless a clumsy sum of Lagrangian densities is viewed as a basic link— although all of them are involved in the weak interaction. According to IT, massive bosons pertain to the same intervalic set and have the same intervalic structure than the family of leptons. This fact can be now easily explained not as a new and perhaps strange family of particles, as it is showed in SM, but as an intermediate state involved in a *change in the structure* of lepton (and other subatomic particles).

Needless to say that the above particles are *logical and directly yielded* starting from a mere change in the system of units and dimensions used —really the only reliable *natural* ones—, but not through a Lagrangian formalism. It is hard to believe that all (and more) results reached by SM through complex and sometimes doubtful mechanisms can be deduced in a much more reliable way by IT using a simple formalism and only starting from new fundamental physical principles. All systems of units and dimensions was supposed to be equivalent in Physics, but clearly they are not since they involves different symmetries of Nature.

AN INTRIGUING COINCIDENCE

As we will understand some chapters later, the most logical result which could be expected from the collision e^+e^- should be all pairs dalino-antidalino allowed by the intervalic structure, and it is sure that above some threshold temperature —which will be defined later— existing in the primordial Universe, this is roughly what happened:

$$D_{+270}D_{-270} \quad 2\,D_{+135}D_{-135} \quad 3\,D_{+90}D_{-90} \quad 5\,D_{+54}D_{-54}$$
$$6\,D_{+45}D_{-45} \quad 9\,D_{+30}D_{-30} \quad 10\,D_{+27}D_{-27} \quad 15\,D_{+18}D_{-18}$$
$$18\,D_{+15}D_{-15} \quad 27\,D_{+10}D_{-10} \quad 30\,D_{+9}D_{-9} \quad 45\,D_{+6}D_{-6}$$
$$54\,D_{+5}D_{-5} \quad 90\,D_{+3}D_{-3} \quad 135\,D_{+2}D_{-2} \quad 270\,D_{+1}D_{-1}$$

Obviously here we have the pairs of leptons-massive bosons:

$D_{+270}D_{-270} = e^+e^-$
$6\,D_{+45}D_{-45} = \mu^+\mu^-$
$15\,D_{+18}D_{-18} = \tau^+\tau^-$
$45\,D_{+6}D_{-6} = Z^+Z^-\ (Z^0)$
$54\,D_{+5}D_{-5} = W^+W^-\ (W^0)$
$90\,D_{+3}D_{-3} = Y^+Y^-\ (Y^0)$
$135\,D_{+2}D_{-2} = X^+X^-\ (X^0)$

But in addition to the Z^0 bosons and the remaining heavy massive bosons, taking a look at the main peaks which appears in the well known graphic of the annihilation cross section ratio ($e^+e^- \rightarrow$ hadrons / $e^+e^- \rightarrow \mu^+\mu^-$) we find the extraordinary coincidences (with masses expressed according to electron's structural energy ratio):

$L_2 = 2\, G_5 = 2 \cdot (5\, D_{54}) = 127.74977 \to \pi^0$ meson (135)
$L_2 = 2\, G_9 = 2 \cdot (9\, D_{30}) = 745.03666 \to \rho, \omega$ mesons (770, 783)
$L_2 = 2\, G_{10} = 2 \cdot (10\, D_{27}) = 1{,}021.9982 \to \Phi$ meson (1,020)
$L_2 = 2\, G_{15} = 2 \cdot (15\, D_{18}) = 3{,}449.2440 \to X_0$ meson (3,415)

This means that the principal peaks of the graphic of the annihilation cross section ratio ($e^+e^- \to$ hadrons / $e^+e^- \to \mu^+\mu^-$) can be simply explained as assembled pairs of elementary gaudinos, having got the intervalic structure of lisztino 2. Such simplicity is quite disconcerting since the last four mesons are fully explained inside IT as having got a quarkic structure. Therefore, the existence of another alternative intervalic structure for these mesons which also fit reasonably well with the experimental data is an intriguing coincidence which is probably meaningless, but that could don't be so.

The challenge is: why should be made in an e^+e^- annihilation both pairs of leptons-massive bosons —which are *lisztinos 2*— and some determined pairs of quarks —just those which are *lisztinos 3 and 4* (as we will see when describing the intervalic principal sequence of quarks)—?

By now we will maintain the quarkic structure for these mesons, but keeping in mind the existence of that alternative structure which could become meaningful.

THE INTERMEDIATE MASSIVE BOSON FAMILY

It is clear that the intervalic structure allows to do a lot of predictions about unknown subatomic particles. We can reasonably think that the minimal threshold energy for the changeful —weak— intervalic interaction is that corresponding to usual massive bosons, since in other case, lesser massive bosons would have been detected in laboratory. On the other hand, there is no reason to believe that it does not exists more massive bosons beyond the actually known. In this way, if they exist, they will necessary be the next allowed lisztinos 1 and 2 starting from the last known. The first of them is the already viewed W^0 boson: $L_2 = 2\, G_{54} = 2 \cdot (54\, D_5) = 160{,}927.93$ (MeV/c^2). The next is the gaudino: $L_1 = G_{90} = 90\, D_3$ and its corresponding lisztino 2: $L_2 = 2\, G_{90} = 2 \cdot (90\, D_3)$, say respectively $Y^\pm$ and Y^0 bosons. And the following allowed massive bosons are only the two remaining intervalic structures to the end, say $X^\pm$ and X^0, and $I^\pm$ and I^0 bosons. This last pair of possible bosons is named *intervalon*, which are the most massive bosons allowed, although their existence is really not postulated. These bosons should not be confounded with the infamous supersymmetric bosons postulated by SUSY: two charged Higgs scalars ($H^\pm$), two neutral scalars (H_1^0, H_2^0) and one neutral pseudoscalar (H_3^0), which are entirely wrong since there is no such a fantastic thing as a Higgs field —really another medieval *ether*— which gives just its mass to all subatomic particles. Recapitulating:

$L_1 = G_{45} = 45\, D_6 = 46{,}564.794 \to Z^\pm$ boson
$L_2 = 2\, G_{45} = 2 \cdot (45\, D_6) = 93{,}129.588 \to Z^0$ boson
$L_1 = G_{54} = 54\, D_5 = 80{,}463.964 \to W^\pm$ boson
$L_2 = 2\, G_{54} = 2 \cdot (54\, D_5) = 160{,}927.93 \to W^0$ boson
$L_1 = G_{90} = 90\, D_3 = 372{,}518.33 \to Y^\pm$ boson
$L_2 = 2\, G_{90} = 2 \cdot (90\, D_3) = 745{,}036.66 \to Y^0$ boson
$L_1 = G_{135} = 135\, D_2 = 1{,}257{,}249.4 \to X^\pm$ boson
$L_2 = 2\, G_{135} = 2 \cdot (135\, D_2) = 2{,}514{,}498.8 \to X^0$ boson
$L_1 = G_{270} = 270\, D_1 = 5{,}621{,}244.5 \to I^\pm$ boson
$L_2 = 2\, G_{270} = 2 \cdot (270\, D_1) = 11{,}242{,}489 \to I^0$ boson

When we will describe later the changeful —weak— intervalic interaction we will give a detailed prediction of the massive bosons whose existence is not only allowed at the Big Bang, but it is postulated to be detected in our cold Universe.

It may be pointed out that W^0 pertains to the {D5} symmetry, the same one as $W^\pm$. However, $Y^\pm$ boson and the two following heavier ones will be the first particles yielded by the intervalic symmetries {D3}, {D2} and {D1}, since no one particle is known in those symmetries, as all of them can yield only intervalic structures with very big masses.

These precise predictions, given by the theory at first sight and without any physical development yet, makes that IT in Particle Physics becomes easily falsable or verifiable, since no one boson is allowed out of this set. Moreover, the magnitudes postulated for leptons-massive bosons follows a fully determined and finite intervalic sequence of really disparate magnitudes which hardly can be reached or yielded by other means. This makes a remarkable contrast with SM, which is not falsable as it can incorporate practically infinite number of bosons and particles of any mass and condition by means of adjusting by hand —or by chance— their several corresponding parameters. It is wonder that such a non falsable model could be formally considered upon a time a scientific theory.

As can be seem, to obtain a little bit of phenomenology in quantum field theory is a highly exceptional case (and in String Theory —or String Lucubration— appears to be forbidden); on the contrary, to obtain a lot of phenomenology in IT is the norm and a starting point, as it was supposed to be in any *scientific theory*.

Need not to say that the above particles are logical and directly yielded starting from a mere change in the system of units and dimensions used —really the only reliable natural ones—, but not through a Lagrangian formalism. It is hard to believe that all (and much more) results reached by SM through complex and sometimes doubtful mechanisms can be deduced in a much more reliable way by IT using a simple formalism and only starting from the intervalic symmetries and applying a very few fundamental physical principles —mainly the spin-statistics theorem and the laws derived from the intervalic energy equation—, but this is an undeniable fact. All systems of units and dimensions was supposed to be equivalent in Physics, but clearly they are not since they involves different symmetries of Nature.

Chapter 14

INTERVALIC ZERO CHARGED MASSIVE BOSON

INTERVALIC Z^0 BOSON

The intervalic structure of Z^0 boson is:

$$Z^0 \rightarrow L_2 = 2\, G_{45} = 90\, D_6 = 540\, I$$

INTERVALIC ENERGY OF Z^0 BOSON

The intervalic Z^0 boson mass is due to the contribution of the intervalic and electromagnetic energies. The first is:

At lisztinian level: $I(Z^0)_{L2} = 0$
At gaudinar level: $I(Z^0)_{G45} = 2\, (c^{\pm 2}\hbar\, (270 q_I)^{-2})$
At dalinar level: $I(Z^0)_{D6} = 90\, (c^{\pm 2}\hbar\, (6 q_I)^{-2})$

Therefore the total intervalic energy of Z^0 boson will be:

$$I(Z^0) = \Sigma(c^{\pm 2}\hbar Q^{-2}) = 2\, c^{\pm 2}\hbar\, e^{-2} + 90\, (c^{\pm 2}\hbar\, (6\mathbf{q_I})^{-2}) =$$
$$= 2 \cdot 270^{-2} c^{-1} + 90 \cdot 6^{-2} c^{-1} = 8.339193892 \cdot 10^{-9}\,(J) =$$
$$= 52{,}049.13174\,(MeV/c^2)$$

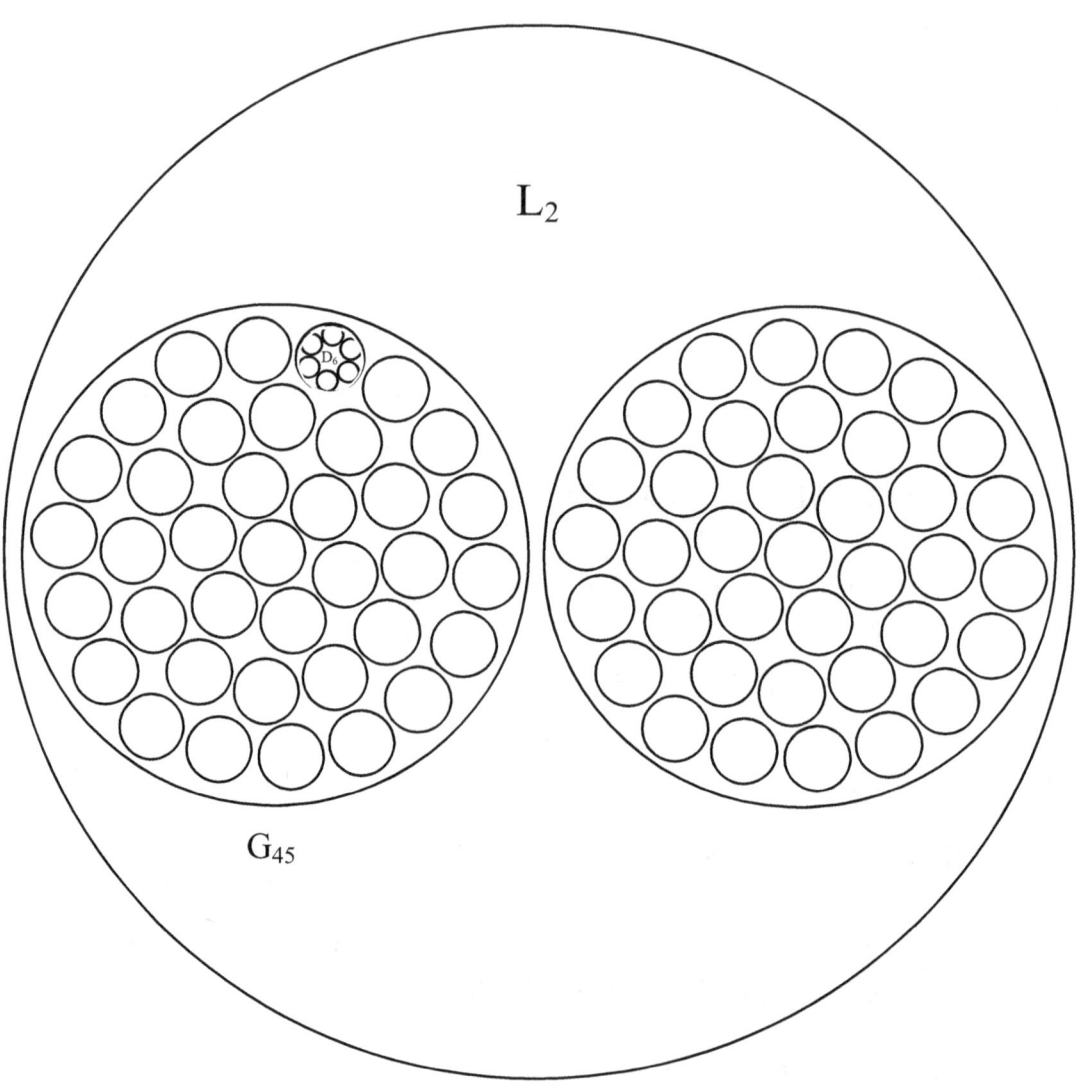

Figured intervalic structure of Z^0 boson:
$L_2 = 2\ G_{45} = 90\ D_6 = 540\ \mathbf{I}$

ELECTROMAGNETIC ENERGY OF Z^0 BOSON

The electromagnetic mass of Z^0 boson is the difference between the total mass, $1.460987054 \cdot 10^{-8}$ (J) = 91,187.60016 (MeV/c²), and the intervalic mass:

$U(Z^0) = E(Z^0)_{tot} - I(Z^0) = 6.270676648 \cdot 10^{-9}$ (J) = 39,138.4682 (MeV/c²)

The electromagnetic mass of the Z^0 boson at the last level of its intervalic structure —the lisztinian level— is zero because the electromagnetic energy at this level is not manifested as *mass* but as *field*. This field is not outwards Z^0 boson because it is zero charged, but inwards, between its two constituent gaudinos 45.

At lisztinian level: $U(Z^0)_{L2} = 0$
At gaudinar level: $U(Z^0)_{G45} = 2\,(1/4\pi\varepsilon_0)\,(270\,\mathbf{q_I})^2 / r_{G45}$
At dalinar level: $U(Z^0)_{D6} = 90\,(1/4\pi\varepsilon_0)\,(6\,\mathbf{q_I})^2 / r_{D6}$

We find the following structural energy ratios between intervalic, electromagnetic and mass energies:

$I(Z^0)/U(Z^0) = 1.330 \approx 4/3$
$I(Z^0)/E(Z^0)_{tot} = 0.571 \approx 4/7$
$U(Z^0)/E(Z^0)_{tot} = 0.429 \approx 3/7$

BOSONIC GAUDINO 45 INTERVALIC, ELECTROMAGNETIC AND SPIN ENERGIES

Since the intervalic and electromagnetic mass energies of Z are just zero at the lisztinian level, we can calculate easily the features of its constituent gaudinos 45. The intervalic and electromagnetic mass of each one of the two constituent gaudinos 45 are necessarily half of the total intervalic and electromagnetic masses of Z^0 boson respectively:

$$I(G_{45})_Z = \tfrac{1}{2} I(Z^0) = 4.169596946 \cdot 10^{-9} \,(J) = 26{,}024.56587 \,(MeV/c^2)$$

$$U(G_{45})_Z = \tfrac{1}{2} U(Z^0) = 3.135338326 \cdot 10^{-9} \,(J) = 19{,}569.23411 \,(MeV/c^2)$$

Its spin energy will be:

$$E_J(G_{45})_Z = I(G_{45})_Z - U(G_{45})_Z = 1.034258622 \cdot 10^{-9} \,(J) = 6{,}455.33177 \,(MeV/c^2)$$

BOSONIC GAUDINO 45 RADIUS

From the electromagnetic energy we can obtain the constituent gaudino 45 radius:

$$r_{G45} = \tfrac{1}{2} (1/4\pi\varepsilon_0)(270\, q_I)^2 / U(G_{45})_Z = 7.358311087 \cdot 10^{-20} \,(m)$$

BOSONIC GAUDINO 45 SPIN ENERGY

Since the intervalic and electromagnetic mass energies of Z are just zero at the lisztinian level, we can calculate easily the features of its constituent gaudinos 45. Its spin energy will be:

$$E_J(G_{45}) = \tfrac{1}{2} I(Z^0) - \tfrac{1}{2} U(Z^0) = 4.169596946 \cdot 10^{-9} - 3.135338326 \cdot 10^{-9} = 1.03425862 \cdot 10^{-9} \,(J) = 6{,}455.33176 \,(MeV/c^2)$$

BOSONIC DALINO 6 RADIUS

We can deduce likewise the magnitude of the constituent bosonic dalino 6 radius, r_{D6}:

$$r_{D6} = 90 \cdot \tfrac{1}{2} (1/4\pi\varepsilon_0) (6\,\mathbf{q}_I)^2 / U(Z^0) = 8.175901202 \cdot 10^{-22}\,(m)$$

This magnitude is in a similar way near the first approximation obtained via electron's ratio: $r(D_6) \approx 7.7892963 \cdot 10^{-22}\,(m)$.

Z^0 BOSON INTERVALIC EXCHANGE FREQUORCE

The overall exchange frequorce due to intervalic interaction among intervalinos inside dalino 6 will be:

$$\varphi(\mathbf{I})_{D6} = I(D_6) / \hbar = 8.777331086 \cdot 10^{23}\,(s^{-1})$$

The related maximum linear velocity of intervalinos inside dalino 6 will be, known the bosonic dalino 6 radius:

$$v(\mathbf{I})_{D6} = 2\pi\,r_{D6}\,\varphi(\mathbf{I})_{D6} = 4{,}508.976622\,(m\,s^{-1})$$

The overall exchange frequorce of dalinos 6 inside each one of the two constituent gaudinos G_{45} of the Z^0 boson will be:

$$\varphi(D_6)_{G45} = I(e) / \hbar = 4.334484487 \cdot 10^{20}\,(s^{-1})$$

And the corresponding maximum linear velocity of dalinos 6 inside each gaudino G_{45} of the Z^0 boson is, known the G_{45} radius:

$$v(D_6)_{G45} = 2\pi\,r_{G45}\,\varphi(D_6)_{G45} = 200.3989611\,(m\,s^{-1})$$

Z^0 BOSON SPIN ENERGY

The Z^0 boson spin energy, $E_j(Z^0)$, is the difference between intervalic and electromagnetic energies, which is *manifested* not as *mass* but as *spin*:

$$E_j(Z^0)_{GD} = I(Z^0)_{GD} - U(Z^0)_{GD} = 2.0684257 \cdot 10^{-9} \text{ (J)} = 12{,}910.09279 \text{ (MeV}/c^2)$$

However this result does not include the Z^0 boson spin energy at the exterior level, but only at the gaudinar and dalinar levels. As we have seen the contribution to *mass* of the intervalic and electromagnetic energies of Z^0 boson at lisztinian level is zero. Therefore the dynamic energies at the exterior level of Z^0 boson can not be calculated through the potential energy at lisztinian level because of the constituent unlike charges, but through the dynamic electromagnetic interaction between gaudinos 45, where the minus or plus sign of the magnitude only indicates the way — attraction or repulsion— of the related frequorce according to conventional use. The simplest assumption on the distance between the two constituent gaudinos 45 is $d_{G45} = 2\, r_{G45}$. Following this way we have got:

$$E_q(Z^0)_L = E_q(G_{45}^{(+)} G_{45}^{(-)})_G = [(+1) \cdot (-1)] \, (1/4\pi\varepsilon_0) \, e^2 / d_{G45} =$$
$$= -1.567669217 \cdot 10^{-9} \text{ (J)} = -9{,}784.617392 \text{ (MeV}/c^2)$$

Therefore the balance dynamical equation of Z^0 boson at lisztinian level becomes:

$$E_j(Z^0)_L + E_q(Z^0)_L = 0$$

And the total spin energy of Z^0 boson is:

$$\Sigma E_j(Z^0)_{LGD} = E_j(Z^0)_{GD} + E_j(Z^0)_L = 3.636095016 \cdot 10^{-9} \text{ (J)} = 22{,}694.71018 \text{ (MeV}/c^2)$$

Z^0 BOSON INTERVALIC KINEMATICS

From here it can be deduced some dynamic features of Z^0 boson by means of the already well known equation $E(J) = m\, r^2 \omega_J^2$, deduced previously from the Virial theorem, where ω_J is the angular velocity due to spin. Supposing likewise that $r_Z = d_{G45} = 1.471662217 \cdot 10^{-19}$ (m) we have:

$$\omega_J(Z^0) = (E_J(Z^0)_L / m_Z r_Z^2)^{1/2} = 6.672930386 \cdot 10^{26}\ (s^{-1})$$

The linear velocity on surface due to spin is:

$$v_J(Z^0) = \omega_J(Z^0)\, r_Z = 9.820299526 \cdot 10^7\ (m\ s^{-1}) = 0.327569932\ c$$

And the acceleration on Z^0 boson's surface due to spin will be:

$$a_J(Z^0) = v_J(Z^0)^2 / r_Z = 6.553017511 \cdot 10^{34}\ (m\ s^{-2})$$

STRUCTURAL ENERGY OF Z^0 BOSON

The structural energy of Z^0 boson is the sum of the energy due to mass plus the energy due to the intrinsic angular momentum which balances mass energy. The structural energy is always just the double of the intervalic energy, so the concept of structural energy does not include the field energy which is not manifested as mass nor its correlated spin energy.

$$A(Z^0) = 2I(Z^0) = E(Z^0)_{mass} + E_J(Z^0)_{GD} = 1.667838778 \cdot 10^{-8}\ (J) = 104{,}098.2635\ (MeV/c^2)$$

Z^0 BOSON INTERVALIC POTENTIAL WELL

The electromagnetic potential energy of Z^0 boson is due to the difference between the electromagnetic energy of Z^0 boson and the electromagnetic energy of its *isolated* constituent dalinos 6, plus the difference between the electromagnetic energy of the dalino 6 and the electromagnetic energy of its *isolated* constituent intervalinos. According to this we have:

$$\Delta U(Z^0) = [U(Z^0)_L - 2U(G_{45})] + 2[U(G_{45}) - 45\ U(D_6)] + 2[45\ U(D_6) - 45\ (6\ U(\mathbf{I}))] = [U(Z^0)_L - 0 = 6.270676648 \cdot 10^{-9} (J) = 39{,}138.4682\ (MeV/c^2)$$

And the electromagnetic potential is:

$$V(Z^0) = \Delta U(Z^0) / m_Z = -3.857531178 \cdot 10^{16}\ (m^2\ s^{-2})$$

In the case of electromagnetic energy (and *long ranged* interactions) 'isolated' means separate by an infinite length. However, in the case of intervalic interaction (and *short ranged* interactions) 'isolated' means *out* of the intervalic structure, an event which does not relies on the macroscopic distance of the separation as in the case of a long ranged interaction. Therefore, the Z^0 boson intervalic potential is determined by the difference on the intervalic energy between Z^0 boson and its constituent disaggregated intervalinos:

$$\Delta I(Z^0) = [I(Z^0)_L - 2I(G_{45})] + 2[I(G_{45}) - 45\ I(D_6)] + 2[45\ I(D_6) - 45\ (6\ I(\mathbf{I}))] = 0 - 540\ (c^{\pm 2}\hbar\ \mathbf{q_I}^{-2}) = -540\ c^{-1} = -1.801246114 \cdot 10^{-6} (J) = -11{,}242{,}489.08\ (MeV/c^2)$$

And the intervalic potential well is:

$$\Phi(Z^0) = \Delta I(Z^0) / m_Z = -2.844390857 \cdot 10^{19}\ (m^2\ s^{-2})$$

INTERVALIC SYMMETRIES OF ZERO CHARGED LISZTINOS

DALINAR SYMMETRY	INTERVALIC STRUCTURE	MASS (MeV/c^2)	PARTICLES WHICH STAY BELOW THE THRESHOLD TEMPERATURE: ZERO CHARGED MASSIVE BOSONS
{D270}	L2G1D270$^{(0)}$	1	
{D135}	L2G2D135$^{(0)}$	8	
{D90}	L2G3D90$^{(0)}$	28	
{D54}	L2G5D54$^{(0)}$	128	
{D45}	L2G6D45$^{(0)}$	221	
{D30}	L2G9D30$^{(0)}$	745	
{D27}	L2G10D27$^{(0)}$	1,022	
{D18}	L2G15D18$^{(0)}$	3,449	
{D15}	L2G18D15$^{(0)}$	5,960	
{D10}	L2G27D10$^{(0)}$	20,116	
{D9}	L2G30D9$^{(0)}$	27,594	
{D6}	L2G45D6$^{(0)}$	91,188	Z^0 boson
{D5}	L2G54D5$^{(0)}$	157,572	W^0 boson
{D3}	L2G90D3$^{(0)}$	729,501	Y^0 boson
{D2}	L2G135D2$^{(0)}$	2,462,065	X^0 boson
{D1}	L2G270D1$^{(0)}$	11,008,055	

Chapter 15

INTERVALIC FRACTIONAL LISZTINO: QUARKS

We are going to describe the *origin of quarks masses*, the *intervalic structures of quarks* and its corresponding intervalic families from a general point of view, and finally the *structural energies of nucleonic quarks* at all levels of its intervalic structure.

As in the description of intervalic leptons and massive bosons, the introduction of the intervalic structure of quarks makes still much more dramatic changes in the number and families of quarks traditionally supposed. It must be noted that the infamous *six+four parameters* of the Standard Model corresponding to quarks masses and the KM matrix, are now irrelevant. Moreover, IT makes useless quantum chromodynamics.

It can be said that the Standard Model *finishes* lamentably without explaining masses, structures and features of fundamental particles and interactions through an overloaded formalism which relies on ~18 or more arbitrary parameters; whilst on the contrary, the Intervalic Theory *begins* describing with astonishing exactness the masses, intervalic structures and features of all fundamental particles, subparticles and interactions, *deduced directly from the intervalic system of dimensions and units*, without using any mathematical formalism, but only applying the new intervalic physics principles and symmetries, and relying on *no one* parameter, with insulting simplicity... This fact involves a powerful logical relation between IT and the very last foundations of Physics, since the intervalic symmetries only can be logically derived under intervalic units and dimensions.

INTERVALIC STRUCTURE OF QUARKS AT THE LISZTINIAN LEVEL

We define a *quark* as a lisztino with fractional charge ⅓ or ⅔ of the elementary charge. As we have seen, Nature choose ever the sets with lowest energy and higher number of symmetries, which a little surprisingly is not $(n/2)e$ but $(n/3)e$. This have been easily explained because ½ e has only 8 symmetries while ⅔ e and ⅓ e has each one 12 symmetries. Even worst, ¾ e and ¼ e are not allowed states because they involve a partition of the intervalino, which is clearly forbidden: ¾ e = 202.5 q_I and ¼ e = 67.5 q_I.

It must be noted that lisztinos based on dalinar structures 270, 135, 54 and 27 cannot yield quarks by no means because those symmetries of compositeness can't make the fractional charges of quarks in any combination.

Now we can list all possible lisztinian structures of quarks in IT. We are going to suppose that we only know the intervalic / electromagnetic energies ratio of the nucleonic quarks, which is greater than the lepton's ratio (roughly 3/2 > 5/4). Although it is sure that the ratio of most of the remaining quarks will not be smaller than the lepton's ratio, we are going to list the predicted masses of quarks between both ratios. It is expected that quarks masses will be very close to the nucleonic quarks' ratio, listed in first place.

There is a difficult related with the nomenclature of quarks because we have seven quarks by each allowed dalinar symmetry. In total 7 x 7 = 49 intervalic structures of quarks. However, only a few of them have till now got a proper name —the six supposed flavoured quarks—. Since the traditional names of quarks are some eccentric and bizarre, we propose to name them according its intervalic structure, as we are going to see. Describing systematically quarks according to the constituent dalinar structure, we have the following possible sets, by order (masses are always in MeV/c^2).

On the contrary, it is necessary to advise that the three families of quarks deplorably intended by SM do not go anyway and the supposed naïve parallelism postulated *ad hoc* between leptons and quarks is doubtful because there are much more than six quarks. Indeed, the observed quarks are only the low energy quarks, but all the 16 intervalic symmetries could exist at the Big Bang, allowing dozens of quarks, although obviously they decayed immediately to yield the presently known particles. There are a lot of intervalic symmetries among all particles, but they are totally different from the ridiculous ones supposed by SM. Of course, there are quarks families focused on the same dalinar symmetries than every lepton-massive boson, but this is due to the features of the own constituent intervalic symmetries, and not to a mystic "trinity" of particles as lamentably prays SM in order to neutralize two errors of infiniteness between themselves.

QUARKS BASED ON {D270} INTERVALIC SYMMETRY (EXISTENCE NOT POSTULATED)

The {D270} symmetry can not make quarks by no means because it can yield any intervalic structure whose electric charge was a fraction of the elementary one.

QUARKS BASED ON {D135} INTERVALIC SYMMETRY (EXISTENCE NOT POSTULATED)

In a similar way, the dalinar symmetry {D135} only can make particles with electric charge e, ½ e or 0. Therefore, it neither can yield quarks.

QUARKS BASED ON {D90} INTERVALIC SYMMETRY (EXISTENCE NOT POSTULATED)

The intervalic symmetry {D90} appears to do not exist in Nature at low energies due to its poor possibilities to produce fractional charges. An even number of dalinos 90 only can make a particle with resultant charge ±⅔ e, and an odd number a particle charged ±⅓ e:

$L_{1/3} = ⅓ \, G_3 = 1 \, D_{90} = 4.3171932 - 4.5989918 \rightarrow$ quark $L⅓D90^{(⅓)}$
$L_{2/3} = ⅔ \, G_3 = 2 \, D_{90} = 8.6343864 - 9.1979836 \rightarrow$ quark $L⅔D90^{(⅔)}$
$L_1 = 1 \, G_3 = 3 \, D_{90} = 12.951579 - 13.796975 \rightarrow$ quark $L1D90^{(⅓)}$
$L_2 = 2 \, G_3 = 6 \, D_{90} = 25.903158 - 27.593950 \rightarrow$ quark $L2D90^{(⅔)}$
$L_3 = 3 \, G_3 = 9 \, D_{90} = 38.854737 - 41.390925 \rightarrow$ quark $L3D90^{(⅓)}$
$L_4 = 4 \, G_3 = 12 \, D_{90} = 51.806316 - 55.187900 \rightarrow$ quark $L4D90^{(⅔)}$
$L_5 = 5 \, G_3 = 15 \, D_{90} = 64.757896 - 68.984875 \rightarrow$ quark $L5D90^{(⅓)}$
$L_6 = 6 \, G_3 = 18 \, D_{90} = 77.709475 - 82.781850$

Nevertheless it have to be noted the close relation in the mass of the quark L4-D90$^{(⅔)}$ of this set —pertaining to the *principal intervalic sequence of quarks* (see below)— with the mass of the stable quark in the fundamental set {D45}, which is just the double:

$L_4 = 4 \, G_3 = 12 \, D_{90} = 51.806316 - 55.187900 \rightarrow$ quark $L4D90^{(⅔)}$
$L_1 = 1 \, G_6 = 6 \, D_{45} = 103.61263 - 110.37580 \rightarrow$ quark $L1D45^{(⅓, ⅔)}$

By this reason we can't discard at first sight that some of these exotic structures could exist in some excited states of quarks, although we personally do not postulate them.

QUARKS BASED ON {D54} INTERVALIC SYMMETRY (EXISTENCE NOT POSTULATED)

The exotic {D54} symmetry can not make quarks by no means, as it is clear that 180/54 and 90/54 do not give integer numbers.

QUARKS BASED ON {D45} INTERVALIC SYMMETRY

In this symmetry set we already find the formerly named quarks u, d, s and a lot of its supposed free masses which, obviously, are not free. As can be seen, the vast majority of quarks with different masses have different intervalic structures and are, therefore, totally different particles, as the intervalic structures of the several supposed quarks up and down dramatically show. The exponent between brackets in the quarks means its fractional electric charges *allowed* by the intervalic structure. As can be seen, most of quarks can have both charges $\pm\frac{2}{3}$ and $\pm\frac{1}{3}$.

$L_{\frac{1}{3}} = \frac{1}{3}\,G_6 = 2\,D_{45} = 34.537545 - 36.791934 \to$ quark $L\frac{1}{3}D45^{(\frac{1}{3})}$
$L_{\frac{2}{3}} = \frac{2}{3}\,G_6 = 4\,D_{45} = 69.075090 - 73.583868 \to$ quark $L\frac{2}{3}D45^{(\frac{1}{3},\frac{2}{3})}$
$L_1 = 1\,G_6 = 6\,D_{45} = 103.61263 - 110.37580 \to$ quark $L1D45^{(\frac{1}{3},\frac{2}{3})}$
$L_2 = 2\,G_6 = 12\,D_{45} = 207.22526 - 220.75160 \to$ quark $L2D45^{(\frac{1}{3},\frac{2}{3})}$
$L_3 = 3\,G_6 = 18\,D_{45} = 310.83790 - 331.12740 \to$ quark $L3D45^{(\frac{1}{3},\frac{2}{3})}$
$L_4 = 4\,G_6 = 24\,D_{45} = 414.45053 - 441.50320 \to$ quark $L4D45^{(\frac{1}{3},\frac{2}{3})}$
$L_5 = 5\,G_6 = 30\,D_{45} = 518.06316 - 551.87900 \to$ quark $L5D45^{(\frac{1}{3},\frac{2}{3})}$
$L_6 = 6\,G_6 = 36\,D_{45} = 621.67579 - 662.25480$

Quarks with two allowed electric charges are named *isoquarks*. For example, quarks L3D45$^{(2/3)}$ and L3D45$^{(1/3)}$ —former quarks *up* and *down*— are isoquarks. This degree of freedom of quarks will be named *isocharge*. As we are going to see, the only symmetries which have an isocharge doublet are {D45}, {D5}, {D3}, {D2} and {D1}.

Remarkable particles in this set are the two *nucleonic quarks*: 3D45 $^{(1/3, 2/3)}$; the two constituent quarks of the π *meson*, which can be any pair adequately chosen between ⅓D45$^{(1/3)}$, ⅔D45$^{(1/3, 2/3)}$ and 1D45$^{(1/3, 2/3)}$; and a intervalic structure of the formerly named quark *strange*: 5D45$^{(1/3, 2/3)}$. The *even* lisztinos can be "fatter" states of all these quarks. It must be noted that quarks up and down have *identical* intervalic structures, that is to say, they are the same quark (with can have different charge) in IT, whilst on the contrary, they are different quarks —with different flavour— in SM. We can also remember that the intervalic structure of *muon* is L1D45$^{(1)}$ → μ.

The total mass energy of the intervalic symmetry {D45} is the sum of all its constituent quarks: 16 L_1 = 1,657.8020—1,766.0128 (MeV/c^2).

QUARKS BASED ON {D30} INTERVALIC SYMMETRY

Due to obvious reason the quarks of this set only can show one fractional charge: ±⅓ *or* ±⅔, but not both. This maybe the reason why although {D30} seems to be a very powerful symmetry, it is really less rich than {D45}. This fact also could explain why this is the only dalinar structure of quarks which is not shared with leptons and intermediate massive bosons of weak intervalic interaction.

$L_{1/3}$ = ⅓ G_9 = 3 D_{30} = 116.56421—124.17278 → quark L⅓D30$^{(1/3)}$
$L_{2/3}$ = ⅔ G_9 = 6 D_{30} = 233.12843—248.34556 → quark L⅔D30$^{(2/3)}$
L_1 = 1 G_9 = 9 D_{30} = 349.69264—372.51833 → quark L1D30$^{(1/3)}$
L_2 = 2 G_9 = 18 D_{30} = 699.38528—745.03666 → quark L2D30$^{(2/3)}$

$L_3 = 3\ G_9 = 27\ D_{30} = 1{,}049.0779 - 1{,}117.5550 \rightarrow$ quark $L3D30^{(\frac{1}{3})}$
$L_4 = 4\ G_9 = 36\ D_{30} = 1{,}398.7705 - 1{,}490.0733 \rightarrow$ quark $L4D30^{(\frac{2}{3})}$
$L_5 = 5\ G_9 = 45\ D_{30} = 1{,}748.4631 - 1{,}862.5916 \rightarrow$ quark $L5D30^{(\frac{1}{3})}$
$L_6 = 6\ G_9 = 54\ D_{30} = 2{,}098.1557 - 2{,}235.1099$

Remarkable particles in this set are the former quark *charm*: L4D30 $^{(\frac{2}{3})}$ as well as one of its possible "fatter" states: L5D30$^{(\frac{1}{3})}$. The last one would involve a mutual interchange in the constituent quarks charges of some hadrons. The first three structures of the set can be seen as excited states of the corresponding intervalic structures of {D45}, as well as the possible quark L3D30$^{(\frac{1}{3})}$.

The total mass energy of the intervalic symmetry {D30} is the sum of all its constituent quarks: $16\ L_1 = 5{,}595.0822 - 5{,}960.2938\ (MeV/c^2)$.

QUARKS BASED ON {D27} INTERVALIC SYMMETRY (EXISTENCE NOT POSTULATED)

As like as {D54}, the exotic {D27} symmetry can not make quarks inasmuch as 180/54 and 90/54 do not give integer numbers.

QUARKS BASED ON {D18} INTERVALIC SYMMETRY

In the following set find mainly quarks with fractional charge $\pm\frac{1}{3}$ since to yield the $\pm\frac{2}{3}$ charge it is necessary to get a number *even* of constituent dalinos.

$L_{\frac{1}{3}} = \frac{1}{3}\ G_{15} = 5\ D_{18} = 539.64917 - 574.87400 \rightarrow$ quark $L\frac{1}{3}D18^{(\frac{1}{3})}$
$L_{\frac{2}{3}} = \frac{2}{3}\ G_{15} = 10\ D_{18} = 1{,}079.2983 - 1{,}149.7480 \rightarrow$ quark $L\frac{2}{3}D18^{(\frac{2}{3})}$
$L_1 = 1\ G_{15} = 15\ D_{18} = 1{,}618.9475 - 1{,}724.6220 \rightarrow$ quark $L1D18^{(\frac{1}{3})}$
$L_2 = 2\ G_{15} = 30\ D_{18} = 3{,}237.8950 - 3{,}449.2440 \rightarrow$ quark $L2D18^{(\frac{2}{3})}$

$L_3 = 3\ G_{15} = 45\ D_{18} = 4{,}856.8425 — 5{,}173.8660 \to$ quark $L3D18^{(\frac{1}{3})}$
$L_4 = 4\ G_{15} = 60\ D_{18} = 6{,}475.7900 — 6{,}898.4880 \to$ quark $L4D18^{(\frac{2}{3})}$
$L_5 = 5\ G_{15} = 75\ D_{18} = 8{,}094.7375 — 8{,}623.1100 \to$ quark $L5D18^{(\frac{1}{3})}$
$L_6 = 6\ G_{15} = 90\ D_{18} = 9{,}713.6850 — 10{,}347.732$

Up to date, quarks of this symmetry with charge ⅓ are well established: we find the former quark *strange*: $L\frac{1}{3}D18^{(\frac{1}{3})}$ and the quark *bottom*: $L3D18^{(\frac{1}{3})}$. And once more we find another possible structure for the interchanged quark *charm*: $L1D18^{(\frac{1}{3})}$. At first sight it appears to exist lesser evidence for the other quarks with charge ⅔, but this is not the case as we will see later, since all deductions for the existence of quarks *inside* SM are doubtfully based on indirect evidences and setting by hand the masses of quarks; even worst, setting by hand the number of quarks through choosing by hand the number of flavours.

Finally, we can remember that the intervalic structure of *tau* is $L1D18^{(1)} \to \tau$.

The total mass energy of the intervalic symmetry {D18} is the sum of all its constituent quarks: $16\ L_1 = 25{,}903.160 — 27{,}593.954\ (MeV/c^2)$.

THE INTERVALIC QUARK LISZTINIAN SEQUENCE

Till now we have developed systematically the intervalic structures at the lisztinian level yielded unavoidably by the symmetries {D45}, {D30} and {D18}. It must be noted that this development of the intervalic compositeness —which makes a rich structurefulness— follows strictly and exclusively the logical rules imposed by the constraint derived from the intervalic units. There is neither any human or subjective *choosing* through all the process, nor has been set by hand any *parameter* or *constant* —really, there is *no one* parameter or constant to be set by hand in all IT—. The only intervention along the process has been to stop at any place the lisztinian sequence in all sets of symmetries: $L_{\frac{1}{3}}$, $L_{\frac{2}{3}}$, L_1, L_2,

$L_3, L_4, L_5, L_6,...$ since it is not a *finite* sequence, as in previous set of symmetries, but it can continue indefinitely. Therefore, we have looked at phenomenology and have seen that there appear to do not exist evidence of particles from lisztino 6 and onwards. Since it is clear that an *infinite* sequence of intervalic structures cannot be postulated, we will therefore stop at lisztino 5 and will check whether such sequence, yielded by Nature in a logical and systematic way, does match or not with the experimental data. Really it may seem strange that lisztino 6 was not made since the hexagonal symmetry is clearly favoured by Nature among any others. On the contrary, the symmetry chosen by Nature at this step to make the most stable particles is the {L3}, which is the highest symmetry which stay necessarily into a bidimensional plane. We will name the systematic sequence $L_{1/3}, L_{2/3}, L_1, L_2, L_3, L_4, L_5$ as the *intervalic quark lisztinian sequence*, and will prove that it is the way in which Nature works. But our partial or incomplete knowledge of any aspect of Nature does not means that God plays dice, as it is sure that *She-He* —or better, *IT*— does not play dice.

QUARKS BASED ON {D15} INTERVALIC SYMMETRY (EXISTENCE NOT POSTULATED)

$L_{1/3} = \frac{1}{3} G_{18} = 6 D_{15} = 932.51359 - 993.38220 \rightarrow$ quark $L\frac{1}{3}D15^{(1/3)}$
$L_{2/3} = \frac{2}{3} G_{18} = 12 D_{15} = 1,865.0271 - 1,986.7644 \rightarrow$ quark $L\frac{2}{3}D15^{(1/3, 2/3)}$
$L_1 = 1 G_{18} = 18 D_{15} = 2,797.5407 - 2,980.1466 \rightarrow$ quark $L1D15^{(1/3, 2/3)}$
$L_2 = 2 G_{18} = 36 D_{15} = 5,595.0815 - 5,960.2932 \rightarrow$ quark $L2D15^{(1/3, 2/3)}$
$L_3 = 3 G_{18} = 54 D_{15} = 8,392.6223 - 8,940.4398 \rightarrow$ quark $L3D15^{(1/3, 2/3)}$
$L_4 = 4 G_{18} = 72 D_{15} = 11,190.162 - 11,920.586 \rightarrow$ quark $L4D15^{(1/3, 2/3)}$
$L_5 = 5 G_{18} = 90 D_{15} = 13,987.703 - 14,900.733 \rightarrow$ quark $L5D15^{(1/3, 2/3)}$
$L_6 = 6 G_{18} = 108 D_{15} = 16,785.244 - 17,880.879 \rightarrow$

QUARKS BASED ON {D10} INTERVALIC SYMMETRY (EXISTENCE NOT POSTULATED)

$L_{1/3} = 1/3\ G_{27} = 9\ D_{10} = 3{,}147.2334 - 3{,}352.6650 \rightarrow$ quark $L1/3D10^{(1/3)}$
$L_{2/3} = 2/3\ G_{27} = 18\ D_{10} = 6{,}294.4668 - 6{,}705.3299 \rightarrow$ quark $L2/3D10^{(2/3)}$
$L_1 = 1\ G_{27} = 27\ D_{10} = 9{,}441.7704 - 10{,}057.995 \rightarrow$ quark $L1D10^{(1/3)}$
$L_2 = 2\ G_{27} = 54\ D_{10} = 18{,}883.40 - 20{,}115.990 \rightarrow$ quark $L2D10^{(2/3)}$
$L_3 = 3\ G_{27} = 81\ D_{10} = 28{,}325.101 - 30{,}173.985 \rightarrow$ quark $L3D10^{(1/3)}$
$L_4 = 4\ G_{27} = 108\ D_{10} = 37{,}766.801 - 40{,}231.98 \rightarrow$ quark $L4D10^{(2/3)}$
$L_5 = 5\ G_{27} = 135\ D_{10} = 47{,}490.119 - 50{,}589.975 \rightarrow$ quark $L5D10^{(1/3)}$
$L_6 = 6\ G_{27} = 162\ D_{10} = 56{,}650.202 - 60{,}347.97$

QUARKS BASED ON {D9} INTERVALIC SYMMETRY (EXISTENCE NOT POSTULATED)

$L_{1/3} = 1/3\ G_{30} = 10\ D_9 = 4{,}317.1927 - 4{,}598.9918 \rightarrow$ quark $L1/3D9^{(1/3)}$
$L_{2/3} = 2/3\ G_{30} = 20\ D_9 = 8{,}634.3854 - 9{,}197.9836 \rightarrow$ quark $L2/3D9^{(1/3,\ 2/3)}$
$L_1 = 1\ G_{30} = 30\ D_9 = 12{,}951.577 - 13{,}796.975 \rightarrow$ quark $L1D9^{(1/3,\ 2/3)}$
$L_2 = 2\ G_{30} = 60\ D_9 = 25{,}903.155 - 27{,}593.950 \rightarrow$ quark $L2D9^{(1/3,\ 2/3)}$
$L_3 = 3\ G_{30} = 90\ D_9 = 38{,}854.733 - 41{,}390.925 \rightarrow$ quark $L3D9^{(1/3,\ 2/3)}$
$L_4 = 4\ G_{30} = 120\ D_9 = 51{,}806.311 - 55{,}187.90 \rightarrow$ quark $L4D9^{(1/3,\ 2/3)}$
$L_5 = 5\ G_{30} = 150\ D_9 = 64{,}757.888 - 68{,}984.875 \rightarrow$ quark $L5D9^{(1/3,\ 2/3)}$
$L_6 = 6\ G_{30} = 180\ D_9 = 77{,}709.466 - 82{,}781.85$

QUARKS BASED ON {D6} INTERVALIC SYMMETRY

From now on we find the heaviest particles, since as smaller is the dalinar structure greater is the related intervalic energy.

$L_{1/3} = 1/3\ G_{45} = 15\ D_6 = 14{,}570.527 - 15{,}521.598 \rightarrow$ quark $L1/3D6^{(1/3)}$
$L_{2/3} = 2/3\ G_{45} = 30\ D_6 = 29{,}141.055 - 31{,}043.196 \rightarrow$ quark $L2/3D6^{(2/3)}$

$L_1 = 1\ G_{45} = 45\ D_6 = 43{,}711.582 — 46{,}564.794 \rightarrow$ quark $L1D6^{(\frac{1}{3})}$
$L_2 = 2\ G_{45} = 90\ D_6 = 87{,}423.165 — 93{,}129.588 \rightarrow$ quark $L2D6^{(\frac{2}{3})}$
$L_3 = 3\ G_{45} = 135\ D_6 = 131{,}134.74 — 139{,}694.38 \rightarrow$ quark $L3D6^{(\frac{1}{3})}$
$L_4 = 4\ G_{45} = 180\ D_6 = 174{,}846.32 — 186{,}259.17 \rightarrow$ quark $L4D6^{(\frac{2}{3})}$
$L_5 = 5\ G_{45} = 225\ D_6 = 218{,}557.91 — 232{,}823.97 \rightarrow$ quark $L5D6^{(\frac{1}{3})}$
$L_6 = 6\ G_{45} = 270\ D_6 = 262{,}269.49 — 279{,}388.76$

The most remarkable quark in this set is the formerly named quark *top*: $L4D6^{(\frac{2}{3})}$, and it can be also remembered that the intervalic structure of the Z^0 *massive boson* is: $L2D6^{(0)} \rightarrow Z^0$. It can be expected that the remaining quarks of this set will be detected as constituents of heavy baryons with {D6} symmetry when higher energies become available, since it appears to be clear that Nature does not merge quarks pertaining to different symmetries when assembling baryons or mesons. Therefore the threshold energy to make such baryons is much greater that the masses of the lightest quarks of the set, which explains why they have not been detected yet.

The total mass energy of the intervalic symmetry {D6} is the sum of all its constituent quarks: $16\ L_1 = 699{,}385.31 — 745{,}036.76$ (MeV/c^2).

QUARKS BASED ON {D5} INTERVALIC SYMMETRY

This set is the last one —following by order the sequence imposed by the dalinar structure— where any subatomic particle had been detected experimentally (the $W^\pm$ massive boson).

$L_{\frac{1}{3}} = \frac{1}{3}\ G_{54} = 18\ D_5 = 25{,}177.871 — 26{,}821.321 \rightarrow$ quark $L\frac{1}{3}D5^{(\frac{1}{3})}$
$L_{\frac{2}{3}} = \frac{2}{3}\ G_{54} = 36\ D_5 = 50{,}355.742 — 53{,}642.642 \rightarrow$ quark $L\frac{2}{3}D5^{(\frac{1}{3},\frac{2}{3})}$
$L_1 = 1\ G_{54} = 54\ D_5 = 75{,}533.615 — 80{,}463.964 \rightarrow$ quark $L1D5^{(\frac{1}{3},\frac{2}{3})}$
$L_2 = 2\ G_{54} = 108\ D_5 = 151{,}067.23 — 160{,}927.93 \rightarrow$ quark $L2D5^{(\frac{1}{3},\frac{2}{3})}$
$L_3 = 3\ G_{54} = 162\ D_5 = 226{,}600.84 — 241{,}391.89 \rightarrow$ quark $L3D5^{(\frac{1}{3},\frac{2}{3})}$
$L_4 = 4\ G_{54} = 216\ D_5 = 302{,}134.45 — 321{,}855.85 \rightarrow$ quark $L4D5^{(\frac{1}{3},\frac{2}{3})}$

$L_5 = 5\ G_{54} = 270\ D_5 = 377{,}668.07 - 402{,}319.82 \rightarrow$ quark L5D5$^{(\frac{1}{3},\frac{2}{3})}$
$L_6 = 6\ G_{54} = 324\ D_5 = 453{,}201.68 - 482{,}783.78$

In this set we predict a new heavy quark: L3D5$^{(\frac{1}{3},\frac{2}{3})}$ with mass 226,600.84—241,391.89 (MeV/c²), based on the principal intervalic sequence of quarks. It can be remembered that the intervalic structure of *W± massive boson* also pertain to this set: L1D5$^{(1)} \rightarrow$ W±. As in the preceding symmetry set, {D6}, there is no reason to expect that the allowed quarks of this symmetry will not be detected when higher energies become available.

The total mass energy of the intervalic symmetry {D5} is the sum of all its constituent quarks: $16\ L_1 = 1{,}208{,}537.8 - 1{,}287{,}423.5$ (MeV/c²).

PRINCIPAL INTERVALIC SEQUENCE OF QUARKS

It can be seen that exists a naïve recurrent formula which alternates lisztinos 3 and 4 at each consecutive dalinar level of structure, and which coincides approximately with the principal peaks that appear in the graphic of the annihilation cross section ratio ($e^+e^- \rightarrow$ hadrons / $e^+e^- \rightarrow \mu^+\mu^-$):

ρ, ω mesons (770, 783)
Φ meson (1,020)
J/ψ meson (3,097)
Y meson (9,460)
Z^0 boson (91,187.6)

Historically, this set of particles was the most important "evidence" to intend the existence of the next quarks above the nucleonic ones. As we have pointed out, the most simple assumption would be to assign to all these mesons an intervalic structure of lisztino 2, just like for the Z^0 boson, since the original interacting particles are equally two lisztinos

INTERVALIC THEORY:
The Intervalic Structures of Subatomic Particles and the Last Foundations of Physics

1 —e^+e^-—. However, according to the usual view of SM it would correspond to have got intervalic structures of alternating lisztinos 3 and lisztinos 4:

- quark L3D45$^{(⅓, ⅔)}$ → former quark up, down
- quark L4D30$^{(⅔)}$ → former quark charm
- quark L3D18$^{(⅓)}$ → former quark bottom
- quark L4D6$^{(⅔)}$ → former quark top
- quark L3D5$^{(⅓, ⅔)}$ → predicted new heavy quark
- quark L4D3$^{(⅓, ⅔)}$ → predicted new heavy quark
- quark L3D2$^{(⅓, ⅔)}$ → predicted new heavy quark

This sequence seems likely to be, compared for example with the stars, as a *principal intervalic sequence* of quarks.

The quark *strange*, to honour its name, is the only one among traditional quarks that does not fit in the recurrent formula.

It is curious that the quarks pertaining to this principal intervalic sequence correspond just with the supposed flavours postulated by SM. This also could mean that such intervalic structures are more stable that the others, although it could also be a pure coincidence. Other possible explanation is that such quarks may be related with the *threshold energy* of each symmetry. It appears to be experimentally clear that no one quark of a determine symmetry is made until there have been made all the quarks of the preceding symmetry, until the preceding symmetry is completed. Therefore the threshold energy of the intervalic symmetries will be equal or greater than the total mass energy of its preceding symmetry. Supposed that the exotic symmetries do not intervene, they would be:

$E_{th}\{D45\} \geq$
$E_{th}\{D30\} \geq 1,657.8020 — 1,766.0128$
$E_{th}\{D18\} \geq 5,595.0822 — 5,960.2938$
$E_{th}\{D6\} \geq 25,903.160 — 27,593.954$
$E_{th}\{D5\} \geq 699,385.31 — 745,036.76$
$E_{th}\{D3\} \geq 1,208,537.8 — 1,287,423.5$
$E_{th}\{D2\} \geq 5,595,082.2 — 5,960,293.8$

But if the exotic symmetries which can made quarks do intervene — namely {D90}, {D15}, {D10} and {D9}, since {D270}, {D135}, {D54} and {D27} can not made fractional charges by no means—, they would be the following:

$E_{th}\{D45\} \geq 207.22526-220.75160$
$E_{th}\{D30\} \geq 1,657.8020-1,766.0128$
$E_{th}\{D18\} \geq 5,595.0822-5,960.2938$
$E_{th}\{D6\} \geq 207,225.26-220,751.60$
$E_{th}\{D5\} \geq 699,385.31-745,036.76$
$E_{th}\{D3\} \geq 1,208,537.8-1,287,423.5$
$E_{th}\{D2\} \geq 5,595,082.2-5,960,293.8$

Of course, at first sight the most simple assumption would be to postulate the *universality* principle in the synthesis of quarks (and disregarding exotic symmetries), so any quark is not made until there is enough energy to make all the 7 quarks of its family. In this case the threshold energy would be the sum of the mass energy of each family of quarks:

$E_{th}\{D45\} = \Sigma(L_nD_{45}) = 1,657.8020-1,766.0128$
$E_{th}\{D30\} = \Sigma(L_nD_{30}) = 5,595.0822-5,960.2938$
$E_{th}\{D18\} = \Sigma(L_nD_{18}) = 25,903.160-27,593.954$
$E_{th}\{D6\} = \Sigma(L_nD_6) = 699,385.31-745,036.76$
$E_{th}\{D5\} = \Sigma(L_nD_5) = 1,208,537.8-1,287,423.5$
$E_{th}\{D3\} = \Sigma(L_nD_3) = 5,595,082.2-5,960,293.8$
$E_{th}\{D2\} = \Sigma(L_nD_2) = 18,883,403-20,115,992$

Whatever will be the threshold energy, what would appear to be most simple is that the skipping from one dalinar symmetry to the next in order to make quarks was not made until the preceding symmetry is not completed.

However, as we will explain when studying the *intervalic decay* of subatomic particles, the rule is to be leaded by the *masses* of the intervalic structures involved, disregarding the dalinar symmetry to which they pertain.

This behaviour can be easily explained because it happens *below* the threshold temperature and there is a limited energy available. On the contrary, *above* the threshold temperature the constraint of the mass becomes useless and inoperative if the availability of energy is endless, as it is the case between the Big Crunch and the next Big Bang. This beautiful state, never seen by human eyes, is leaded completely by the intervalic structures derived from the intervalic symmetries of Nature, and will be described at due course.

QUARKS BASED ON {D3} INTERVALIC SYMMETRY

Although their energies are not available yet, we give the intervalic structures and masses allowed for the heavier quarks.

$L_{⅓} = ⅓\ G_{90} = 30\ D_3 = 116{,}564.21 - 124{,}172.78 \rightarrow$ quark $L⅓D3^{(⅓)}$
$L_{⅔} = ⅔\ G_{90} = 60\ D_3 = 233{,}128.43 - 248{,}345.56 \rightarrow$ quark $L⅔D3^{(⅓,\ ⅔)}$
$L_1 = 1\ G_{90} = 90\ D_3 = 349{,}692.64 - 372{,}518.33 \rightarrow$ quark $L1D3^{(⅓,\ ⅔)}$
$L_2 = 2\ G_{90} = 180\ D_3 = 699{,}385.28 - 745{,}036.66 \rightarrow$ quark $L2D3^{(⅓,\ ⅔)}$
$L_3 = 3\ G_{90} = 270\ D_3 = 1{,}049{,}077.9 - 1{,}117{,}555.0 \rightarrow$ quark $L3D3^{(⅓,\ ⅔)}$
$L_4 = 4\ G_{90} = 360\ D_3 = 1{,}398{,}770.5 - 1{,}490{,}073.3 \rightarrow$ quark $L4D3^{(⅓,\ ⅔)}$
$L_5 = 5\ G_{90} = 450\ D_3 = 1{,}748{,}463.2 - 1{,}862{,}591.7 \rightarrow$ quark $L5D3^{(⅓,\ ⅔)}$
$L_6 = 6\ G_{90} = 540\ D_3 = 2{,}098{,}155.9 - 2{,}235{,}110.0$

The quark $L4D3^{(⅓,\ ⅔)}$ would be the next one of the *principal sequence*.

The total mass energy of the intervalic symmetry {D3} is the sum of all its constituent quarks: $16\ L_1 = 5{,}595{,}082.2 - 5{,}960{,}293.8\ (MeV/c^2)$.

QUARKS AND ISOCHARGES ALLOWED BY THE INTERVALIC SYMMETRIES

DALINAR SYMMETRY	ALLOWED NUMBER OF QUARKIC INTERVALIC STRUCTURES	ISOCHARGE ALLOWED
{D270}	0	-
{D135}	0	-
{D90}	7	no
{D54}	0	-
{D45}	7	yes
{D30}	7	no
{D27}	0	-
{D18}	7	no
{D15}	7	yes
{D10}	7	no
{D9}	7	yes
{D6}	7	no
{D5}	7	yes
{D3}	7	yes
{D2}	7	yes
{D1}	-	-

TABLE OF INTERVALIC QUARKS
—Fractional lisztinos—

According to its dalinar symmetry and electric charge: 25 **uniquarks** (quarks with one allowed charge: $\frac{1}{3}$ or $\frac{2}{3}$) and 24 **isoquarks** (quarks with two allowed charges: $\frac{1}{3}$ and $\frac{2}{3}$). Mass in (MeV/c²).

Dalinar symmetry		{D270}	{D135}	{D90}	{D54}	{D45}	{D30}	{D27}	{D18}	{D15}	{D10}
UNI-QUARKS	Charge $\frac{1}{3}$					Quark $L_0.\frac{1}{3}G_02D_5^{(\frac{1}{3})}$ (35) *last radiant decaying quark*	Quark $L_0.\frac{1}{3}G_63D_{30}^{(\frac{1}{3})}$ (117) Quark $L_11G_69D_{30}^{(\frac{1}{3})}$ (350) Quark $L_23G_627D_{30}^{(\frac{1}{3})}$ (1,049) Quark $L_55G_645D_{30}^{(\frac{1}{3})}$ (1,748)		Quark $L_0.\frac{1}{3}G_{15}5D_{18}^{(\frac{1}{3})}$ (540) Quark $L_11G_{15}15D_{18}^{(\frac{1}{3})}$ (1,619) Quark $L_23G_{15}45D_{18}^{(\frac{1}{3})}$ (4,857) *former quark bottom* Quark $L_55G_{15}75D_{18}^{(\frac{1}{3})}$ (8,095)		
	Charge $\frac{2}{3}$						Quark $L_0.\frac{2}{3}G_66D_{30}^{(\frac{2}{3})}$ (233) Quark $L_22G_618D_{30}^{(\frac{2}{3})}$ (699) Quark $L_44G_636D_{30}^{(\frac{2}{3})}$ (1,399) *former quark charm*		Quark $L_0.\frac{2}{3}G_{15}10D_{18}^{(\frac{2}{3})}$ (1,079) Quark $L_22G_{15}30D_{18}^{(\frac{2}{3})}$ (3,238) Quark $L_44G_{15}60D_{18}^{(\frac{2}{3})}$ (6,476)		
ISOQUARKS	Charge $\frac{1}{3}$ and $\frac{2}{3}$					Quark $L_0.\frac{2}{3}G_64D_{45}^{(\frac{1}{3},\frac{2}{3})}$ (69) *constituent quark of π meson* Quark $L_11G_66D_{45}^{(\frac{1}{3},\frac{2}{3})}$ (104) Quark $L_22G_612D_{45}^{(\frac{1}{3},\frac{2}{3})}$ (207) Quark $L_33G_618D_{45}^{(\frac{1}{3},\frac{2}{3})}$ (311) *former quarks up, down* Quark $L_44G_224D_{45}^{(\frac{1}{3},\frac{2}{3})}$ (414) Quark $L_55G_630D_{45}^{(\frac{1}{3},\frac{2}{3})}$ (518) *former quark strange*					

Dalinar symmetry		{D9}	{D6}	{D5}	{D3}	{D2}	{D1}
UNI-QUARKS	Charge $\frac{1}{3}$		Quark $L_0.\frac{1}{3}G_{45}15D_6^{(\frac{1}{3})}$ (14,571) Quark $L_11G_{45}45D_6^{(\frac{1}{3})}$ (43,712) Quark $L_33G_{45}135D_6^{(\frac{1}{3})}$ (131,135) Quark $L_55G_{45}225D_6^{(\frac{1}{3})}$ (218,558)	Quark $L_0.\frac{1}{3}G_{45}18D_5^{(\frac{1}{3})}$ (25,178)	Quark $L_0.\frac{1}{3}G_{90}30D_3^{(\frac{1}{3})}$ (116,564)	Quark $L_0.\frac{1}{3}G_{135}45D_2^{(\frac{1}{3})}$ (393,404)	
	Charge $\frac{2}{3}$		Quark $L_0.\frac{2}{3}G_{45}30D_6^{(\frac{2}{3})}$ (29,141) Quark $L_22G_{45}90D_6^{(\frac{2}{3})}$ (87,426) Quark $L_44G_{45}180D_6^{(\frac{2}{3})}$ (174,846) *former quark top*				
ISO-QUARKS	Charge $\frac{1}{3}$ and $\frac{2}{3}$			Quark $L_0.\frac{2}{3}G_{54}36D_6^{(\frac{1}{3},\frac{2}{3})}$ (50,356) Quark $L_11G_{90}54D_5^{(\frac{1}{3},\frac{2}{3})}$ (75,534) Quark $L_22G_{54}108D_5^{(\frac{1}{3},\frac{2}{3})}$ (151,068) Quark $L_33G_{54}162D_5^{(\frac{1}{3},\frac{2}{3})}$ (226,601) Quark $L_44G_{54}216D_5^{(\frac{1}{3},\frac{2}{3})}$ (302,134) Quark $L_55G_{54}270D_5^{(\frac{1}{3},\frac{2}{3})}$ (377,668)	Quark $L_0.\frac{2}{3}G_{90}60D_3^{(\frac{1}{3},\frac{2}{3})}$ (233,128) Quark $L_11G_{90}90D_3^{(\frac{1}{3},\frac{2}{3})}$ (349,693) Quark $L_22G_{90}180D_3^{(\frac{1}{3},\frac{2}{3})}$ (699,384) Quark $L_33G_{90}270D_3^{(\frac{1}{3},\frac{2}{3})}$ (1,049,078) Quark $L_44G_{90}360D_3^{(\frac{1}{3},\frac{2}{3})}$ (1,398,771) Quark $L_55G_{90}450D_3^{(\frac{1}{3},\frac{2}{3})}$ (1,748,463)		Quark $L_0.\frac{2}{3}G_{135}90D_1^{(\frac{1}{3},\frac{2}{3})}$ (786,808) Quark $L_11G_{135}135D_1^{(\frac{1}{3},\frac{2}{3})}$ (1,180,213) Quark $L_22G_{135}270D_1^{(\frac{1}{3},\frac{2}{3})}$ (2,360,424) Quark $L_33G_{135}405D_2^{(\frac{1}{3},\frac{2}{3})}$ (3,540,638) Quark $L_44G_{135}540D_2^{(\frac{1}{3},\frac{2}{3})}$ (4,720,850) Quark $L_55G_{135}675D_2^{(\frac{1}{3},\frac{2}{3})}$ (5,901,063)

INTERVALIC STRUCTURE OF THE COMPLETE FAMILY OF INTERVALIC QUARKS

Intervalic structure levels: 1 Intervalic String (S), 2 Photon (γ), 3 Intervalino (I), 4 Dalino (D), 5 Gaudino (G), 6 Lisztino (L), 7 Monteverdino (M), 8 Palestrino (P).
Dalinar symmetries: {270, 135, 90, 54, 45, 30, 27, 18, 15, 10, 9, 6, 5, 3, 2, 1}, of which only {45, 30, 18, 6, 5, 3, 2, 1} are allowed for lisztinos. (Mass in MeV/c^2)

Dal. Sym.	Lisztinian structure L$_{\frac{1}{2}}$ (quarks composed by 90 intervalinos)	Lisztinian structure L$_{\frac{2}{3}}$ (quarks composed by 180 intervalinos)	Lisztinian structure L$_1$ (quarks composed by 270 intervalinos)	Lisztinian structure L$_2$ (quarks composed by 540 intervalinos)	Lisztinian structure L$_3$ (quarks composed by 810 intervalinos)	Lisztinian structure L$_4$ (quarks composed by 1080 intervalinos)	Lisztinian structure L$_5$ (quarks composed by 1350 intervalinos)
{D45}	$\frac{1}{2}$G$_6$ D$_{45}$ Last radiant decay quark (35) Quark L$_{\frac{1}{2}}\frac{1}{2}G_6$2D$_{45}$ $^{(\frac{1}{2},\frac{2}{3})}$ = 90 I = 180 γ = 360 S	$\frac{2}{3}$G$_6$ D$_{45}$ Constituent quark of π meson (69) Quark L$_{\frac{2}{3}}\frac{2}{3}G_6$4D$_{45}$ $^{(\frac{1}{2},\frac{2}{3})}$ = 180 I = 360 γ = 720 S	G$_6$ D$_{45}$ (104) Quark L$_1$G$_6$6D$_{45}$ $^{(\frac{1}{2},\frac{2}{3})}$ = 270 I = 540 γ = 1080 S	G$_6$ D$_{45}$ L$_2$ (207) Quark L$_2$2G$_6$12D$_{45}$ $^{(\frac{1}{2},\frac{2}{3})}$ = 540 I = 1080 γ = 2160 S	G$_6$ D$_{45}$ L$_3$ Former quarks up, down (311) Quark L$_3$3G$_6$18D$_{45}$ $^{(\frac{1}{2},\frac{2}{3})}$ = 810 I = 1620 γ = 3240 S	G$_6$ D$_{45}$ L$_4$ (414) Quark L$_4$4G$_6$24D$_{45}$ $^{(\frac{1}{2},\frac{2}{3})}$ = 1080 I = 2160 γ = 4320 S	G$_6$ D$_{45}$ L$_5$ Former quark strange (518) Quark L$_5$5G$_6$30D$_{45}$ $^{(\frac{1}{2},\frac{2}{3})}$ = 1350 I = 2700 γ = 5400 S
{D30}	$\frac{1}{2}$G$_9$ D$_{30}$ (117) Quark L$_{\frac{1}{2}}\frac{1}{2}G_9$3D$_{30}$ $^{(\frac{2}{3})}$ = 90 I = 180 γ = 360 S	$\frac{2}{3}$G$_9$ D$_{30}$ (233) Quark L$_{\frac{2}{3}}\frac{2}{3}G_9$6D$_{30}$ $^{(\frac{2}{3})}$ = 180 I = 360 γ = 720 S	G$_9$ D$_{30}$ (350) Quark L$_1$G$_9$9D$_{30}$ $^{(\frac{2}{3})}$ = 270 I = 540 γ = 1080 S	G$_9$ D$_{30}$ L$_2$ (699) Quark L$_2$2G$_9$18D$_{30}$ $^{(\frac{2}{3})}$ = 540 I = 1080 γ = 2160 S	G$_9$ D$_{30}$ L$_3$ (1,049) Quark L$_3$3G$_9$27D$_{30}$ $^{(\frac{2}{3})}$ = 810 I = 1620 γ = 3240 S	G$_9$ D$_{30}$ L$_4$ Former quark charm (1,399) Quark L$_4$4G$_9$36D$_{30}$ $^{(\frac{2}{3})}$ = 1080 I = 2160 γ = 4320 S	G$_9$ D$_{30}$ L$_5$ (1,748) Quark L$_5$5G$_9$45D$_{30}$ $^{(\frac{2}{3})}$ = 1350 I = 2700 γ = 5400 S
{D18}	$\frac{1}{2}$G$_{15}$ D$_{18}$ (540) Quark L$_{\frac{1}{2}}\frac{1}{2}G_{15}$5D$_{18}$ $^{(\frac{2}{3})}$ = 90 I = 180 γ = 360 S	$\frac{2}{3}$G$_{15}$ D$_{18}$ (1,079) Quark L$_{\frac{2}{3}}\frac{2}{3}G_{15}$10D$_{18}$ $^{(\frac{2}{3})}$ = 180 I = 360 γ = 720 S	G$_{15}$ D$_{18}$ (1,619) Quark L$_1$1G$_{15}$15D$_{18}$ $^{(\frac{2}{3})}$ = 270 I = 540 γ = 1080 S	G$_{15}$ D$_{18}$ L$_2$ (3,238) Quark L$_2$2G$_{15}$30D$_{18}$ $^{(\frac{2}{3})}$ = 540 I = 1080 γ = 2160 S	G$_{15}$ D$_{18}$ L$_3$ Former quark bottom (4,857) Quark L$_3$3G$_{15}$45D$_{18}$ $^{(\frac{2}{3})}$ = 810 I = 1620 γ = 3240 S	G$_{15}$ D$_{18}$ L$_4$ (6,476) Quark L$_4$4G$_{15}$60D$_{18}$ $^{(\frac{2}{3})}$ = 1080 I = 2160 γ = 4320 S	G$_{15}$ D$_{18}$ L$_5$ (8,095) Quark L$_5$5G$_{15}$75D$_{18}$ $^{(\frac{2}{3})}$ = 1350 I = 2700 γ = 5400 S

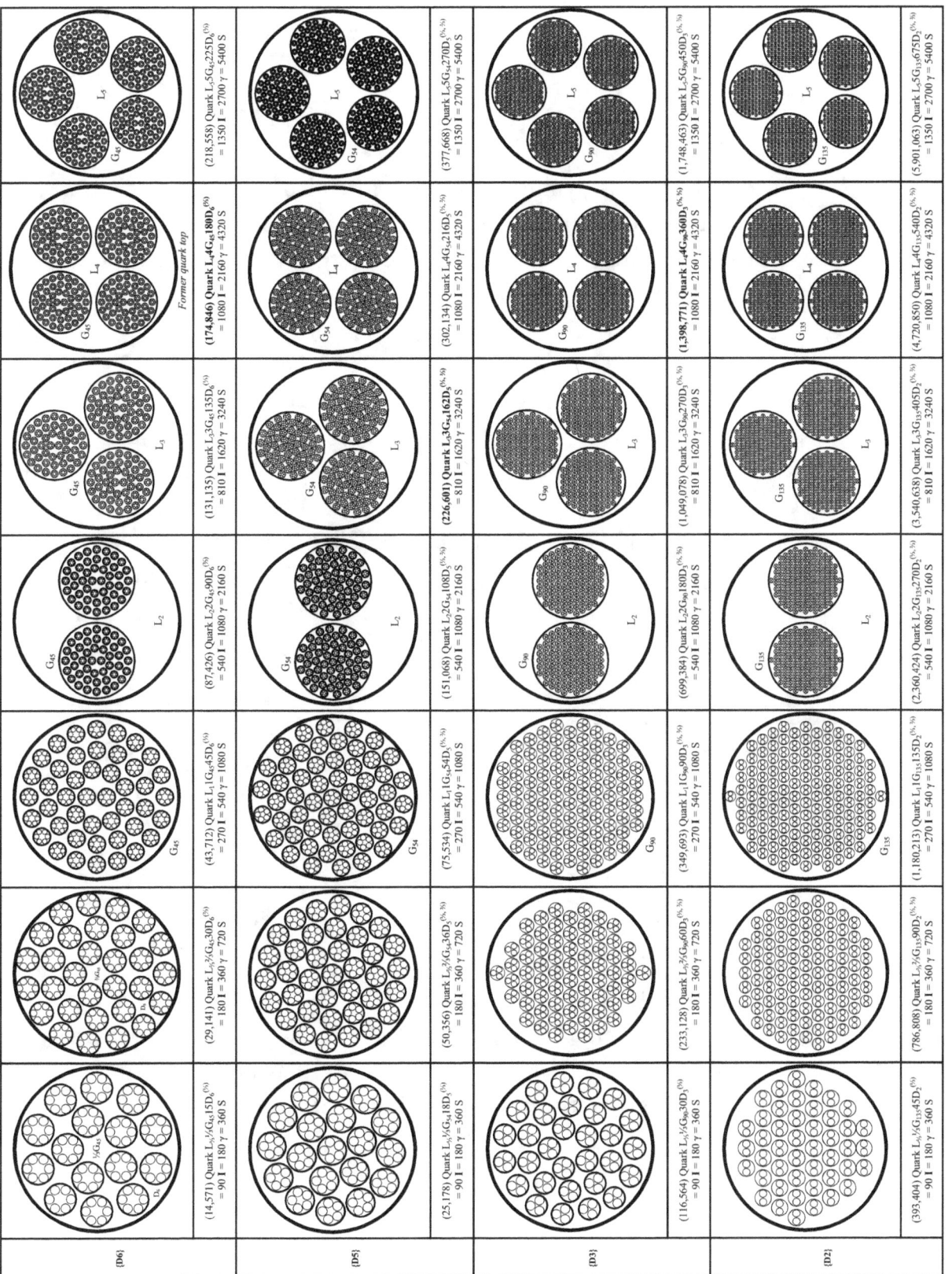

QUARKS BASED ON {D2} INTERVALIC SYMMETRY

And this is the last symmetry which is intended to make quarks, as it is supposed that the symmetry {D1} does not yield any quarks since its energy balance is totally different from the energy balance usually shown by quarks and leptons-massive bosons.

$L_{⅓} = ⅓\ G_{135} = 45\ D_2 = 393{,}404.23 — 419{,}083.13 →$ quark $L⅓D2^{(⅓)}$
$L_{⅔} = ⅔\ G_{135} = 90\ D_2 = 786{,}808.46 — 838{,}166.26 →$ quark $L⅔D2^{(⅓,\ ⅔)}$
$L_1 = 1\ G_{135} = 135\ D_2 = 1{,}180{,}212.7 — 1{,}257{,}249.4 →$ quark $L1D2^{(⅓,\ ⅔)}$
$L_2 = 2\ G_{135} = 270\ D_2 = 2{,}360{,}425.4 — 2{,}514{,}498.8 →$ quark $L2D2^{(⅓,\ ⅔)}$
$L_3 = 3\ G_{135} = 405\ D_2 = 3{,}540{,}638.1 — 3{,}771{,}748.2 →$ quark $L3D2^{(⅓,\ ⅔)}$
$L_4 = 4\ G_{135} = 540\ D_2 = 4{,}720{,}850.8 — 5{,}028{,}997.6 →$ quark $L4D2^{(⅓,\ ⅔)}$
$L_5 = 5\ G_{135} = 675\ D_2 = 5{,}901{,}063.5 — 6{,}286{,}247.0 →$ quark $L5D2^{(⅓,\ ⅔)}$
$L_6 = 6\ G_{135} = 810\ D_2 = 7{,}081{,}276.2 — 7{,}543{,}496.4$

The quark $L3D2^{(⅓,\ ⅔)}$ would be the last quark pertaining to the principal intervalic sequence.

The total mass energy of the intervalic symmetry {D2} is the sum of all its constituent quarks: $16\ L_1 = 18{,}883{,}403 — 20{,}115{,}992\ (MeV/c^2)$.

QUARKS BASED ON {D1} INTERVALIC SYMMETRY

As {D1} is a limit symmetry since the dalino 1 is the subatomic particle closest to the intervalino, it only can make particles above the threshold temperature. Therefore this symmetry has not importance at low energies, but only at the intervalic primordial aggregation.

QUARKS FLAVOURS AND INTERVALIC SYMMETRIES

It is really difficult to understand how have been possible that it have been generally accepted some obviously wrong affirmations of SM. For example, regarding the constituent quarks of, say, the SU(3) nonet of mesons with spin 0, it is clear the constituent light quarks of the π meson can't be, by no means, the same quarks *up* and *down* of nucleons... unless it was stupidly postulated that sometimes some quarks can have "free masses", which however they are not really "free" because the mass of the π meson is precisely always the same (!). The intervalic structure of quarks involves dramatic charges in quantum chromodynamics because colour and gluon fields become unnecessary and superfluous assumptions in IT, as I have explained in other sites. The existence of colour and gluons is based on some features and results, like those of the e^-e^+ annihilation, the electromagnetic decay of π^0 meson, the hadronic decay of $W^\pm$ and Z^0, the *jets* events in e^-e^+ annihilations, the symmetry of the baryon states, the strong interaction, etc., but all of them can be explained in a much more fundamental way through the intervalic symmetries in spite of the fantastic absurdity about the 3 colours and the 8 gluon fields postulated by SM.

In a similar way as colour, *flavour* becomes irrelevant since it is substituted by the *intervalic structure* of quarks. It is clear that the traditional model of quarks is, at least, incomplete because it does no explain the hundreds of particles detected experimentally, but only a few of them. This model does not lead to any place. Moreover, its naïve combinations of quarks, *mixing flavours* of all kinds and conditions, is absolutely misleading although Lie groups may look mathematically pretty to somebody, since it is not based on any reliable fundamental feature, but only in the trust of the existence of underlying symmetries in Nature —which of course ever can be *partially* found through very different ways, inclusive through wrong means, as is the case of SM—. Nevertheless, we are not satisfied with some *partial* symmetries yet, but we wish to find the whole reliable underlying symmetries of Nature. And it is sure that they are not those ones derived miserably from the flavours, colours, arbitrary mixing angles and other disastrous features postulated by SM.

The most economic assumption in IT regarding the aggregation of quarks (and also any other particles) postulates a homogeneity of symmetries, which states that there is *no mixing of intervalic symmetries* in the assembly of quarks. It can be shown that all the detected baryons and meson can be fully explained as aggregations of quarks pertaining to the same family of intervalic symmetries.

Therefore, the assembly of hadrons, postulated by SM to be composed by any mixture containing several of the supposed strange, charm, beauty or top quarks, is absolutely false. Such hadrons are monteverdinos composed by *lisztinos pertaining exclusively to a unique family of intervalic symmetries*, for example:

- Hadrons with up, down, strange → Monteverdino of {D45} symmetry
- Hadrons with charm → Monteverdino of {D30} intervalic symmetry
- Hadrons with beauty → Monteverdino of {D18} intervalic symmetry
- Hadrons with top → Monteverdino of {D6} intervalic symmetry

The logical economy and elegance of the intervalic postulate on the homogeneity of the intervalic symmetries in the assembly of subatomic particles speaks by itself. Needless to say that according to IT the three traditional quarks families and their half a dozen of flavours postulated by SM, or in other words, the full QCD and the traditional model of flavoured quarks become irrelevant. Fortunately, all experimental data remain, and only need to be interpreted in a radical new way, according to the new intervalic symmetries. As the concepts which handles SM are likely to be a bag of mistakes, it is sometimes awkward to establish a meaningful correspondence between those erroneous concepts and the right ones introduced by IT. From now on we are not going to try it unless it can help to clarify some physical feature. SM can not be repaired by no means because it is wrong from the most superficial up to the very last foundations. Therefore it sinks noisily anyway, and can only be discarded completely, quickly and at once, before to suffer a resounding defeat. In resume, right after the postulation of IT, the Standard Model is perhaps not yet a whimsical discipline of *Science*, but of *History*.

LISTING THE INTERVALIC STRUCTURES OF QUARKS

Ordering in a list all these symmetries which have been logical and systematically yielded from the intervalic units, we have got an impressive set of intervalic structurefulness, a total of 49 allowed intervalic structures of quarks which fit superbly with experimental data in an astonishing way and can make a lot of predictions of all kinds, easily falsable or verifiable. And most important, all this have been reached without having introduced any constant or parameter at every step, in the whole theory.

$L_{1/3} = 1/3\ G_6 = 2\ D_{45} = 34.537545 - 36.791934 \rightarrow$ quark $L{1/3}D45^{(1/3)}$
$L_{2/3} = 2/3\ G_6 = 4\ D_{45} = 69.075090 - 73.583868 \rightarrow$ quark $L{2/3}D45^{(1/3, 2/3)}$
$L_1 = 1\ G_6 = 6\ D_{45} = 103.61263 - 110.37580 \rightarrow$ quark $L1D45^{(1/3, 2/3)}$
$L_2 = 2\ G_6 = 12\ D_{45} = 207.22526 - 220.75160 \rightarrow$ quark $L2D45^{(1/3, 2/3)}$
$L_3 = 3\ G_6 = 18\ D_{45} = 310.83790 - 331.12740 \rightarrow$ quark $L3D45^{(1/3, 2/3)}$
$L_4 = 4\ G_6 = 24\ D_{45} = 414.45053 - 441.50320 \rightarrow$ quark $L4D45^{(1/3, 2/3)}$
$L_5 = 5\ G_6 = 30\ D_{45} = 518.06316 - 551.87900 \rightarrow$ quark $L5D45^{(1/3, 2/3)}$

$L_{1/3} = 1/3\ G_9 = 3\ D_{30} = 116.56421 - 124.17278 \rightarrow$ quark $L{1/3}D30^{(1/3)}$
$L_{2/3} = 2/3\ G_9 = 6\ D_{30} = 233.12843 - 248.34556 \rightarrow$ quark $L{2/3}D30^{(2/3)}$
$L_1 = 1\ G_9 = 9\ D_{30} = 349.69264 - 372.51833 \rightarrow$ quark $L1D30^{(1/3)}$
$L_2 = 2\ G_9 = 18\ D_{30} = 699.38520 - 745.03666 \rightarrow$ quark $L2D30^{(2/3)}$
$L_3 = 3\ G_9 = 27\ D_{30} = 1{,}049.0779 - 1{,}117.5550 \rightarrow$ quark $L3D30^{(1/3)}$
$L_4 = 4\ G_9 = 36\ D_{30} = 1{,}398.7705 - 1{,}490.0733 \rightarrow$ quark $L4D30^{(2/3)}$
$L_5 = 5\ G_9 = 45\ D_{30} = 1{,}748.4631 - 1{,}862.5916 \rightarrow$ quark $L5D30^{(1/3)}$

$L_{1/3} = 1/3\ G_{15} = 5\ D_{18} = 539.64917 - 574.87400 \rightarrow$ quark $L{1/3}D18^{(1/3)}$
$L_{2/3} = 2/3\ G_{15} = 10\ D_{18} = 1{,}079.2982 - 1{,}149.7480 \rightarrow$ quark $L{2/3}D18^{(2/3)}$
$L_1 = 1\ G_{15} = 15\ D_{18} = 1{,}618.9475 - 1{,}724.6220 \rightarrow$ quark $L1D18^{(1/3)}$
$L_2 = 2\ G_{15} = 30\ D_{18} = 3{,}237.8950 - 3{,}449.2443 \rightarrow$ quark $L2D18^{(2/3)}$
$L_3 = 3\ G_{15} = 45\ D_{18} = 4{,}856.8425 - 5{,}173.8660 \rightarrow$ quark $L3D18^{(1/3)}$

$L_4 = 4\ G_{15} = 60\ D_{18} = 6{,}475.7900 — 6{,}898.4886 \rightarrow$ quark $L4D18^{(2/3)}$
$L_5 = 5\ G_{15} = 75\ D_{18} = 8{,}094.7375 — 8{,}623.1100 \rightarrow$ quark $L5D18^{(1/3)}$

$L_{1/3} = 1/3\ G_{45} = 15\ D_6 = 14{,}570.527 — 15{,}521.598 \rightarrow$ quark $L1/3D6^{(1/3)}$
$L_{2/3} = 2/3\ G_{45} = 30\ D_6 = 29{,}141.055 — 31{,}043.196 \rightarrow$ quark $L2/3D6^{(2/3)}$
$L_1 = 1\ G_{45} = 45\ D_6 = 43{,}711.582 — 46{,}564.794 \rightarrow$ quark $L1D6^{(1/3)}$
$L_2 = 2\ G_{45} = 90\ D_6 = 87{,}423.165 — 93{,}129.588 \rightarrow$ quark $L2D6^{(2/3)}$
$L_3 = 3\ G_{45} = 135\ D_6 = 131{,}134.74 — 139{,}694.38 \rightarrow$ quark $L3D6^{(1/3)}$
$L_4 = 4\ G_{45} = 180\ D_6 = 174{,}846.32 — 186{,}259.17 \rightarrow$ quark $L4D6^{(2/3)}$
$L_5 = 5\ G_{45} = 225\ D_6 = 218{,}557.91 — 232{,}823.97 \rightarrow$ quark $L5D6^{(1/3)}$

$L_{1/3} = 1/3\ G_{54} = 18\ D_5 = 25{,}177.871 — 26{,}821.321 \rightarrow$ quark $L1/3D5^{(1/3)}$
$L_{2/3} = 2/3\ G_{54} = 36\ D_5 = 50{,}355.742 — 53{,}642.642 \rightarrow$ quark $L2/3D5^{(1/3,\ 2/3)}$
$L_1 = 1\ G_{54} = 54\ D_5 = 75{,}533.615 — 80{,}463.964 \rightarrow$ quark $L1D5^{(1/3,\ 2/3)}$
$L_2 = 2\ G_{54} = 108\ D_5 = 151{,}067.23 — 160{,}927.93 \rightarrow$ quark $L2D5^{(1/3,\ 2/3)}$
$L_3 = 3\ G_{54} = 162\ D_5 = 226{,}600.84 — 241{,}391.89 \rightarrow$ quark $L3D5^{(1/3,\ 2/3)}$
$L_4 = 4\ G_{54} = 216\ D_5 = 302{,}134.45 — 321{,}855.85 \rightarrow$ quark $L4D5^{(1/3,\ 2/3)}$
$L_5 = 5\ G_{54} = 270\ D_5 = 377{,}668.07 — 402{,}319.82 \rightarrow$ quark $L5D5^{(1/3,\ 2/3)}$

$L_{1/3} = 1/3\ G_{90} = 30\ D_3 = 116{,}564.21 — 124{,}172.78 \rightarrow$ quark $L1/3D3^{(1/3)}$
$L_{2/3} = 2/3\ G_{90} = 60\ D_3 = 233{,}128.43 — 248{,}345.56 \rightarrow$ quark $L2/3D3^{(1/3,\ 2/3)}$
$L_1 = 1\ G_{90} = 90\ D_3 = 349{,}692.64 — 372{,}518.33 \rightarrow$ quark $L1D3^{(1/3,\ 2/3)}$
$L_2 = 2\ G_{90} = 180\ D_3 = 699{,}385.28 — 745{,}036.66 \rightarrow$ quark $L2D3^{(1/3,\ 2/3)}$
$L_3 = 3\ G_{90} = 270\ D_3 = 1{,}049{,}077.9 — 1{,}117{,}555.0 \rightarrow$ quark $L3D3^{(1/3,\ 2/3)}$
$L_4 = 4\ G_{90} = 360\ D_3 = 1{,}398{,}770.5 — 1{,}490{,}073.3 \rightarrow$ quark $L4D3^{(1/3,\ 2/3)}$
$L_5 = 5\ G_{90} = 450\ D_3 = 1{,}748{,}463.2 — 1{,}862{,}591.7 \rightarrow$ quark $L5D3^{(1/3,\ 2/3)}$

$L_{1/3} = 1/3\ G_{135} = 45\ D_2 = 393{,}404.23 — 419{,}083.13 \rightarrow$ quark $L1/3D2^{(1/3)}$
$L_{2/3} = 2/3\ G_{135} = 90\ D_2 = 786{,}808.46 — 838{,}166.26 \rightarrow$ quark $L2/3D2^{(1/3,\ 2/3)}$
$L_1 = 1\ G_{135} = 135\ D_2 = 1{,}180{,}212.7 — 1{,}257{,}249.4 \rightarrow$ quark $L1D2^{(1/3,\ 2/3)}$
$L_2 = 2\ G_{135} = 270\ D_2 = 2{,}360{,}425.4 — 2{,}514{,}498.8 \rightarrow$ quark $L2D2^{(1/3,\ 2/3)}$
$L_3 = 3\ G_{135} = 405\ D_2 = 3{,}540{,}638.1 — 3{,}771{,}748.2 \rightarrow$ quark $L3D2^{(1/3,\ 2/3)}$
$L_4 = 4\ G_{135} = 540\ D_2 = 4{,}720{,}850.8 — 5{,}028{,}997.6 \rightarrow$ quark $L4D2^{(1/3,\ 2/3)}$
$L_5 = 5\ G_{135} = 675\ D_2 = 5{,}901{,}063.5 — 6{,}286{,}247.0 \rightarrow$ quark $L5D2^{(1/3,\ 2/3)}$

INTERVALIC STRUCTURES OF FUNDAMENTAL PARTICLES ALLOWED BY THE INTERVALIC SYMMETRIES BELOW THE THRESHOLD TEMPERATURE ACCORDING TO THE 16 DALINAR SYMMETRIES

DALINAR SYMMETRY	LEPTONS-MASSIVE BOSONS	QUARKS
{D270}	G1D270$^{\pm}$ lepton (0.5) = e	
{D135}		
{D90}		
{D54}		
{D45}	G6D45$^{\pm}$ lepton (106) = μ	L⅓D45$^{(⅓)}$ (35) L⅔D45$^{(⅓, ⅔)}$ (69) L1D45$^{(⅓, ⅔)}$ (104) L2D45$^{(⅓, ⅔)}$ (207) L3D45$^{(⅓, ⅔)}$ (311) L4D45$^{(⅓, ⅔)}$ (414) L5D45$^{(⅓, ⅔)}$ (518)
{D30}	G9D30$^{\pm}$ lepton (373)	L⅓D30$^{(⅓)}$ (117) L⅔D30$^{(⅔)}$ (233) L1D30$^{(⅓)}$ (350) L2D30$^{(⅔)}$ (699) L3D30$^{(⅓)}$ (1,049) L4D30$^{(⅔)}$ (1,399) L5D30$^{(⅓)}$ (1,748)
{D27}		
{D18}	G15D18$^{\pm}$ lepton (1,777) = τ	L⅓D18$^{(⅓)}$ (540) L⅔D18$^{(⅔)}$ (1,079) L1D18$^{(⅓)}$ (1,619) L2D18$^{(⅔)}$ (3,238) L3D18$^{(⅓)}$ (4,857) L4D18$^{(⅔)}$ (6,476) L5D18$^{(⅓)}$ (8,095)

INTERVALIC STRUCTURES OF FUNDAMENTAL PARTICLES ALLOWED BY THE 16 DALINAR SYMMETRIES BELOW THE THRESHOLD TEMPERATURE ACCORDING TO THE 16 DALINAR SYMMETRIES

DALINAR SYMMETRY	LEPTONS-MASSIVE BOSONS	QUARKS
{D15}		
{D10}		
{D9}		
{D6}	$G45D6^{\pm}$ (46,565) = $Z^{\pm}$ $L2G45D6^{0}$ (91,188) = Z^{0}	$L\frac{1}{3}D6^{(1/3)}$ (14,571) $L\frac{2}{3}D6^{(2/3)}$ (29,141) $L1D6^{(1/3)}$ (43,712) $L2D6^{(2/3)}$ (87,426) $L3D6^{(1/3)}$ (131,135) $L4D6^{(2/3)}$ (174,846) $L5D6^{(1/3)}$ (218,558)
{D5}	$G54D5^{\pm}$ (80,423) = $W^{\pm}$ $L2G54D5^{0}$ (160,928) = W^{0}	$L\frac{1}{3}D5^{(1/3)}$ (25,178) $L\frac{2}{3}D5^{(1/3, 2/3)}$ (50,356) $L1D5^{(1/3, 2/3)}$ (75,534) $L2D5^{(1/3, 2/3)}$ (151,068) $L3D5^{(1/3, 2/3)}$ (226,601) $L4D5^{(1/3, 2/3)}$ (302,134) $L5D5^{(1/3, 2/3)}$ (377,668)
{D3}	$G90D3^{\pm}$ (372,518) = $Y^{\pm}$ $L2G90D3^{0}$ (745,037) = Y^{0}	$L\frac{1}{3}D3^{(1/3)}$ (116,564) $L\frac{2}{3}D3^{(1/3, 2/3)}$ (233,128) $L1D3^{(1/3, 2/3)}$ (349,693) $L2D3^{(1/3, 2/3)}$ (699,384) $L3D3^{(1/3, 2/3)}$ (1,049,078) $L4D3^{(1/3, 2/3)}$ (1,398,771) $L5D3^{(1/3, 2/3)}$ (1,748,463)
{D2}	$G135D2^{\pm}$ (1,257,249) = $X^{\pm}$ $L2G135D2^{0}$ (2,514,499) = X^{0}	$L\frac{1}{3}D2^{(1/3)}$ (393,404) $L\frac{2}{3}D2^{(1/3, 2/3)}$ (786,808) $L1D2^{(1/3, 2/3)}$ (1,180,213) $L2D2^{(1/3, 2/3)}$ (2,360,424) $L3D2^{(1/3, 2/3)}$ (3,540,638) $L4D2^{(1/3, 2/3)}$ (4,720,850) $L5D2^{(1/3, 2/3)}$ (5,901,063)
{D1}		

INTERVALIC STRUCTURES
OF SUBATOMIC PARTICLES
ALLOWED BY THE INTERVALIC SYMMETRIES (mass in MeV)

DALINAR SYMMETRY	LEPTONS-MASSIVE BOSONS AND NEUTRINOS	QUARKS (FRACTIONAL LISZTINOS)
{D270}	$G_1D_{270}^{(\pm)}$ (0.5) = $e^{\pm}$ electron $v_{D270} = v_e$ neutrino (1.1833119 · 10^{-14})	-
{D135}	-	-
{D90}	-	-
{D54}	-	-
{D45}	$G_6D_{45}^{(\pm)}$ (106) = $\mu^{\pm}$ muon $v_{D45} = v_{\mu}$ neutrino (2.0005108 · 10^{-7})	$L_{1/5}\frac{1}{3}G_62D_{45}^{(1/3)}$ (35) *last radiant decay quark* $L_{2/5}\frac{2}{3}G_64D_{45}^{(2/3)}$ (69) *constituent quark of π meson* $L_1 1G_66D_{45}^{(1/3, 2/3)}$ (104) $L_2 2G_612D_{45}^{(1/3, 2/3)}$ (207) **$L_3 3G_618D_{45}^{(1/3, 2/3)}$ (311)** *former quarks up, down* $L_4 4G_624D_{45}^{(1/3, 2/3)}$ (414) $L_5 5G_630D_{45}^{(1/3, 2/3)}$ (518) *former quark strange*
{D30}	$G_9D_{30}^{(\pm)}$ (373) - v_{D30} -	$L_{1/5}\frac{1}{3}G_93D_{30}^{(1/3)}$ (117) $L_{2/5}\frac{2}{3}G_96D_{30}^{(2/3)}$ (233) $L_1 1G_99D_{30}^{(1/3)}$ (350) $L_2 2G_918D_{30}^{(2/3)}$ (699) $L_3 3G_927D_{30}^{(1/3)}$ (1,049) **$L_4 4G_936D_{30}^{(2/3)}$ (1,399)** *former quark charm* $L_5 5G_945D_{30}^{(1/3)}$ (1,748)
{D27}	-	-
{D18}	$G_{15}D_{18}^{(\pm)}$ (1,777) = $\tau^{\pm}$ tau $v_{D18} = v_{\tau}$ neutrino (2.6777745 · 10^{-4})	$L_{1/5}\frac{1}{3}G_{15}5D_{18}^{(1/3)}$ (540) $L_{2/5}\frac{2}{3}G_{15}10D_{18}^{(2/3)}$ (1,079) $L_1 1G_{15}15D_{18}^{(1/3)}$ (1,619) $L_2 2G_{15}30D_{18}^{(2/3)}$ (3,238) **$L_3 3G_{15}45D_{18}^{(1/3)}$ (4,857)** *former quark bottom* $L_4 4G_{15}60D_{18}^{(2/3)}$ (6,476) $L_5 5G_{15}75D_{18}^{(1/3)}$ (8,095)
{D15}		
{D10}		
{D9}		
{D6}	$G_{45}45D_6^{(\pm)}$ (46,565) $Z^{\pm}$ massive boson $L_2 2G_{45}90D_6^{(0)}$ (91,188) Z^0 **massive boson** v_{D6} neutrino	$L_{1/5}\frac{1}{3}G_{45}15D_6^{(1/3)}$ (14,571) $L_{2/5}\frac{2}{3}G_{45}30D_6^{(2/3)}$ (29,141) $L_1 1G_{45}45D_6^{(1/3)}$ (43,712) $L_2 2G_{45}90D_6^{(2/3)}$ (87,426) $L_3 3G_{45}135D_6^{(1/3)}$ (131,135) **$L_4 4G_{45}180D_6^{(2/3)}$ (174,846)** *former quark top* $L_5 5G_{45}225D_6^{(1/3)}$ (218,558)
{D5}	$G_{54}54D_5^{(\pm)}$ (80,423) **$W^{\pm}$ massive boson** $L_2 2G_{54}108D_5^{(0)}$ (160,928) W^0 **massive boson** v_{D5} neutrino	$L_{1/5}\frac{1}{3}G_{54}18D_5^{(1/3)}$ (25,178) $L_{2/5}\frac{2}{3}G_{54}36D_5^{(1/3, 2/3)}$ (50,356) $L_1 1G_{54}54D_5^{(1/3, 2/3)}$ (75,534) $L_2 2G_{54}108D_5^{(1/3, 2/3)}$ (151,068) **$L_3 3G_{54}162D_5^{(1/3, 2/3)}$ (226,601)** $L_4 4G_{54}216D_5^{(1/3, 2/3)}$ (302,134) $L_5 5G_{54}270D_5^{(1/3, 2/3)}$ (377,668)
{D3}	$G_{90}90D_3^{(\pm)}$ (372,518) $Y^{\pm}$ massive boson $L_2 2G_{90}180D_3^{(0)}$ (745,037) Y^0 **massive boson** v_{D3} neutrino	$L_{1/5}\frac{1}{3}G_{90}30D_3^{(1/3)}$ (116,564) $L_{2/5}\frac{2}{3}G_{90}60D_3^{(1/3, 2/3)}$ (233,128) $L_1 1G_{90}90D_3^{(1/3, 2/3)}$ (349,693) $L_2 2G_{90}180D_3^{(1/3, 2/3)}$ (699,384) $L_3 3G_{90}270D_3^{(1/3, 2/3)}$ (1,049,078) **$L_4 4G_{90}360D_3^{(1/3, 2/3)}$ (1,398,771)** $L_5 5G_{90}450D_3^{(1/3, 2/3)}$ (1,748,463)
{D2}	$G_{135}135D_2^{(\pm)}$ (1,257,249) **$X^{\pm}$ massive boson** $L_2 2G_{135}270D_2^{(0)}$ (2,514,499) X^0 massive boson v_{D2} neutrino	$L_{1/5}\frac{1}{3}G_{135}45D_2^{(1/3)}$ (393,404) $L_{2/5}\frac{2}{3}G_{135}90D_2^{(1/3, 2/3)}$ (786,808) $L_1 1G_{135}135D_2^{(1/3, 2/3)}$ (1,180,213) $L_2 2G_{135}270D_2^{(1/3, 2/3)}$ (2,360,424) $L_3 3G_{135}405D_2^{(1/3, 2/3)}$ (3,540,638) $L_4 4G_{135}540D_2^{(1/3, 2/3)}$ (4,720,850) $L_5 5G_{135}675D_2^{(1/3, 2/3)}$ (5,901,063)
{D1}	-	-

Lisztinian structure $L_{1/5}$
(quarks composed by 90 intervalinos)

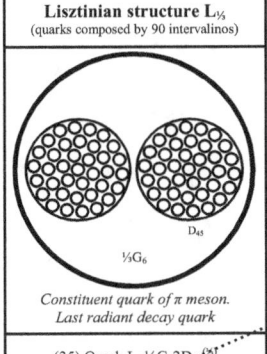

Constituent quark of π meson. Last radiant decay quark

(35) Quark $L_{1/5}\frac{1}{3}G_62D_{45}^{(1/3)}$
= 90 I = 180 γ = 360 S

Lisztinian structure L_3
(quarks composed by 810 intervalinos)

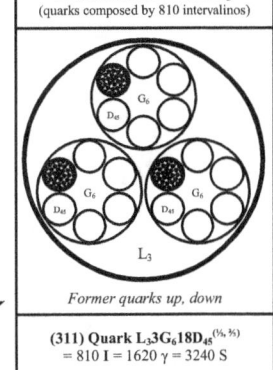

Former quarks up, down

(311) Quark $L_3 3G_6 18D_{45}^{(1/3, 2/3)}$
= 810 I = 1620 γ = 3240 S

Lisztinian structure L_5
(quarks composed by 1350 intervalinos)

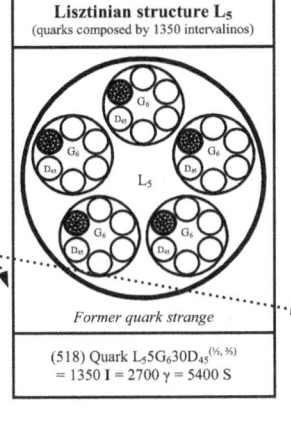

Former quark strange

(518) Quark $L_5 5G_6 30D_{45}^{(1/3, 2/3)}$
= 1350 I = 2700 γ = 5400 S

Lisztinian structure L_4
(quarks composed by 1080 intervalinos)

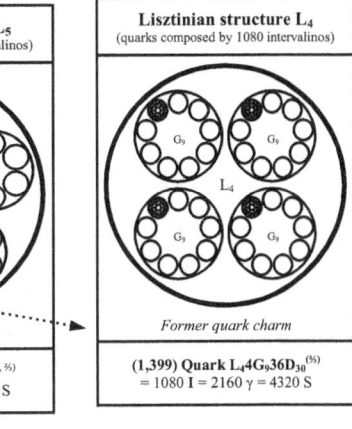

Former quark charm

(1,399) Quark $L_4 4G_9 36D_{30}^{(2/3)}$
= 1080 I = 2160 γ = 4320 S

Lisztinian structure L_3
(quarks composed by 810 intervalinos)

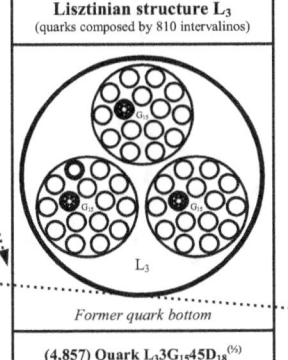

Former quark bottom

(4,857) Quark $L_3 3G_{15} 45D_{18}^{(1/3)}$
= 810 I = 1620 γ = 3240 S

Lisztinian structure L_4
(quarks composed by 1080 intervalinos)

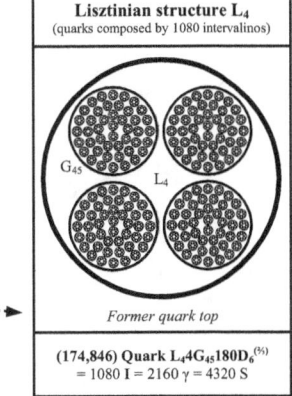

Former quark top

(174,846) Quark $L_4 4G_{45} 180D_6^{(2/3)}$
= 1080 I = 2160 γ = 4320 S

Lisztinian structure L_3
(quarks composed by 810 intervalinos)

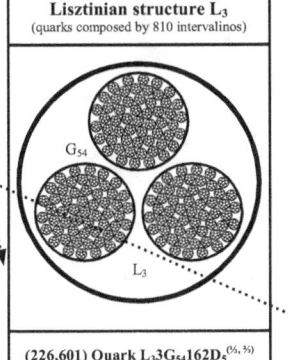

(226,601) Quark $L_3 3G_{54} 162D_5^{(1/3, 2/3)}$
= 810 I = 1620 γ = 3240 S

Lisztinian structure L_4
(quarks composed by 1080 intervalinos)

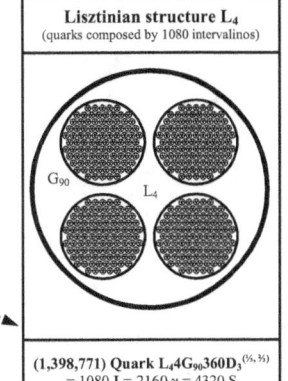

(1,398,771) Quark $L_4 4G_{90} 360D_3^{(1/3, 2/3)}$
= 1080 I = 2160 γ = 4320 S

Chapter 16

INTERVALIC QUARKS UP AND DOWN

NUCLEONIC QUARKS: GAUDINAR STRUCTURE

Let us recapitulate the gaudinar structure of nucleonic quarks. The primordial aggregation of intervalinos to make dalinos, and of dalinos to make gaudinos was symmetric under interchange and therefore follows Bose-Einstein statistics. On the contrary, the primordial aggregation of gaudinos made symmetric and antisymmetric states under interchange: the first is found in zero charge massive bosons and the second is found in lisztinos with fractional charge, that is to say, in *quarks*.

Regarding stable quarks constituent of nucleons, there are only seven modes to make those aggregation of dalinos 45 to make gaudinos 6:

$$6\,(D_{+45}) + 0\,(D_{-45}) = G_6^+$$
$$5\,(D_{+45}) + 1\,(D_{-45}) = G_6^{+2/3}$$
$$4\,(D_{+45}) + 2\,(D_{-45}) = G_6^{+1/3}$$

$$3\,(D_{+45}) + 3\,(D_{-45}) = G_6^{\,0}$$
$$2\,(D_{+45}) + 4\,(D_{-45}) = G_6^{\,-\frac{1}{3}}$$
$$1\,(D_{+45}) + 5\,(D_{-45}) = G_6^{\,-\frac{2}{3}}$$
$$0\,(D_{+45}) + 6\,(D_{-45}) = G_6^{\,-}$$

Since quarks have fractional charges, the only possible aggregations of gaudinos (G_6) to compose a quark are:

$$q(+\tfrac{2}{3}) = G_6^{+} + G_6^{-} + G_6^{+\frac{2}{3}}$$
$$q(+\tfrac{2}{3}) = G_6^{+} + G_6^{-\frac{2}{3}} + G_6^{+\frac{1}{3}}$$
$$q(+\tfrac{2}{3}) = G_6^{+\frac{2}{3}} + G_6^{-\frac{2}{3}} + G_6^{+\frac{2}{3}}$$
$$q(+\tfrac{2}{3}) = G_6^{+} + G_6^{-\frac{1}{3}} + G_6^{0}$$
$$q(+\tfrac{2}{3}) = G_6^{+\frac{2}{3}} + G_6^{-\frac{1}{3}} + G_6^{+\frac{1}{3}}$$
$$q(+\tfrac{2}{3}) = G_6^{+\frac{2}{3}} + G_6^{0} + G_6^{0}$$
$$q(+\tfrac{2}{3}) = G_6^{+\frac{1}{3}} + G_6^{0} + G_6^{+\frac{1}{3}}$$

$$q(-\tfrac{1}{3}) = G_6^{+} + G_6^{-} + G_6^{-\frac{1}{3}}$$
$$q(-\tfrac{1}{3}) = G_6^{+} + G_6^{-\frac{2}{3}} + G_6^{-\frac{2}{3}}$$
$$q(-\tfrac{1}{3}) = G_6^{+\frac{2}{3}} + G_6^{-} + G_6^{0}$$
$$q(-\tfrac{1}{3}) = G_6^{+\frac{1}{3}} + G_6^{-} + G_6^{+\frac{1}{3}}$$
$$q(-\tfrac{1}{3}) = G_6^{+\frac{2}{3}} + G_6^{-\frac{2}{3}} + G_6^{-\frac{1}{3}}$$
$$q(-\tfrac{1}{3}) = G_6^{+\frac{1}{3}} + G_6^{-\frac{2}{3}} + G_6^{0}$$
$$q(-\tfrac{1}{3}) = G_6^{+\frac{1}{3}} + G_6^{-\frac{1}{3}} + G_6^{-\frac{1}{3}}$$
$$q(-\tfrac{1}{3}) = G_6^{-\frac{1}{3}} + G_6^{0} + G_6^{0}$$

It can be supposed that states containing a gaudino 6 with zero charge would make unstable structures because they do not have intervalic nor electromagnetic energy at the last structure level, and therefore these states are not allowed. This clear constraint eliminates six possible combinations among gaudinos from the above lists.

On the other hand, it can be supposed that the constituent charges of these gaudinos will not be the elementary charge but fractional charges. If this assumption was held the only remaining combinations would be:

$$q(+\tfrac{2}{3}) = G_6^{+\tfrac{2}{3}} + G_6^{-\tfrac{2}{3}} + G_6^{+\tfrac{2}{3}}$$
$$q(+\tfrac{2}{3}) = G_6^{+\tfrac{2}{3}} + G_6^{-\tfrac{1}{3}} + G_6^{+\tfrac{1}{3}}$$

$$q(-\tfrac{1}{3}) = G_6^{+\tfrac{2}{3}} + G_6^{-\tfrac{2}{3}} + G_6^{-\tfrac{1}{3}}$$
$$q(-\tfrac{1}{3}) = G_6^{+\tfrac{1}{3}} + G_6^{-\tfrac{1}{3}} + G_6^{-\tfrac{1}{3}}$$

Since the intervalic energy at the gaudinar structure level should be greater than the intervalic energy at the lisztinian level, the first two intervalic structures of both quarks must be eliminated. Therefore, the intervalic structure of nucleonic quarks will be:

$$q \to L_3 = 3\, G_6 = 18\, D_{45} = 810\, I$$

$$u^{+\tfrac{2}{3}} \to L_3 = G_6^{+\tfrac{2}{3}} + G_6^{-\tfrac{1}{3}} + G_6^{+\tfrac{1}{3}} = 18\, D_{45} = 810\, I$$
$$d^{-\tfrac{1}{3}} \to L_3 = G_6^{+\tfrac{1}{3}} + G_6^{-\tfrac{1}{3}} + G_6^{-\tfrac{1}{3}} = 18\, D_{45} = 810\, I$$

We will make use in advance the astonishing results obtained in the deduction of the nucleons masses (cfr. Intervalic Nucleon) with null error made in our study of the intervalic nucleon. There were obtained the following nucleons and quarks masses:

$$m(p) = 2\, m(u) + m(d) + c^{\pm 2}\, I(e) = 938.2723027\ (MeV/c^2)$$
$$m(n) = m(u) + 2\, m(d) = 939.5656372\ (MeV/c^2)$$

$$m(u) = m(u)_{com} + c^{\pm 2}\, I(u) + c^{\pm 2}\, U(u) = 312.1359302\ (MeV/c^2)$$
$$m(d) = m(d)_{com} + c^{\pm 2}\, I(d) + c^{\pm 2}\, U(d) = 313.7148535\ (MeV/c^2)$$

Needless to say that some fantastic assumptions introduced in SM, like the acrobatic "free masses" of quarks, the elusive creation of mass through an unexplained spin-spin interaction —however single nucleus do not loose dramatically its mass—, and the mysterious Higgs mechanism (which looks like a certain medieval ether) are, from the point of

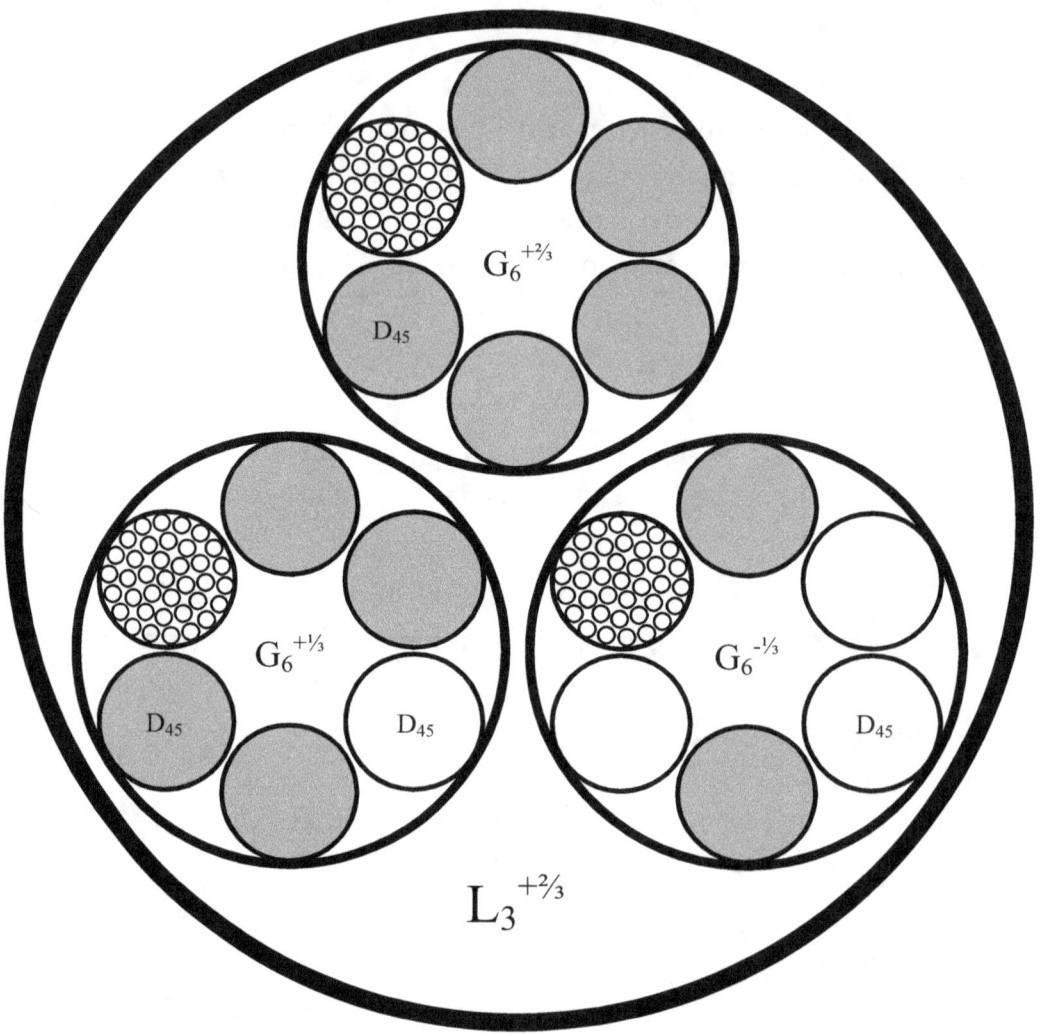

Figured intervalic structure of *nucleonic isoquark* $+^2/_3$ *(up)*:
$L_3^{+2/3} = 3\ G_6\ (G_6^{+2/3},\ G_6^{-1/3},\ G_6^{+1/3}) = 18\ D_{45}\ (11\ D_{+45},\ 7\ D_{-45}) = 810\ \mathbf{I}$

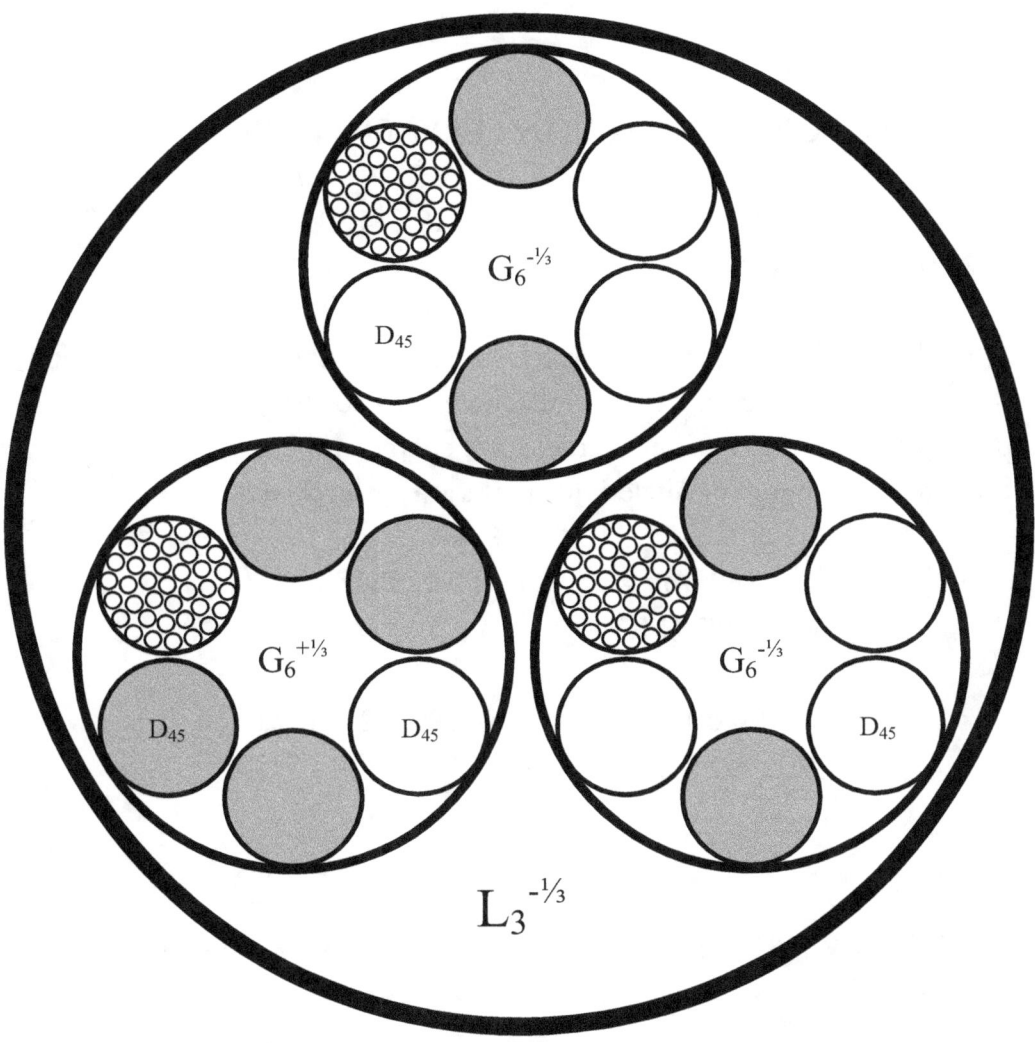

Figured intervalic structure of *nucleonic isoquark -⅓ (down)*:
$L_3^{+2/3} = 3\ G_6\ (G_6^{-1/3},\ G_6^{+1/3},\ G_6^{-1/3}) = 18\ D_{45}\ (8\ D_{+45},\ 10\ D_{-45}) = 810\ \mathbf{I}$

view of IT, naïve and clumsy assumptions which have become to be unnecessary and irrelevant, as we have discussed in other sites.

NUCLEONIC QUARKS INTERVALIC ENERGY

INTERVALIC ENERGY AT LISZTINIAN STRUCTURE LEVEL: L_3

The intervalic energy of quarks at the last structure level is simply the intervalic energy of its overall charge. According to the intervalic principle of equivalence between electric charge and energy, $I = c^{\pm 2} \hbar\, Q^{-2}$, the intervalic energy of nucleonic quarks at this level is:

$$I(u)_L = c^{\pm 2} \hbar\, (180\, q_I)^{-2} = c^{-1}\, 180^{-2} = 1.0295188 \cdot 10^{-13}\, (J) =$$
$$= 0.64257482\, (MeV/c^2)$$

$$I(d)_L = c^{\pm 2} \hbar\, (90\, q_I)^{-2} = c^{-1}\, 90^{-2} = 4.1180752 \cdot 10^{-13}\, (J) =$$
$$= 2.5702993\, (MeV/c^2)$$

INTERVALIC ENERGY AT GAUDINAR STRUCTURE LEVEL: $3\, G_6$

Each quark —lisztino 3— is composed by three fractional charged gaudinos 6, each of them has its proper intervalic energy according to its overall electric charge. The electric charge of gaudinos 6 can take in general the values: 0, $\pm 90 q_I$, $\pm 180 q_I$ and $\pm 270 q_I$. Its corresponding gaudinar intervalic energy will be respectively:

$I(G_6^0) = 0$
$I(G_6^{\pm 1/3}) = c^{\pm 2} \hbar \, (1/3 \, e)^{-2} = 90^{-2} \, c^{-1} = 2.5702993 \, (MeV/c^2)$
$I(G_6^{\pm 2/3}) = c^{\pm 2} \hbar \, (2/3 \, e)^{-2} = 180^{-2} \, c^{-1} = 0.64257482 \, (MeV/c^2)$
$I(G_6^{\pm 1}) = c^{\pm 2} \hbar \, e^{-2} = 270^{-2} \, c^{-1} = 0.28558881 \, (MeV/c^2)$

The total intervalic energy of the quark at the gaudinar structure level is the sum of the intervalic energy of each one of its three constituent gaudinos.

$I(u)_G = I(G_6^{+2/3}) + 2 \, I(G_6^{\pm 1/3}) = 5.78317342 \, (MeV/c^2)$
$I(d)_G = I(G_6^{-1/3}) + 2 \, I(G_6^{\pm 1/3}) = 7.71089790 \, (MeV/c^2)$

INTERVALIC ENERGY AT DALINAR STRUCTURE LEVEL: 18 D_{45}

Each gaudino 6 is composed by six dalinos 45. The intervalic energy of one dalino 45 is:

$I(D_{45}) = c^{\pm 2} \hbar \, (45 \, q_I)^{-2} = c^{\pm 2} \hbar \, [45 \sqrt{-(c^{-1}\hbar)}]^{-2} = 45^{-2} \, c^{-1} =$
$= 1.64723010 \cdot 10^{-12} \, (J) = 10.28119715 \, (MeV/c^2)$

Therefore the dalinar intervalic energy of the gaudino 6 is:

$I(G_6)_D = 6 \cdot I(D_{45}) = 9.8833806 \cdot 10^{-12} \, (J) = 61.687184 \, (MeV/c^2)$

And the dalinar intervalic energy of the lisztino 3 —quark— is:

$I(L_3)_D = 3 \cdot I(G_6) = 2.9650142 \cdot 10^{-11} \, (J) = 185.06155 \, (MeV/c^2)$

Or in other words: $I(L_3)_{D45} = 18 \cdot I(D_{45})$.

NUCLEONIC QUARKS TOTAL INTERVALIC ENERGY

The total intervalic energy of nucleonic quarks is the sum of the intervalic energy at all levels of its intervalic structure:

$$I(u)_t = I(L_3^{+2/3}) + I(G_6^{+2/3}) + 2I(G_6^{\pm 1/3}) + 18I(D_{45}) = 191.4872982 \; (MeV/c^2)$$

$$I(d)_t = I(L_3^{-1/3}) + I(G_6^{-1/3}) + 2I(G_6^{\pm 1/3}) + 18I(D_{45}) = 195.3427472 \; (MeV/c^2)$$

NUCLEONIC QUARKS ELECTROMAGNETIC ENERGY

NUCLEONIC QUARKS MATERIAL ENERGY

From the nucleonic quark mass we can deduce the *material* energy of quarks —of electromagnetic origin—, which is *by definition* the difference between the total and the intervalic energies:

$$U(u)_t = c^{\pm 2}m(u) - I(u)_t = 1.93300503 \cdot 10^{-11} (J) = 120.6486320 \; (MeV/c^2)$$

$$U(d)_t = c^{\pm 2}m(d) - I(d)_t = 1.89653105 \cdot 10^{-11} (J) = 118.3721063 \; (MeV/c^2)$$

Now we have got the total intervalic / electromagnetic energy ratios of nucleonic quarks:

$I(u)/U(u) = 1.587148524$
$I(u)/E(u)_t = 0.613474065$
$U(u)/E(u)_t = 0.386525934$

$I(d)/U(d) = 1.650243062$
$I(d)/E(d)_t = 0.622676118$
$U(d)/E(d)_t = 0.377323881$

ELECTROMAGNETIC ENERGY AT LISZTINIAN STRUCTURE LEVEL: L_3

Using the already known magnitude in IT of the intervalic nucleonic quark radius $r_q = 6.88054386 \cdot 10^{-16}$ (m), the electromagnetic energy in the last structure level of nucleonic quarks is:

$U(u)_L = U(L_3^{+2/3}) = \frac{1}{2} (1/4\pi\varepsilon_0) (180\ q_I)^2 / r_q = 7.451218056 \cdot 10^{-14}$ (J)
$= 0.465068249$ (MeV/c²)

$U(d)_L = U(L_3^{-1/3}) = \frac{1}{2} (1/4\pi\varepsilon_0) (90\ q_I)^2 / r_q = 1.862804514 \cdot 10^{-14}$ (J) =
$= 0.116267062$ (MeV/c²)

DISTANCE BETWEEN GAUDINOS 6

The electromagnetic energy at the gaudinar level is the sum of the energies of the three constituent gaudinos 6:

$U(u)_G = U(G_6^{+2/3}) + 2\ U(G_6^{\pm 1/3})$
$U(d)_G = U(G_6^{-1/3}) + 2\ U(G_6^{\pm 1/3})$

In a traditional view we can suppose that the difference in the material mass of quarks is due to the electromagnetic interaction between constituent gaudinos 6 at gaudinar level. To make this calculus we have to eliminate the other involved energy levels: the lisztinian and the dalinar electromagnetic contributions to mass:

$$U(u)_G = U(u)_t - U(u)_L - U(u)_D = 120.1835638 - U(u)_D$$
$$U(d)_G = U(d)_t - U(d)_L - U(d)_D = 118.2558392 - U(d)_D$$

Combining the preceding expressions we have:

$$U(u)_G - U(d)_G = 1.927724562 \ (MeV/c^2) = 3.088556592 \cdot 10^{-13} \ (J)$$
$$= E_q(G_6^{+\frac{2}{3}} G_6^{+\frac{1}{3}} G_6^{-\frac{1}{3}}) - E_q(G_6^{-\frac{1}{3}} G_6^{+\frac{1}{3}} G_6^{-\frac{1}{3}}) = \{[(\tfrac{2}{3} \cdot \tfrac{1}{3}) + (\tfrac{2}{3} \cdot \tfrac{1}{3}) + (\tfrac{1}{3})^2] -$$
$$- [(\tfrac{1}{3})^2 + (\tfrac{1}{3})^2 + (\tfrac{1}{3})^2]\} (1/4\pi\varepsilon_0) \ e^2 / d_{G6} = (2/9) (1/4\pi\varepsilon_0) \ e^2 / d_{G6}$$

From here we can deduce the distance between gaudinos 6 inside quark:

$$d_{G6} = 1.659948039 \cdot 10^{-16} \ (m)$$

QUARKIC GAUDINO 6 RADIUS

Immediately we can deduce a magnitude of the quarkic gaudino 6 radius suited for the calculation of the electromagnetic energy:

$$r_{G6} = d_{G6} / 2 = 8.299740193 \cdot 10^{-17} \ (m)$$

ELECTROMAGNETIC ENERGY AT GAUDINAR STRUCTURE LEVEL: 3 G_6

Henceforth, the electromagnetic energy of quarkic gaudinos 6 is:

$$U(G_6^{2/3}) = \tfrac{1}{2}(1/4\pi\varepsilon_0)(180\,q_I)^2 / r_{G6} = 6.177112952 \cdot 10^{-13}\,(J) =$$
$$= 3.85544898\,(MeV/c^2)$$

$$U(G_6^{1/3}) = \tfrac{1}{2}(1/4\pi\varepsilon_0)(90\,q_I)^2 / r_{G6} = 1.544278238 \cdot 10^{-13}\,(J) =$$
$$= 0.963862244\,(MeV/c^2)$$

And the total electromagnetic energy of quarks at gaudinar level is:

$$U(u)_G = U(G_6^{2/3}) + 2\,U(G_6^{1/3}) = 5.783173468\,(MeV/c^2)$$
$$U(d)_G = 3\,U(G_6^{1/3}) = 2.891586735\,(MeV/c^2)$$

ELECTROMAGNETIC ENERGY AT DALINAR STRUCTURE LEVEL: 18 D_{45}

Since we know the electromagnetic energy at all the preceding energy levels of the intervalic structure we have immediately:

$$U(u)_D = U(u)_t - U(u)_L - U(u)_G = 114.4003903\,(MeV/c^2)$$
$$U(d)_D = U(d)_t - U(d)_L - U(d)_G = 115.3642525\,(MeV/c^2)$$

Therefore the electromagnetic energy of the constituent gaudinos 6 at the dalinar level is:

$$U(G_6^{\pm 1/3})_D = \tfrac{1}{3}\,U(d)_D = 38.45475083\,(MeV/c^2)$$
$$U(G_6^{\pm 2/3})_D = U(u)_D - 2\,U(G_6^{\pm 1/3})_D = 37.49088864\,(MeV/c^2)$$

And the electromagnetic energy of each one of the six constituent dalinos 45 at the dalinar level of the intervalic structure is:

$$U_{G6\pm\frac{1}{3}}(D_{45}) = (1/6)\ U(G_6^{\pm\frac{1}{3}})_D = 6.409125138\ (MeV/c^2)$$
$$U_{G6\pm\frac{2}{3}}(D_{45}) = (1/6)\ U(G_6^{\pm\frac{2}{3}})_D = 6.24848144\ (MeV/c^2)$$

QUARKIC DALINO 45 RADIUS

Since the value of the electromagnetic energy of the quarkic dalino 45 is known, we can calculate the magnitude of the *quarkic dalino 45 radius*, r_{D45}:

$$U_{G6\pm\frac{2}{3}}(D_{45}) = \tfrac{1}{2}\ (1/4\pi\varepsilon_0)\ (45\ q_I)^2\ /\ r_{G6\pm\frac{2}{3}}(D_{45}) =$$
$$= 1.001117531 \cdot 10^{-12}\ (J)$$
$$r_{G6\pm\frac{2}{3}}(D_{45}) = 3.200696811 \cdot 10^{-18}\ (m)$$

$$U_{G6\pm\frac{1}{3}}(D_{45}) = \tfrac{1}{2}\ (1/4\pi\varepsilon_0)\ (45\ q_I)^2\ /\ r_{G6\pm\frac{1}{3}}(D_{45}) =$$
$$= 1.02685550 \cdot 10^{-12}\ (J)$$
$$r_{G6\pm\frac{1}{3}}(D_{45}) = 3.12047512 \cdot 10^{-18}\ (m)$$

These magnitudes are slightly greater than the muonic dalino 45 radius, $r_{D45}(\mu) = 2.746832954 \cdot 10^{-18}\ (m)$, as expected.

Due to the hexagonal symmetry of gaudino 6, the distance among dalinos 45 will be approximately:

$$d_{G6\pm\frac{1}{3}}(D_{45}) \approx 15\ (1/4\pi\varepsilon_0)\ (45\ q_I)^2\ /\ U(G_6^{\pm\frac{1}{3}})_D =$$
$$= 1.560235928 \cdot 10^{-17}\ (m)$$
$$d_{G6\pm\frac{2}{3}}(D_{45}) \approx 15\ (1/4\pi\varepsilon_0)\ (45\ q_I)^2\ /\ U(G_6^{\pm\frac{2}{3}})_D =$$
$$= 1.600348405 \cdot 10^{-17}\ (m)$$

This magnitude are likewise slightly greater than the distance among muonic dalinos 45, $d_{D45}(\mu) = 1.373416477 \cdot 10^{-17}$ (m), as expected.

NUCLEONIC QUARKS ENERGY RATIOS

Now we can write the energy ratios of nucleonic quarks at every level of the intervalic structure with such astonishing precision which can never be imagined by SM at its best.

The data of the constituent dalinos 45 are:

$I(D_{45}) = 1.6472301 \cdot 10^{-12}$ (J) $= 10.28119715$ (MeV/c²)
$U_{G6\pm\frac{1}{3}}(D_{45}) = (1/6)\, U(G_6^{\pm\frac{1}{3}})_D = 6.409125138$ (MeV/c²)
$U_{G6\pm\frac{2}{3}}(D_{45}) = (1/6)\, U(G_6^{\pm\frac{2}{3}})_D = 6.24848144$ (MeV/c²)
$m_{G6\pm\frac{1}{3}}(D_{45}) = 2.67408560 \cdot 10^{-12}$ (J) $= 16.69032229$ (MeV/c²)
$m_{G6\pm\frac{2}{3}}(D_{45}) = 2.648347631 \cdot 10^{-12}$ (J) $= 16.52967859$ (MeV/c²)

And their corresponding energy ratios:

$I(D_{45})/U_{G6\pm\frac{2}{3}}(D_{45}) = 1.645381324$
$I(D_{45})/E_{G6\pm\frac{2}{3}}(D_{45})_{tot} = 0.621984093$
$U_{G6\pm\frac{2}{3}}(D_{45})/E_{G6\pm\frac{2}{3}}(D_{45})_{tot} = 0.378015906$

$I(D_{45})/U_{G6\pm\frac{1}{3}}(D_{45}) = 1.604149853$
$I(D_{45})/E_{G6\pm\frac{1}{3}}(D_{45})_{tot} = 0.61599752$
$U_{G6\pm\frac{1}{3}}(D_{45})/E_{G6\pm\frac{1}{3}}(D_{45})_{tot} = 0.384002479$

The data of the constituent gaudinos 6 are:

$I(G_6^{\pm\frac{1}{3}})_{tot} = 1.029518813 \cdot 10^{-11}$ (J) $= 64.2574822$ (MeV/c²)
$I(G_6^{\pm\frac{2}{3}})_{tot} = 9.98633248 \cdot 10^{-12}$ (J) $= 62.32975772$ (MeV/c²)
$U(G_6^{\pm\frac{1}{3}})_{tot} = 6.624416481 \cdot 10^{-12}$ (J) $= 41.34633762$ (MeV/c²)

$U(G_6^{\pm 2/3})_{tot} = 6.315560825 \cdot 10^{-12} \, (J) = 39.41861307 \, (MeV/c^2)$
$m(G_6^{\pm 1/3}) = 1.69196046 \cdot 10^{-11} \, (J) = 105.6038198 \, (MeV/c^2)$
$m(G_6^{\pm 2/3}) = 1.63018933 \cdot 10^{-11} \, (J) = 101.7483708 \, (MeV/c^2)$

And their corresponding energy ratios:

$I(G_6^{\pm 1/3})/U(G_6^{\pm 1/3}) = 1.554127545$
$I(G_6^{\pm 1/3})/E(G_6^{\pm 1/3})_{tot} = 0.608476874$
$U(G_6^{\pm 1/3})/E(G_6^{\pm 1/3})_{tot} = 0.391523125$

$I(G_6^{\pm 2/3})/U(G_6^{\pm 2/3}) = 1.581226554$
$I(G_6^{\pm 2/3})/E(G_6^{\pm 2/3})_{tot} = 0.61258728$
$U(G_6^{\pm 2/3})/E(G_6^{\pm 2/3})_{tot} = 0.387412719$

The data of the constituent lisztinos 3 are:

$I(L_3^{+2/3})_{tot} \quad I(u)_{tot} = 3.067966082 \cdot 10^{-11} \, (J) = 191.4872982 \, (MeV/c^2)$
$I(L_3^{-1/3})_{tot} \quad I(d)_{tot} = 3.129737211 \cdot 10^{-11} \, (J) = 195.3427472 \, (MeV/c^2)$
$U(L_3^{+2/3})_{tot} \quad U(u)_{tot} = 1.93300503 \cdot 10^{-11} \, (J) = 120.6486320 \, (MeV/c^2)$
$U(L_3^{-1/3})_{tot} \quad U(d)_{tot} = 1.89653105 \cdot 10^{-11} \, (J) = 118.3721063 \, (MeV/c^2)$
$m(L_3^{+2/3}) \quad m(u) = 5.000971112 \cdot 10^{-11} \, (J) = 312.1359302 \, (MeV/c^2)$
$m(L_3^{-1/3}) \quad m(d) = 5.026268264 \cdot 10^{-11} \, (J) = 313.7148535 \, (MeV/c^2)$

And their corresponding energy ratios:

$I(u)/U(u) = 1.587148524$
$I(u)/E(u)_{tot} = 0.613474065$
$U(u)/E(u)_{tot} = 0.386525934$

$I(d)/U(d) = 1.650243062$
$I(d)/E(d)_{tot} = 0.622676118$
$U(d)/E(d)_{tot} = 0.377323881$

The average ratios between both quarks is:

$\langle I(d)/U(d) \rangle = 1.618695793 \sim \Phi$
$\langle I(d)/E(d)_{tot} \rangle = 0.618075091 \sim \Phi^{-1}$
$\langle U(d)/E(d)_{tot} \rangle = 0.381924907 \sim 1\text{-}\Phi^{-1}$

The deviation of these ratios from the golden mean, $\Phi = 1.61803398875$, is:

$\Delta[\langle I(d)/U(d) \rangle] = +0.0409018\%$
$\Delta[\langle I(d)/E(d)_{tot} \rangle] = +0.0066505\%$
$\Delta[\langle U(d)/E(d)_{tot} \rangle] = -0.0107624\%$

As we will see when describing the intervalic nucleon, there is a simple and elegant relation between the intervalic and electromagnetic energies of nucleons at the lisztinian —quarks— level:

$I(n)_L / I(p)_L = 1.500000000$
$U(p)_L / U(n)_L = 1.500000000$

Chapter 17

INTERVALIC NUCLEON

INTERVALIC NUCLEON

INTERVALIC, MATERIAL, ELECTROMAGNETIC, AND MASS ENERGIES

The total mass of a particle at rest is the sum of its constituent structural energies which are due: first, to electric charge, which is named *intervalic* energy, $I = c^{\pm 2} \hbar Q^{-2}$; and second, the remaining energy, which is named *material* energy, which, and usually has an *electromagnetic* origin. Thus, we have in the following equation two of the three intervalic principles of equivalence —between electric charge, energy and matter— already described in other site:

$$E = c^{\pm 2} \hbar Q^{-2} + c^{\pm 2} m$$

In not-singular units —that is to say, units with $c \neq 1$ (although the inconsistent and trivial "geometrized units" are naively singular, they are clearly geometryless and do not have any intervalic symmetry)— the equation *looses* a half of its dimensional symmetries and have to be written as:

$$E = c^{-2}\hbar\, Q^{-2} + c^2 m$$

Therefore, the constituent masses of a subatomic particle can be separated according to its different origin. Mass due to the electric charge is named *intervalic* mass, m_{in} (this is the formerly named "intrinsic" mass, a term that described its unknown origin); and the remaining mass is named *material* mass, m_{mat} —and *electromagnetic* mass, m_{em}, if it comes from electromagnetic sources—.

As I have explained in other site, the electromagnetic energy —as all the four supposed "forces" of Nature— is derived from the intervalic energy, being just its *inverse* in intervalic units: $U \quad I^{-1} = c^{\pm 2}(n\hbar)^{-1} Q^2 = c^{\pm 2} Q^2 / r$. In SI units it has to be added to the equation the permeability of vacuum factor, which was conventionally set by definition as $\mu_0/4\pi = 10^{-7}(-1)$, instead of $1\,(-1)$ as in intervalic units. Thus, the total energy manifested as mass, that is to say, the *mass energy* of subatomic particles is due to a sum between two inverse factors, a relation of paramount elegance:

$$E_{mass} = I + I^{-1} = c^{\pm 2}\hbar\, Q^{-2} + c^{\pm 2}(n\hbar)^{-1} Q^2$$

NUCLEON INTERVALIC STRUCTURE

The difference between nucleons masses —treated here, as well as the supposed binding energy of nucleons in the nucleus—, is principally due to the involved structural energies in the last monteverdic level of the intervalic structure.

According to Intervalic Theory in Particle Physics nucleons are composed by an assembly of three lisztinos or quarks $L3D45^{(\frac{1}{3},\frac{2}{3})}$. Its full intervalic structure, showing its compositeness and structurefulness, is:

$$M_3 = 3\, L_3 = 9\, G_6 = 54\, D_{45} = 2430\, I$$

Although this paper is not devoted to the complex intervalic structure of nucleons, which have been explained by separate and with detail in other sites, it means that nucleon is a *monteverdino* 3 composed by the intervalic aggregation of three *lisztinos* 3, each one of them is composed by three *gaudinos* 6, each one of them is formed by six *dalinos* 45, and each of them is composed by forty five *intervalinos*, making a total of 2430 intervalinos. These terms have been taken in honour of the great artists Claudio Monteverdi (1567-1643), Franz Liszt (1811-1886), Antonio Gaudí (1852-1926) and Salvador Dalí (1904-1989) respectively. All these particles are the unavoidable result of the primordial aggregation of intervalinos which made the Big Bang through *intervalic interaction* and according to the *intervalic symmetries* of the electric charge and the fundamental *intervalic principle of energy balance for subatomic particles*, which establish the perfect balance reached by the involved energies in all subatomic particles, which are named *structural energies*. Being $E(J)$ the spin energy:

$$I - I^{-1} - E(J) = 0$$

The first ones, intervalic energy and its inverse the electromagnetic energies, are manifested as *mass*, while the last structural energy is not manifested as mass but as *spin*.

NUCLEON INTERVALIC ENERGY

The magnitudes of the intervalic energy of nucleon at every step of its intervalic structure can be known with remarkable accuracy and with all desired precision. This full knowledge, which is a great achievement of IT, is due mainly to the fact that the intervalic interaction is a short ranged one which does not varies progressively according to the distance of the involved particles, as long ranged interactions do, but only relies on spin and its derived physical quantity: the electric charge. The physical features of the intervalic interaction are completely determined by the intervalic structure and vice versa. They are very different from the electromagnetic features, and will be seen in the suitable chapter later.

INTERVALIC ENERGY AT MONTEVERDIC STRUCTURE LEVEL: M_3

At the last level of its intervalic structure the monteverdic intervalic energy of nucleons is simply the energy of its total overall electric charge. Thus, for proton it is the intervalic energy of the elementary charge, and for neutron is zero:

$$I(M_3^+) = I(p)_{M3} = c^{\pm 2}\, \hbar\, e^{-2} = c^{\pm 2}\, \hbar\, (270\, \mathbf{q_I})^{-2} = c^{\pm 2}\, \hbar\, [270\, \sqrt{-(c^{-1}\hbar)}]^{-2}$$
$$= 270^{-2}\, c^{-1} = 4.57563917 \cdot 10^{-14}\, (J) = 0.28558881\, (MeV/c^2)$$

$$I(M_3^0) = I(n)_{M3} = 0$$

INTERVALIC ENERGY AT LISZTINIAN STRUCTURE LEVEL: $3\, L_3$

The next level of the intervalic structure of nucleons is the lisztinian one where we find three lisztinos, namely, quarks. Since the electric charge of nucleons is composed by the electric charges of its constituent quarks, the proton and neutron intervalic energies are different between them.

According to the intervalic principle of equivalence between electric charge and energy, $I = c^{\pm 2}\hbar\, Q^{-2}$, the lisztinian intervalic energy of quarks *up* and *down*, $I(u)$ and $I(d)$, are respectively 9/4 and 9 times the energy of the elementary charge:

INTERVALIC THEORY:
The Intervalic Structures of Subatomic Particles and the Last Foundations of Physics

$$I(L_3^{+2/3}) = I(u)_{L3} = c^{\pm 2} \hbar \, (\tfrac{2}{3} e)^{-2} = (9/4) \, I(e) = c^{\pm 2} \hbar \, (180 \, q_I)^{-2} =$$
$$= c^{-1} \, 180^{-2} = 1.0295188 \cdot 10^{-13} \, (J) = 0.64257483 \, (MeV/c^2)$$

$$I(L_3^{-1/3}) = I(d)_{L3} = c^{\pm 2} \hbar \, (\tfrac{1}{3} e)^{-2} = 9 \, I(e) = c^{\pm 2} \hbar \, (90 \, q_I)^{-2} =$$
$$= c^{-1} \, 90^{-2} = 4.1180752 \cdot 10^{-13} \, (J) = 2.5702993 \, (MeV/c^2)$$

When joining three quarks in a proton (as in primordial times), there is liberated a considerable amount of intervalic energy in different processes by means of the aggregation of subatomic intervalic electric charges, up to reaching the value of the elementary charge —the state of minimal intervalic energy—. The lisztinian intervalic energy of the three quarks in the proton is 54/4 times the energy of the elementary charge:

$$I(p)_{L3} = I(u+u+d)_{L3} = ((9/4) + (9/4) + 9) \, c^{-2} \hbar \, e^{-2} =$$
$$= (27/2) \, 270^{-2} c^{-1} = 6.1771128 \cdot 10^{-13} \, (J) = 3.8554490 \, (MeV/c^2)$$

In a similar way, we have that the lisztinian intervalic energy of the three quarks that conform the neutron is 81/4 times greater:

$$I(n)_{L3} = I(u+d+d)_{L3} = ((9/4) + 9 + 9) \, c^{-2} \hbar \, e^{-2} =$$
$$= (81/4) \, 270^{-2} c^{-1} = 9.2656692 \cdot 10^{-13} \, (J) = 5.7831734 \, (MeV/c^2)$$

At the lisztinian level of intervalic structure, the proton/electron, neutron/electron and neutron/proton intervalic energy ratios are:

$I(p) / I(e) = 27/2 = 13.500000$
$I(n) / I(e) = 81/4 = 20.250000$
$I(p) / I(n) = (27/2)/(81/4) = 0.66666667$
$I(n) / I(p) = (81/4)/(27/2) = 1.50000000$

INTERVALIC ENERGY AT GAUDINAR STRUCTURE LEVEL: 9 G_6

Each quark —lisztino 3— is composed by three gaudinos 6, each of them has its proper intervalic energy according to its overall electric charge. Due to powerful intervalic symmetries constraints explained in other site the electric charge of gaudinos 6 can only take the values: 0, $\pm 90 q_I$, $\pm 180 q_I$ and $\pm 270 q_I$. Therefore its corresponding gaudinar intervalic energy will be respectively:

$I(G_6^0) = 0$
$I(G_6^{\pm 1/3}) = c^{\pm 2} \hbar\, (\tfrac{1}{3} e)^{-2} = 90^{-2}\, c^{-1} = 2.570299287\ (MeV/c^2)$
$I(G_6^{\pm 2/3}) = c^{\pm 2} \hbar\, (\tfrac{2}{3} e)^{-2} = 180^{-2}\, c^{-1} = 0.642574821\ (MeV/c^2)$
$I(G_6^{\pm 1}) = c^{\pm 2} \hbar\, e^{-2} = 270^{-2}\, c^{-1} = 0.285588809\ (MeV/c^2)$

The total intervalic energy of the nucleon at the gaudinar structure level is the sum of the intervalic energy of each one of its nine constituent gaudinos. Since we postulated that the gaudinar structure of nucleonic quarks is:

$u^{+2/3} \rightarrow L_3 = G_6^{+2/3} + G_6^{-1/3} + G_6^{+1/3}$
$d^{-1/3} \rightarrow L_3 = G_6^{+1/3} + G_6^{-1/3} + G_6^{-1/3}$

The intervalic energy of quarks at the gaudinar structure level will be:

$I(u)_G = I(G_6^{\pm 2/3}) + 2\, I(G_6^{\pm 1/3}) = 5.783173396\ (MeV/c^2)$
$I(d)_G = 3\, I(G_6^{\pm 1/3}) = 7.710897861\ (MeV/c^2)$

And the intervalic energy of nucleons at this level will be:

$I(p)_G = I\,[2(G_6^{\pm 2/3}) + 7(G_6^{\pm 1/3})]_G = 19.27724465\ (MeV/c^2)$
$I(n)_G = I\,[(G_6^{\pm 2/3}) + 8(G_6^{\pm 1/3})]_G = 21.20496912\ (MeV/c^2)$

INTERVALIC ENERGY AT DALINAR STRUCTURE LEVEL: 54 D_{45}

Each G_6 is composed by six D_{45}. The intervalic energy of one D_{45} is:

$$I(D_{45}) = c^{\pm 2} \, \hbar \, (45 \, q_I)^{-2} = c^{\pm 2} \, \hbar \, [45 \, \sqrt{-(c^{-1}\hbar)}]^{-2} = 45^{-2} \, c^{-1} =$$
$$= 1.6472301 \cdot 10^{-12} \, (J) = 10.281197 \, (MeV/c^2)$$

Therefore the dalinar intervalic energy of the gaudino 6 is:

$$I(G_6)_D = 6 \cdot I(D_{45}) = 9.8833806 \cdot 10^{-12} \, (J) = 61.687184 \, (MeV/c^2)$$

And the dalinar intervalic energy of the lisztino 3 —quark— is:

$$I(L_3)_D = 3 \cdot I(G_6) = 2.9650142 \cdot 10^{-11} \, (J) = 185.06155 \, (MeV/c^2)$$

And finally the total dalinar intervalic energy of the nucleon is:

$$I(M_3)_D = 3 \cdot I(G_6) = 8.8950426 \cdot 10^{-11} \, (J) = 555.184650 \, (MeV/c^2)$$

Or in other words: $I(M_3)_D = 54 \cdot I(D_{45})$.

NUCLEON TOTAL INTERVALIC ENERGY

The total intervalic energy of nucleons is simply the sum of its constituent intervalic energies at all levels of its intervalic structure:

$$I(p)_{tot} = I(M_3^+) + 2\, I(L_3^{+2/3}) + I(L_3^{-1/3}) + 2\, I(u)_G + I(d)_G + 54\, I(D_{45}) =$$
$$= 9.270245023 \cdot 10^{-11} \, (J) = 578.602933 \, (MeV/c^2)$$

$$I(n)_{tot} = I(M_3^0) + I(L_3^{+2/3}) + 2\, I(L_3^{-1/3}) + I(u)_G + 2\, I(d)_G + 54\, I(D_{45})$$
$$= 9.327440495 \cdot 10^{-11} \, (J) = 582.172792 \, (MeV/c^2)$$

NUCLEON ELECTROMAGNETIC ENERGY

Contrarily to the short ranged intervalic energy, the electromagnetic energy relies on the distance of the involved particles, or on its radius if it is expressed as potential energy. This means that the electromagnetic mass of an intervalic structure may be different when existing in *isolated state* as when forming part of a further intervalic structure, that is to say, in a *binding state*. The difference of mass between both states is named *binding energy*, although sometimes this might be a misleading concept as we will see when describing the intervalic nuclei. All magnitudes of the electromagnetic energy of nucleon in this chapter are the masses of the intervalic structures as 'binding states' inside nucleon, as we only have got direct experimental data on nucleon mass as a whole. Although it seems to be absolutely impossible to know even approximately the electromagnetic mass of the intervalic structures of nucleon as 'isolated states' — which would lead us to the knowledge of their corresponding binding energies—, we will see opportunely how the magic of IT inclusive reaches that unexpected achievement.

NUCLEON TOTAL ELECTROMAGNETIC ENERGY

Since we already know the magnitude of the total intervalic energy of nucleon, we can deduce immediately its total electromagnetic energy:

$U(p)_t = c^{\pm 2} m(p) - I(p)_t = 5.762528048 \cdot 10^{-11} (J) = 359.6685548$ (MeV/c^2)

$U(n)_t = c^{\pm 2} m(n) - I(n)_t = 5.726054096 \cdot 10^{-11} (J) = 357.3920307$ (MeV/c^2)

Then, the ratios between the intervalic, electromagnetic and total mass energies of nucleon will be:

$I(p)/U(p) = 1.60871148$
$I(p)/E(p)_{mass} = 0.616668452$
$U(p)/E(p)_{mass} = 0.383330671$

$I(n)/U(n) = 1.628947324$
$I(n)/E(n)_{mass} = 0.61961908$
$U(n)/E(n)_{mass} = 0.38038006$

NUCLEON MATERIAL ENERGY

According to the definition of *material* energy of a particle as the difference between its *mass* energy and its *intervalic* energy, it can be easily calculated the nucleons material energy. Of course, this definition of material energy does not affirm anything about its origin. In the present case, since nucleons have a complex intervalic structure with several levels, the concept of material energy in nucleons have to be carefully handled because it is only valid at its own level of structure. It can not be used to deduce nothing about the magnitudes and physical quantities involved in deeper levels of structure. Taking into consideration these precisions, it can be however useful. For example, we may wish to calculate the electromagnetic energy and other physical features of nucleon at the last structure levels as if we did not know its intervalic structure at the gaudinar level and inwards. Really, we are going to do it right now. In that case, we begin calculating the material energy up to the gaudinar level inclusive. It will be, supposed that our knowledge of the structure of nucleon finishes on the traditional model of quarks:

$\Sigma^{9 \cdot G6} E(p)_{mat} = m(p) - I(M_3^+) - I(u+u+d) = 934.13131 \ (MeV/c^2)$
$\Sigma^{9 \cdot G6} E(n)_{mat} = m(n) - I(M_3^0) - I(u+d+d) = 933.78251 \ (MeV/c^2)$

As it can be seen, the *material* mass of proton at this structure level is slightly *greater* than neutron's one.

DISTANCE BETWEEN QUARKS IN NUCLEONS

If it is supposed that the difference between proton and neutron masses is due to only to the constituent electric charges of quarks, as it is usually assumed, their *material* masses must be exactly the same after subtracting the different intervalic and electromagnetic energies among quarks in both nucleons. Besides, since *material* masses of proton and neutron has been deduced by means of the intervalic principles of equivalence, we can *define* that subatomic electromagnetic energy as the *difference* between nucleons material masses. Following that definition, we can calculate with full precision the electromagnetic interaction between quarks although the distance between quarks, d_q, was not known. Moreover, from this *difference of material energy* between nucleons, previously deduced by intervalic geometrical means, it can be deduced the *distance* between quarks inside nucleon, d_q:

$$m(p)_{mat} - m(n)_{mat} = 0.3488012 \, (MeV/c^2) = 5.58841375 \cdot 10^{-14} \, (J) =$$
$$= E_q(uud) - E_q(udd) = \{[(\tfrac{2}{3})^2 + (\tfrac{2}{3} \cdot \tfrac{1}{3}) + (\tfrac{2}{3} \cdot \tfrac{1}{3})] -$$
$$- [(\tfrac{1}{3})^2 + (\tfrac{2}{3} \cdot \tfrac{1}{3}) + (\tfrac{2}{3} \cdot \tfrac{1}{3})]\} (1/4\pi\varepsilon_0) \, e^2 / d_q = \tfrac{1}{3} (1/4\pi\varepsilon_0) \, e^2 / d_q$$

$$d_q = 1.37610877 \cdot 10^{-15} \, (m)$$

This is a meaningful result, as I have explained in other site, because it shows that there is a changeless —strong— intervalic interaction among *quarks* instead of *nucleons,* as usually supposed, since the distance reached by the π meson as intermediate state is: $c^{-1}\hbar / m_\pi = 1.4152474 \cdot 10^{-15}$ (m).

INTERVALIC NUCLEONIC QUARK RADIUS

Immediately, the intervalic nucleonic quark radius, r_q, is, if it is supposed that the three quarks are tangent among themselves in nucleon:

$$r_q = d_q / 2 = 6.88054386 \cdot 10^{-16} \,(m)$$

We can use this value in order to express the electromagnetic mass energy of quarks inside nucleon as a potential energy.

ELECTROMAGNETIC ENERGY AT MONTEVERDIC STRUCTURE LEVEL: M_3

We are tempted to write for the proton:

$$U(M_3^+) = U(p)_M = \tfrac{1}{2}\,(1/4\pi\varepsilon_0)\,(270\,\mathbf{q_I})^2 / r_N = 0.58535170 \,(MeV/c^2)$$

However this electromagnetic energy is just the energy of the electromagnetic *field*, and therefore it is not manifested as *mass* energy. On the contrary, the intervalic energy at this level is the "equivalent energy" of the total charge which is ever manifested as mass energy.

ELECTROMAGNETIC ENERGY AT LISZTINIAN STRUCTURE LEVEL: L_3

The electromagnetic energy of quarks at the lisztinian structure level is:

$$U(L_3^{+\frac{2}{3}}) = U(u)_L = \tfrac{1}{2}(1/4\pi\varepsilon_0)(180\,\mathbf{q_I})^2/r_q = 0.46506825\,(MeV/c^2)$$
$$U(L_3^{-\frac{1}{3}}) = U(d)_L = \tfrac{1}{2}(1/4\pi\varepsilon_0)(90\,\mathbf{q_I})^2/r_q = 0.116267062\,(MeV/c^2)$$

The electromagnetic energy of nucleon at this structure level is:

$$U(p)_L = U(u+u+d)_L = (180^2+180^2+90^2)\,\tfrac{1}{2}(1/4\pi\varepsilon_0)\,\mathbf{q_I}^2/r_q =$$
$$= 1.67652406 \cdot 10^{-13}\,(J) = 1.04640356\,(MeV/c^2)$$

$$U(n)_L = U(u+d+d)_L = (180^2+90^2+90^2)\,\tfrac{1}{2}(1/4\pi\varepsilon_0)\,\mathbf{q_I}^2/r_q =$$
$$= 1.11768271 \cdot 10^{-13}\,(J) = 0.697602374\,(MeV/c^2)$$

Remembering that the *electromagnetic* energy of electron is the difference between its *total* and *intervalic* energies: $U(e) = E(e)_{tot} - I(e) = 3.6114722 \cdot 10^{-14}\,(J) = 0.2254103\,(MeV/c^2)$, the corresponding electromagnetic energy ratios at the lisztinian level of the intervalic structure are:

$$U(p)_L / U(e) = 4.64221715$$
$$U(n)_L / U(e) = 3.09481144$$
$$U(p)_L / U(n)_L = 1.50000000$$
$$U(n)_L / U(p)_L = 0.66666667$$

As can be seen, there is a very simple and elegant relation between the intervalic and electromagnetic energies of nucleons at this level:

$$I(n)_L / I(p)_L = 1.50000000$$
$$U(p)_L / U(n)_L = 1.50000000$$

For our purpose in this chapter we don't need to consider now the electromagnetic energy of nucleon at the complex gaudinar and dalinar structure levels, which have been described in their corresponding sites dedicated to these intervalic structures.

THE DIFFERENCE OF MASSES BETWEEN PROTON AND NEUTRON ACCORDING TO THE INTERVALIC THEORY

Experimentally, $m(n) - m(p) = 1.29332 \pm 0.00028$ (MeV/c²).

If it is postulated a structural similitude between nucleons, this difference of mass must be due to the *difference between its involved intervalic and electromagnetic energies at the last structure level* —monteverdic level for nucleons and lisztinian level for quarks—. Applying the above results on the intervalic and electromagnetic energies of constituent quarks, we have (in MeV/c²):

$m(n) - m(p) = {}_{M3}\Sigma^{3.L3}[I(n) + U(n)] - [I(p) + U(p)] =$
$= [(I(M_3^0)+I(u)+2I(d)) + (U(M_3^0)+U(u)+2U(d))] -$
$- [(I(M_3^+)+2I(u)+I(d)) + (U(M_3^+)+2U(u)+U(d))] =$
$= [(0+0.64257483+5.14059860) +$
$+ (0+0.465068250+0.232534124)] -$
$- [(0.28558880+1.2851497+2.5702993)+$
$+ (0+0.9301365+0.116267062)] =$
$= 5.78317343+0.697602374-4.1410378-1.046403562 =$
$= 1.293334442$

Therefore, the difference between proton and neutron masses is, according to intervalic principles of equivalence, exactly 1.293334442 (MeV/c²). It means a theoretical exactitude up to a thousandth of electron volt, 0.001 eV (!), while the best present experimental values have an error of around 280 eV. (Actually, the result is yielded theoretically with *null error*, since the pointed one of 0.001 eV is due only to the limitation represented by the number of digits that we have chosen to make the calculus from the values of the fundamental constants, c and ℏ).

In concrete, according to IT the present experimental value about the difference of mass between proton and neutron has a very little deviation with respect to the theoretical intervalic value of only 14.44 eV (our experimental physicists have done a really good measurement!). This simple deduction is completely based on intervalic physical principles, which lies only on the intervalic system of units and dimensions. It appears to be the first truthful and largely exact explanation about the difference of masses between proton and neutron.

INTERVALIC NUCLEON COMMON MASS

It can be supposed that the dalinar structure level is equal for both nucleons, and their dalinar masses will be identical, $m_p(54\ D_{45})\ m_n(54\ D_{45})$. At the next gaudinar structure level the constituent electric charges of gaudinos are different in quarks up and down, hence their different masses. Nevertheless, if we subtract the contribution to mass of the monteverdic and lisztinian structure levels of proton and neutron, it will remain a magnitude that we are going to name *intervalic nucleon common mass*, $m(N)_{com}$:

$$m(N)_{com} = m(p) - [I(p)_{M3+L3} + U(p)_{M3+L3}] = 933.0848686\ (MeV/c^2)$$
$$\approx m(N)_{com} = m(n) - [I(n)_{M3+L3} + U(n)_{M3+L3}] = 933.0848542\ (MeV/c^2)$$

The difference of 14.44 eV between both results is due to the deviation in the experimental data already viewed. We will take the middle value between both ones as the intervalic nucleon common mass: $933.0848614\ (MeV/c^2)$.

NUCLEONIC QUARK COMMON MASS

If the deeper levels of intervalic structure would not have influence on the masses of the nucleonic quarks, we could define a nucleonic quark common mass as:

$$m(u)_{com} \approx m(d)_{com} = \tfrac{1}{3} m(N)_{com} = 311.0282871 \text{ (MeV}/c^2)$$

Nevertheless, as the different constituent electric charges at the gaudinar level has influence on quarks masses, this magnitude will have to be interpreted as the average nucleonic quark common mass.

INTERVALIC QUARKS UP AND DOWN MASSES

Following the preceding results it can be deduced the nucleonic quarks up and down total masses:

$$m(u) = m(u)_{com} + I(u)_{L3} + U(u)_{L3} = 312.1359302 \text{ (MeV}/c^2)$$
$$m(d) = m(d)_{com} + I(d)_{L3} + U(d)_{L3} = 313.7148535 \text{ (MeV}/c^2)$$

As can be seen, the contribution of the *intervalic energy* explain at last why the mass of quark *d* is *greater* than the mass of quark *u*, which fully explains the difference of masses between nucleons:

$$I(u)_L = c^{\pm 2} \hbar \, (180 \, q_I)^{-2} = c^{-1} 180^{-2} = 1.0295188 \cdot 10^{-13} \text{ (J)} =$$
$$= 0.64257482 \text{ (MeV}/c^2)$$

$$I(d)_L = c^{\pm 2} \hbar \, (90 \, q_I)^{-2} = c^{-1} 90^{-2} = 4.1180752 \cdot 10^{-13} \text{ (J)} =$$
$$= 2.5702993 \text{ (MeV}/c^2)$$

$$U(u)_L = U(L_3^{+2/3}) = \tfrac{1}{2} \, (1/4\pi\varepsilon_0) \, (180 \, q_I)^2 / r_q =$$
$$= 7.451218056 \cdot 10^{-14} \text{ (J)} = 0.465068249 \text{ (MeV}/c^2)$$

$$U(d)_L = U(L_3^{-1/3}) = \tfrac{1}{2} \, (1/4\pi\varepsilon_0) \, (90 \, q_I)^2 / r_q =$$
$$= 1.862804514 \cdot 10^{-14} \text{ (J)} = 0.116267062 \text{ (MeV}/c^2)$$

Thus, to the value of the common mass of quarks displayed in our tables of intervalic quarks, it have to be added the contribution to mass of the last lisztinian level of quarks, which is roughly ~2.7 and ~1.1 (MeV/c^2) to quarks with charge ±⅓ and ±⅔ respectively, in order to obtain the actual masses of all existing 49 quarks.

INTERVALIC NUCLEONS MASSES

In a similar way, we have immediately the intervalic proton and neutron masses, with *null error*, by means of the simple sum of its constituent intervalic quarks total masses, a result made without adjusting any constant, and directly deduced from the intervalic system of units and dimensions:

$$m(p) = 2\,m(u) + m(d) + I(e) = 938.2723027 \text{ (MeV/}c^2)$$
$$m(n) = m(u) + 2\,m(d) = 939.5656372 \text{ (MeV/}c^2)$$

Needless to say that some fantastic assumptions introduced by SM, like the acrobatic "free masses" of quarks, the elusive creation of mass through a supposed spin-spin interaction, and the mysterious Higgs mechanism (which looks like a certain medieval ether) are, from the point of view of IT, naïve and clumsy assumptions which have become unnecessary and irrelevant.

NUCLEONIC COMMON QUARK ELECTROMAGNETIC ENERGY

From the nucleonic quark common mass we can deduce the common dalinar electromagnetic energy of the three constituent gaudinos of the lisztino 3:

$U(L_3) = E(L_3)_{com} - I(L_3) = 2.0182105 \cdot 10^{-11} (J) = 125.9667371$ (MeV/c²)

The nucleonic common quark ratios of the involved energies are, as expected:

$I(L_3)/U(L_3) = 1.469130298$
$I(L_3)/E(L_3)_{com} = 0.594999097$
$U(L_3)/E(L_3)_{com} = 0.405000902$

Compared with the electron energy ratio, $I(e)/U(e) = 1.266973485$, the electromagnetic energy of quarks is considerably smaller.

INTERVALIC NUCLEON RADIUS

To better understand the nucleon radius we could define two extreme magnitudes for that value, and it is expected that the nucleon radius will be between both values: an interior and an exterior radius.

To determine the first one, based on volume means, we have that the total volume of the three quarks according to the magnitude of the intervalic quark radius is:

$$3 V(q) = 3 \cdot (4/3) \pi r_q^3 = 4.093343275 \cdot 10^{-45} \ (m^3)$$

This volume would correspond to the interior radius:

$$r_{int} = [(3/4)V(q) / \pi]^{1/3} = 9.923461426 \cdot 10^{-16} \ (m)$$

To determine the second one, based on geometric means, we will define an exterior radius tangent to the constituent quarks in nucleon. We can deduce it from the magnitude of the intervalic quark radius after some straightforward algebra:

$$r_{ext} = [1 + (4/3)\sqrt{(3/4)}]\, r_q = 2.1547005\, r_q = 1.48255113 \cdot 10^{-15}\, (m)$$

Curiously, $r_{int} \approx \frac{2}{3}\, r_{ext}$. The experimental magnitude of the nucleon radius, $r_N \sim 1.23 \cdot 10^{-15}$ (m), is between both values, as expected. Moreover, it is just in the middle between the interior and exterior radius: $\frac{1}{2}(r_{int} + r_{ext}) = 1.237448636 \cdot 10^{-15}$ (m) $\approx r_N$.

NUCLEON ENERGY MASS ORGANIZED BY INTERVALIC STRUCTURAL LEVELS

We can organize the preceding results in order to show the mass of nucleons which is added as we go into deeper levels of the intervalic structure. So we have that the nucleon energy mass at the most exterior level, the monteverdic one, is:

$$I(p)_M = 4.575639155 \cdot 10^{-14}\, (J) = 0.285588809\, (MeV/c^2)$$
$$I(n)_M = 0$$

$$U(p)_M = 0$$
$$U(n)_M = 0$$

At the monteverdic and lisztinian level the involved energies are:

$$I(p)_{ML} = I(p)_M + I(u+u+d)_L =$$
$$= 6.63467679 \cdot 10^{-13}\, (J) = 4.14103774\, (MeV/c^2)$$
$$I(n)_{ML} = I(n)_M + I(u+d+d)_L =$$
$$= 9.265669311 \cdot 10^{-13}\, (J) = 5.783173396\, (MeV/c^2)$$

$$U(p)_{ML} = U(p)_M + U(u+u+d)_L =$$
$$= 1.67652406 \cdot 10^{-13}\, (J) = 1.04640356\, (MeV/c^2)$$

$$U(n)_{ML} = U(n)_M + U(u+d+d)_L =$$
$$= 1.11768271 \cdot 10^{-13} \text{ (J)} = 0.697602374 \text{ (MeV/c}^2)$$

At the three most exterior levels of nucleon —monteverdic, lisztinian and gaudinar— the mass energies are:

$$I(p)_{MLG} = I(p)_M + I(u+u+d)_L + I[2(G_6^{\pm 2/3})+7(G_6^{\pm 1/3})]_G =$$
$$= 3.752024115 \cdot 10^{-12} \text{ (J)} = 23.41828239 \text{ (MeV/c}^2)$$
$$I(n)_{MLG} = I(n)_M + I(u+d+d)_L + I[(G_6^{\pm 2/3})+8(G_6^{\pm 1/3})]_G =$$
$$= 4.323979011 \cdot 10^{-12} \text{ (J)} = 26.98814251 \text{ (MeV/c}^2)$$

$$U(p)_{MLG} = U(p)_M + U(u+u+d)_L + U[2(G_6^{\pm 2/3})+7(G_6^{\pm 1/3})]_G =$$
$$= 2.484069763 \cdot 10^{-12} \text{ (J)} = 15.50433723 \text{ (MeV/c}^2)$$
$$U(n)_{MLG} = U(n)_M + U(u+d+d)_L + U[(G_6^{\pm 2/3})+8(G_6^{\pm 1/3})]_G =$$
$$= 1.964902156 \cdot 10^{-12} \text{ (J)} = 12.26394931 \text{ (MeV/c}^2)$$

And finally, we have the intervalic and electromagnetic energies from the monteverdic up to the dalinar levels of the intervalic structure, whose sums yield the total intervalic and electromagnetic masses of nucleon:

$$I(p)_{MLGD} = I(p)_M + I(u+u+d)_L + I[2(G_6^{\pm 2/3})+7(G_6^{\pm 1/3})]_G + 54I(D_{45})_D$$
$$= 9.270245014 \cdot 10^{-11} \text{ (J)} = 578.6029324 \text{ (MeV/c}^2)$$
$$I(n)_{MLGD} = I(n)_M + I(u+d+d)_L + I[(G_6^{\pm 2/3})+8(G_6^{\pm 1/3})]_G + 54I(D_{45})_D =$$
$$= 9.327440504 \cdot 10^{-11} \text{ (J)} = 582.1727926 \text{ (MeV/c}^2)$$

$$U(p)_{MLGD} = U(p)_M + U(u+u+d)_L + U[2(G_6^{\pm 2/3})+7(G_6^{\pm 1/3})]_G +$$
$$+ 54U(D_{45})_D = 5.762541114 \cdot 10^{-11} \text{ (J)} = 359.6693703 \text{ (MeV/c}^2)$$
$$U(n)_{MLGD} = U(n)_M + U(u+d+d)_L + U[(G_6^{\pm 2/3})+8(G_6^{\pm 1/3})]_G +$$
$$+ 54U(D_{45})_D = 5.72606702 \cdot 10^{-11} \text{ (J)} = 357.3928374 \text{ (MeV/c}^2)$$

$$I(p)_{MLGD} + U(p)_{MLGD} = c^{\pm 2} m(p) = 938.2723027 \text{ (MeV/c}^2)$$
$$I(n)_{MLGD} + U(n)_{MLGD} = c^{\pm 2} m(n) = 939.5656300 \text{ (MeV/c}^2)$$

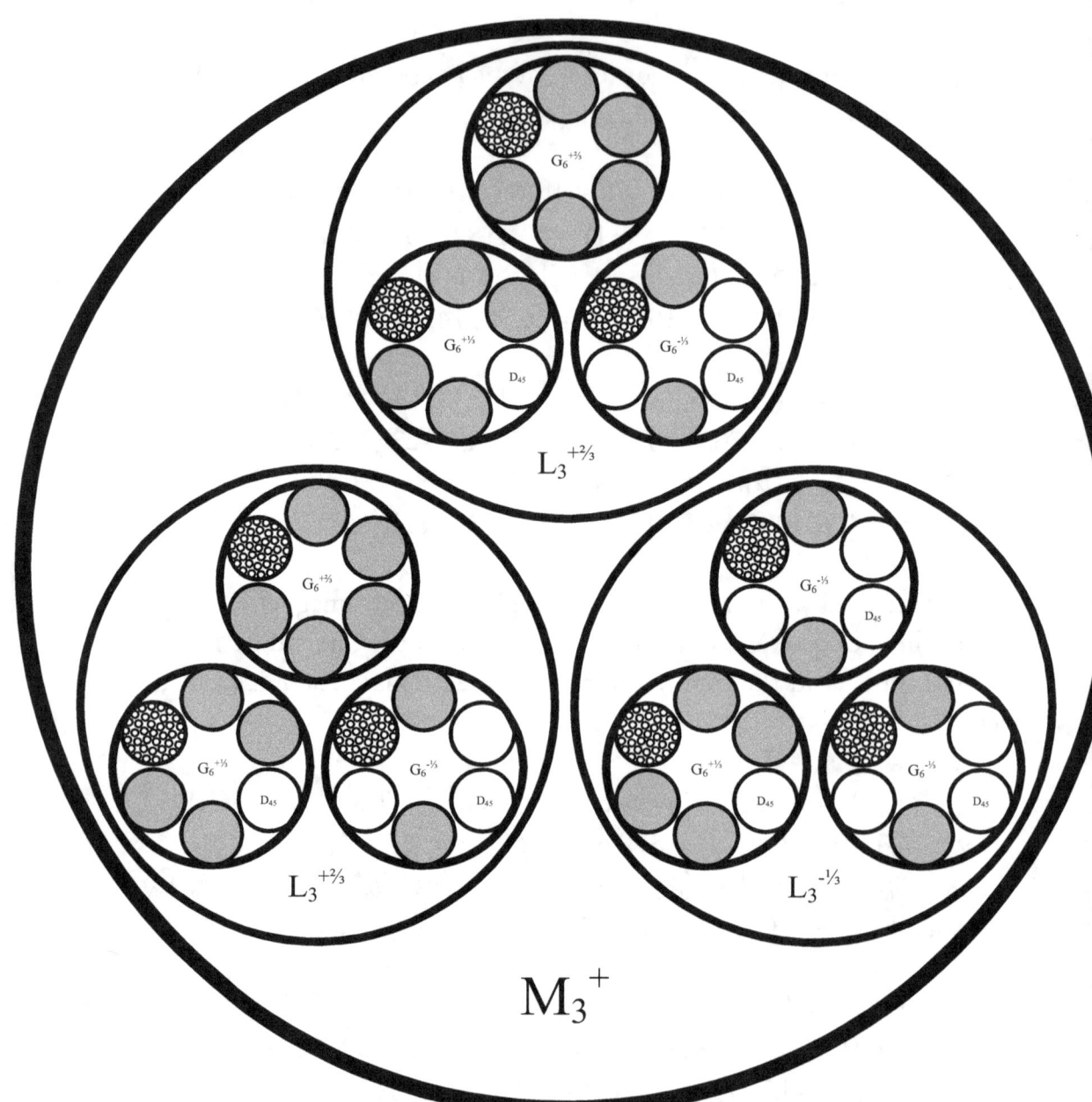

Intervalic structure of *proton* (unlike charges shadowed):
$$M_3^+ = 3\ L_3\ (L_3^{+2/3},\ L_3^{-1/3},\ L_3^{+2/3}) =$$
$$9\ G_6\ (2G_6^{+2/3},\ 4G_6^{-1/3},\ 3G_6^{+1/3}) = 54\ D_{45}\ (30D_{+45},\ 24D_{-45}) = 2430\ \mathbf{I}$$

INTERVALIC THEORY:
The Intervalic Structures of Subatomic Particles and the Last Foundations of Physics

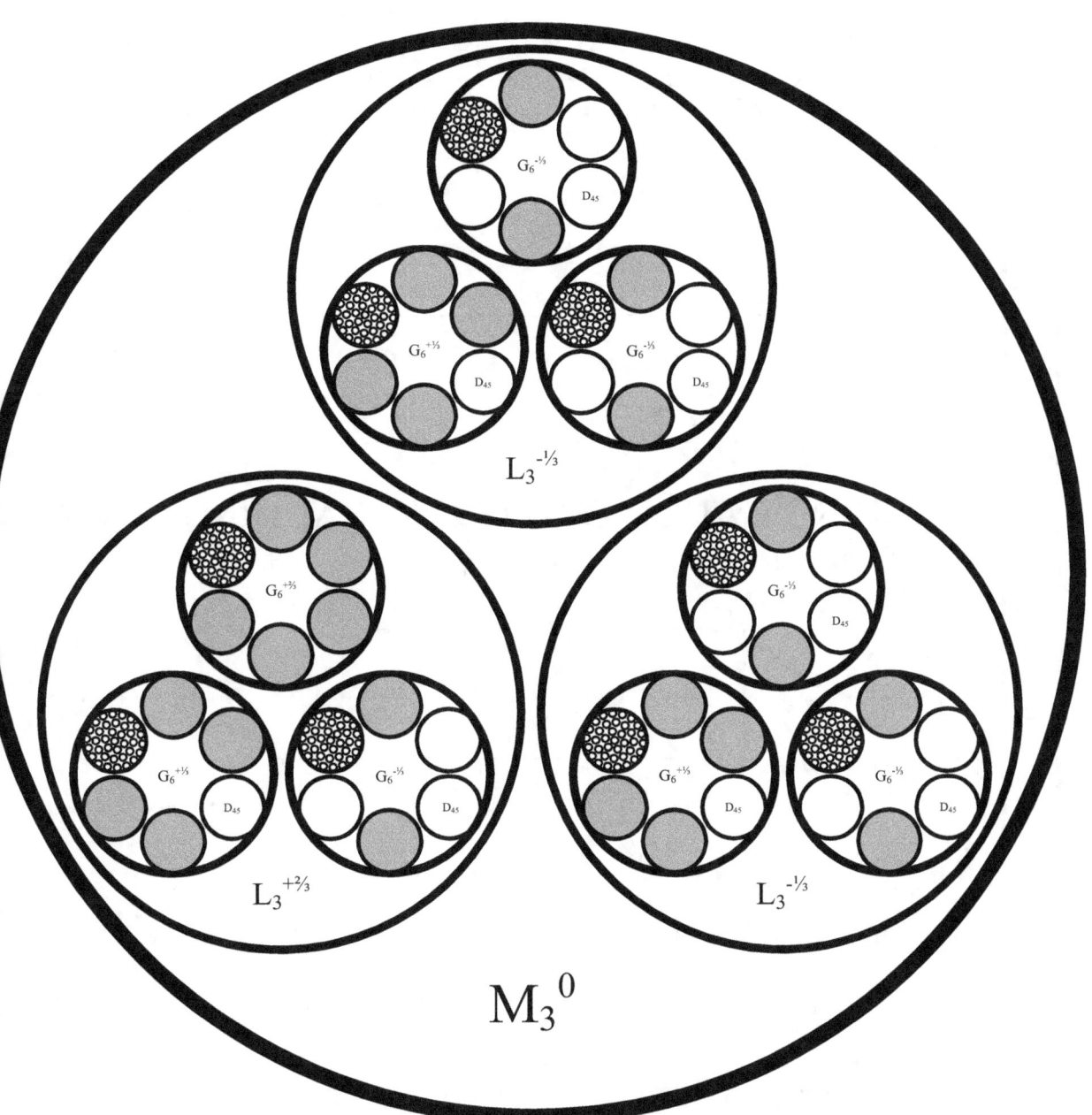

Intervalic structure of *neutron* (unlike charges shadowed):
$$M_3^0 = 3\, L_3\, (L_3^{-1/3},\, L_3^{+2/3},\, L_3^{-1/3}) =$$
$$9\, G_6\, (1G_6^{+2/3},\, 5G_6^{-1/3},\, 3G_6^{+1/3}) = 54\, D_{45}\, (27D_{+45},\, 27D_{-45}) = 2430\, \mathbf{I}$$

CONSTITUENT PARTI

INTERVALIC STRUCTURE OF NUCLE

$p = M_3^+ = udu = 3\ L_3\ (L_3^{+2/3}, L_3^{-1/3}, L_3^{+2/3}) = 9\ G_6\ \{(2G_6^{+2/3}, 4G_6^{-1/3}, 3G_6^{+1/3})\} = (3G_6^{-1},$

$n = M_3^0 = dud = 3\ L_3\ (L_3^{-1/3}, L_3^{+2/3}, L_3^{-1/3}) = 9\ G_6\ \{(1G_6^{+2/3}, 5G_6^{-1/3}, 3G_6^{+1/3})\} = (3G_6^{-1},$

Intervalic structure levels: 0: Point, 1: Intervalic String (S), 2: Photon (γ), 3: Intervalino (I)

Phase	
7	NUCL... M
6	QUARK L_3 | QUA... L_3
5	GAUDINO G_6 | GAUDINO G_6 | GAUDINO G_6 | GAUDINO G_6 | GAUD... G_6
4	D_{45} ...
3	I I I I ...
2	γ γ γ γ ...
1	s s s s ...

CLES OF NUCLEON

:ON —PROTON AND NEUTRON—

$\{3G_6^{+1}, 2G_6^{+2/3}, 1G_6^{-1/3})\} = 54\ D_{45}\ (30D_{+45}, 24D_{-45}) = 2430\ I = 4860\ \gamma = 9720\ S$

$\{3G_6^{+1}, 1G_6^{+2/3}, 2G_6^{-1/3})\} = 54\ D_{45}\ (27D_{+45}, 27D_{-45}) = 2430\ I = 4860\ \gamma = 9720\ S$

, 4: Dalino (D), 5: Gaudino (G), 6: Lisztino (L), 7: Monteverdino (M), 8: Palestrino (P)

	Number of constituyent particles
:EON ... 3	1 monte-verdino
RK / QUARK L_3	3 quarks
INO / GAUDINO G_6 / GAUDINO G_6 / GAUDINO G_6 / GAUDINO G_6	9 gaudinos
D_{45} ... D_{45}	54 dalinos
I I I ... I	2430 inter-valinos
γ γ γ ... γ	4860 photons
S S S ... S	9720 intervalic strings

INTERVALIC STRUCTURE OF NUCLEON

$p = M_3^+ = udu = 3\,L_3\,(L_3^{+2/3}, L_3^{-1/3}, L_3^{+2/3}) = 9\,G_6\,\{(2G_6^{+2/3}, 4G_6^{-1/3}, 3G_6^{+1/3}) = (3G_6^{-1}, 3G_6^{+1}, 2G_6^{+2/3}, 1G_6^{-1/3})\} = 54\,D_{45}\,(30D_{+45}, 24D_{-45}) = 2430\,\mathbf{I} = 4860\,\gamma = 9720\,S$

$n = M_3^0 = dud = 3\,L_3\,(L_3^{-1/3}, L_3^{+2/3}, L_3^{-1/3}) = 9\,G_6\,\{(1G_6^{+2/3}, 5G_6^{-1/3}, 3G_6^{+1/3}) = (3G_6^{-1}, 3G_6^{+1}, 1G_6^{+2/3}, 2G_6^{-1/3})\} = 54\,D_{45}\,(27D_{+45}, 27D_{-45}) = 2430\,\mathbf{I} = 4860\,\gamma = 9720\,S$

Intervalic structure levels: 0: Point, 1: Intervalic String (S), 2: Photon (γ), 3: Intervalino (**I**), 4: Dalino (D), 5: Gaudino (G), 6: Lisztino(L), 7: Monteverdino (M), 8: Palestrino (P)

PHOTON: Light

Photon intervalic structure: $\gamma = 2\,S$

Photon state: $\gamma = |\underline{S}\,\underline{S}|_{sym} = \{\,|\uparrow\uparrow|,\ 2^{-1/2}\,(\,|\uparrow\downarrow| + |\downarrow\uparrow|\,),\ |\downarrow\downarrow|\,\}$

Photon radius: $r_\gamma = \hbar = 1.0556363 \cdot 10^{-34}$ (m)

Photon spin: $J_\gamma = \hbar = 1.0556363 \cdot 10^{-34}$ (m)

Photon total length: $l_\gamma = 2\pi\,\hbar$

Frequorce of primordial photon:

$\varphi(\gamma_{pr}) = \boldsymbol{\varphi}_\mathbf{I} = c\,\hbar^{-1} = 2.839921837 \cdot 10^{42}\,(s^{-1})$

Temperature of primordial photon:

$\Theta(\gamma_{pr}) = \boldsymbol{\Theta}_\mathbf{I} = c\,k_B^{-1} = 2.17138589 \cdot 10^{31}$ (K)

Timeless Universe limit (Intervalic-relativistic transformations of time regarding temperature):

$t = t_0 / \sqrt{(1-[\Theta/\boldsymbol{\Theta}_\mathbf{I}]^2)} = \infty$

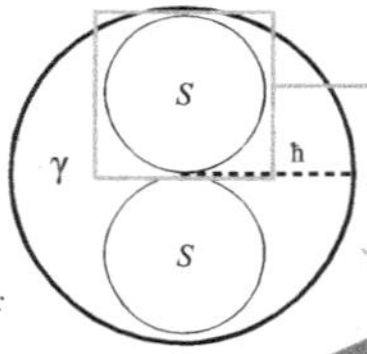

INTERVALIC STRING: Consciousness

Irreducible features of the Intervalic String: Sat—Chit—Anand ≈

≈ Point: being — *Interval:* space — *Intervalic string:* consciousness

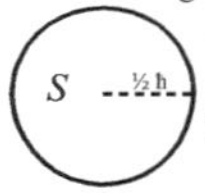

Intervalic String state: $S = \{\uparrow, \downarrow\}$

Intervalic String radius: $r_S = ½\,\hbar$

Intervalic String spin: $J_S = ½\,\hbar$

Intervalic String total length: $l_S = \pi\,\hbar$

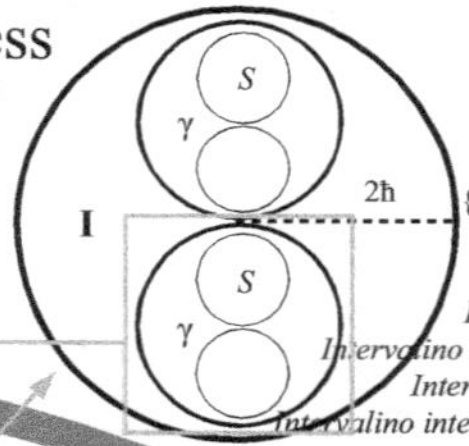

INTERVALINO: Mass

Intervalino intervalic structure: $\mathbf{I} = 2\,\gamma = 4\,S$

Intervalino state: $\mathbf{I} = |\gamma\,\underline{\gamma}|_{ant} = 2^{-1/2}\,(\,|\gamma\underline{\gamma}| - |\underline{\gamma}\gamma|\,) =$

$= 2^{-1/2}\,(\,|\{\,|\uparrow\uparrow|,\ 2^{-1/2}\,(\,|\uparrow\downarrow| + |\downarrow\uparrow|\,),\ |\downarrow\downarrow|\,\}$

$\{\,|\downarrow\downarrow|,\ 2^{-1/2}\,(\,|\downarrow\uparrow| + |\uparrow\downarrow|\,),\ |\uparrow\uparrow|\,\}| - |\{\,|\downarrow\downarrow|,\ 2^{-1/2}\,(\,|\downarrow\uparrow| + |\uparrow\downarrow|\,),\ |\uparrow\uparrow|\,\}$

$\{\,|\uparrow\uparrow|,\ 2^{-1/2}\,(\,|\uparrow\downarrow| + |\downarrow\uparrow|\,),\ |\downarrow\downarrow|\,\}|)$

Intervalino radius: $r_\mathbf{I} = c / \omega_\mathbf{I} = 2\,\hbar = 2.1112726 \cdot 10^{-34}$ (m)

Intervalino spin: $J_\mathbf{I} = 0\,\hbar$ — *Intervalino total length:* $l_\mathbf{I} = 4\pi\,\hbar$

Intervalino electric charge: $Q(\mathbf{I}) = \mathbf{q}_\mathbf{I} = \sqrt{-(c^{-1}\hbar)} = 5.93398995 \cdot 10^{-22}$ (C)

Intervalino structural energy balance: $[\,c^{\pm 2}\,\hbar\,\mathbf{q}_\mathbf{I}^{-2}\,] - m_\mathbf{I}\,\omega_\mathbf{I}^2\,r_\mathbf{I}^2 = 0$

Intervalino intervalic energy: $\mathbf{I}(\mathbf{I}) = c^{\pm 2}\,\hbar\,\mathbf{q}_\mathbf{I}^{-2} = c^{-1} = 20{,}819.42423\ (MeV/c^2)$

Intervalino electromagnetic energy: $U(\mathbf{I}) = 0$

Intervalino spin energy: $E_J(\mathbf{I}) = \mathbf{I}(\mathbf{I}) - U(\mathbf{I}) = c^{-1} = 20{,}819.42423\ (MeV/c^2)$

Intervalino mass: $m(\mathbf{I}) = \mathbf{I}(\mathbf{I}) = c^{-1} = 20{,}819.42423\ (MeV/c^2)$

Intervalino linear velocity due to spin on surface: $v(\mathbf{I}) = c$

Intervalino angular velocity due to spin:

$\omega_J(\mathbf{I}) = c / r_\mathbf{I} = ½\,c\,\hbar^{-1} = 1.419960918 \cdot 10^{42}\,(s^{-1})$

Intervalino coupling temperature:

$\Theta_{cp} = 1/(4\pi\,ck_B) = 1.922575127 \cdot 10^{13}$ (K)

Intervalino coupling frequorce:

$\varphi_{cp} = 1/(4\pi\,c\hbar) = 2.51452013 \cdot 10^{24}\,(s^{-1})$

Intervalino structural energy ratios:

$\mathbf{I}(\mathbf{I})/E(\mathbf{I})_{mass} = 1$

$U(\mathbf{I})/E(\mathbf{I})_{mass} = 0$

$E_J(\mathbf{I})/E(\mathbf{I})_{mass} = 1$

DALINO: Electric Charge

Energy released by the assembly of dalinos at the intervalic Primordial Assembly:

$E_B(D_{270}) = 270m(\mathbf{I})\text{-}m(D_{270}) = 5{,}621{,}244.136\ (MeV/c^2)$

$E_B(D_{45}) = 45m(\mathbf{I})\text{-}m(D_{45}) = 936{,}855.694\ (MeV/c^2)$

Assemblies of intervalinos in symmetric and antisymmetric state under interchange:

$D^{(\pm)} = |\mathbf{I}\,\underline{\mathbf{I}}|_{sym} =$

$= \{\,|\mathbf{I}\,\mathbf{I}|,\ 2^{-1/2}\,(\,|\mathbf{I}\,\underline{\mathbf{I}}| + |\underline{\mathbf{I}}\,\mathbf{I}|\,),\ |\underline{\mathbf{I}}\,\underline{\mathbf{I}}|\,\}$

$D^{(0)} = |\mathbf{I}\,\underline{\mathbf{I}}|_{ant} = 2^{-1/2}\,(\,|\mathbf{I}\,\underline{\mathbf{I}}| - |\underline{\mathbf{I}}\,\mathbf{I}|\,)$

Origin of the electric charge from the primordial assembly of intervalinos:

$|\mathbf{I}\,\mathbf{I}| = (+)$ charge

$2^{-1/2}\,(\,|\mathbf{I}\,\underline{\mathbf{I}}| \pm |\underline{\mathbf{I}}\,\mathbf{I}|\,) =$

= zero charge

$|\underline{\mathbf{I}}\,\underline{\mathbf{I}}| = (-)$ charge

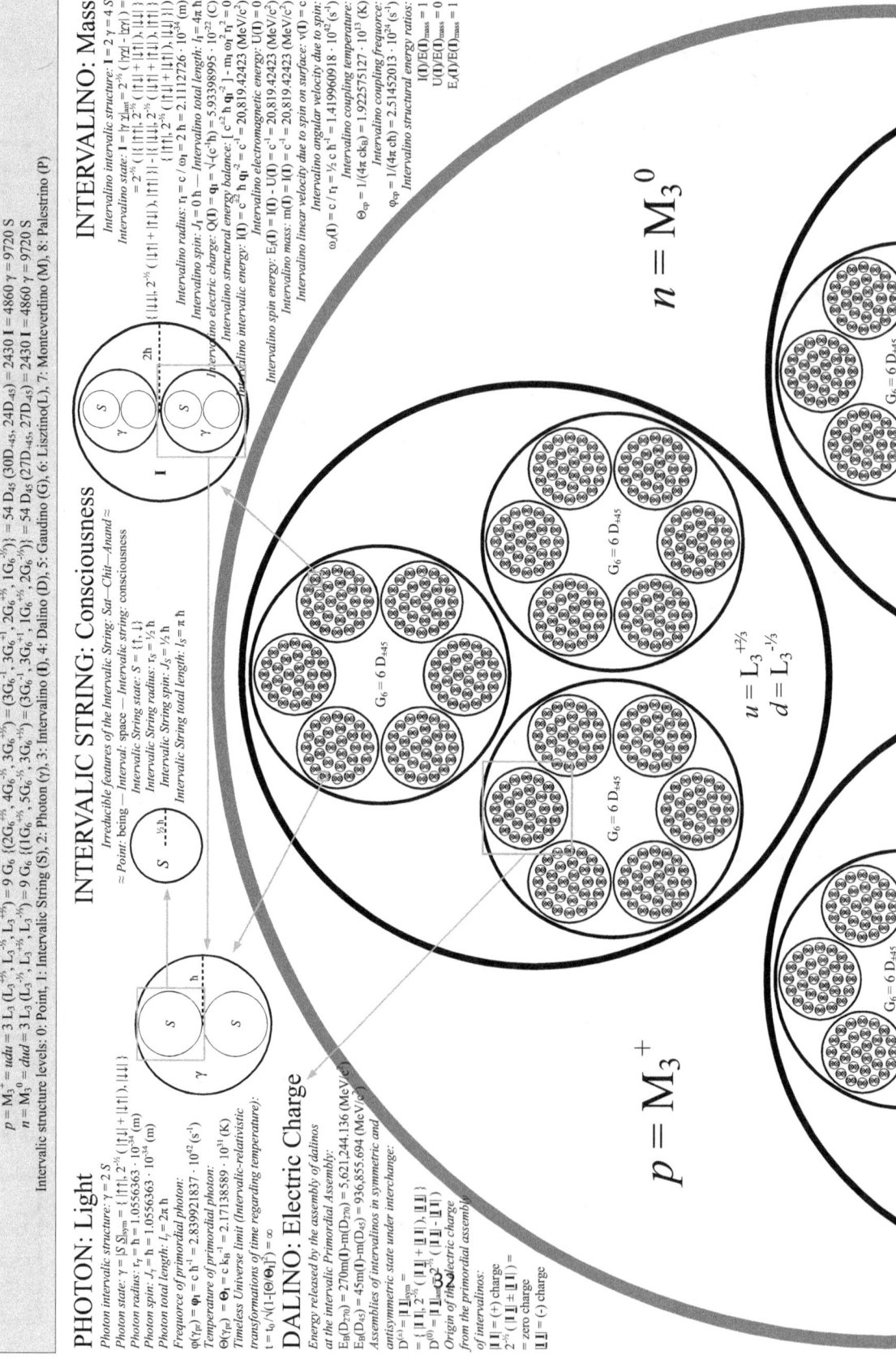

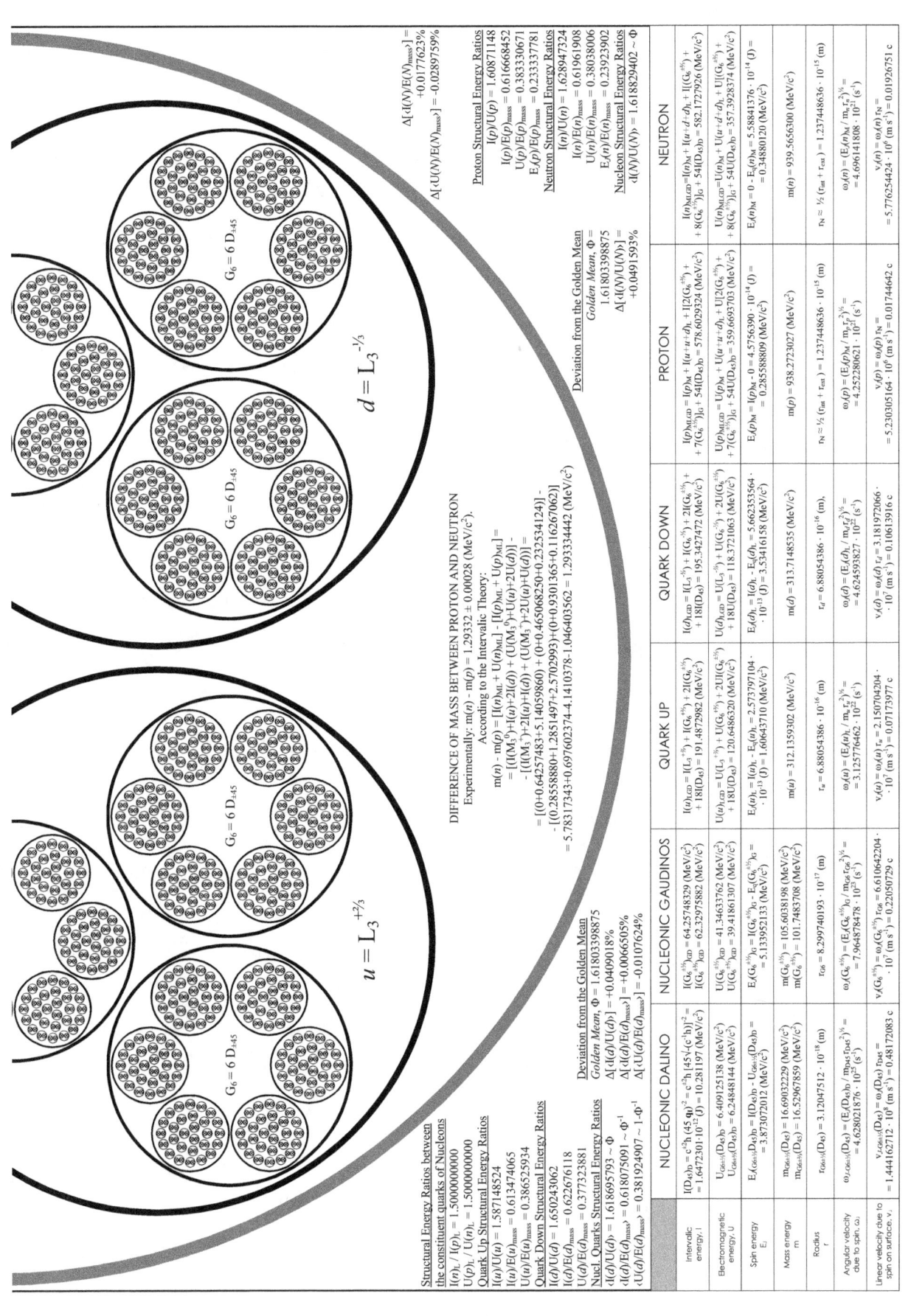

Chapter 18

INTERVALIC BARYON

INTERVALIC MONTEVERDINO: BARYON

THE NEW STATUS OF BARYONS

While nucleons have been supposed to be the fundamental constituents of nucleus, the study and research of baryonic states has become of great importance for Physics. However, after IT has postulated that the constituents of nucleus may not be nucleons, but deeper intervalic structures below the monteverdic level —quarks or gaudinos—, the early importance about baryons has decreased considerably. As baryons only can exist as a single particle, isolated from any other baryons, they are not the usual state of matter, but an extraordinary one (being the only exception to that rule the nucleus of ^{2}H, ^{3}H, and ^{3}He). Therefore, the unique stable baryons of Nature are proton and neutron, which would be only found in ^{1}H, ^{2}H, ^{3}H and ^{3}He nucleus. In any case, the identification of all baryons are not already a main filed by itself, but only another medium to reach a better understand of the intervalic symmetries. However, the development of further devices for the detection of quarks in those processes involving *universality*, such as the decays of massive bosons, appear to be a better way to research the full spectrum of the intervalic symmetries of fundamental particles, as the quarks production is not induced "artificially" as in a collision, but the quarks pairs are produced in a systematic and "natural" way in the decay.

Starting from the intervalic structure of quarks, we have got a new set of physical concepts and quantum numbers to combine for making a new intervalic model of quarks which yield 2-quarks systems (mesons) and 3-quarks systems (baryons). We are going to see that the old model of quarks is, say, to grope about in the dark, and therefore it can match some reveille by chance, as the majority of its quantum numbers in SM are partial or wrongly defined in some way. For example, IT postulated that Nature does not mix intervalic structures —in this case, quarks— pertaining to different symmetry families ({D45}, {D30}, {D18}, {D6}, {D5}, etc.), whilst the misleading SM merely supposes naïve combinations among flavours. But such intended flavours are really nothing but a fantastic name for some intervalic structures of quarks whose approximate masses have been experimentally detected through a purely empirical way. As those supposed flavoured quarks pertains to several symmetry families, the only allowed combinations among them are precisely those ones which does not mix different intervalic symmetries, and event which may or may not happen. That is to say, pure chance, as several of the fundamental principles that govern the assembly of monteverdinos —baryons and mesons— are not right and completely understood inside SM. Fortunately, things go in a different way in IT.

INTERVALIC BARYONS WITH {D45} SYMMETRY

As we have seen opportunely, every set of quark symmetries is composed by 7 intervalic structures. Those corresponding to {D45} symmetry are the following:

$L_{⅓} = ⅓\,G_6 = 2\,D_{45} = 34.537545—36.791934 \rightarrow$ quark $L⅓D45^{(⅓)}$
$L_{⅔} = ⅔\,G_6 = 4\,D_{45} = 69.075090—73.583868 \rightarrow$ quark $L⅔D45^{(⅓, ⅔)}$
$L_1 = 1\,G_6 = 6\,D_{45} = 103.61263—110.37580 \rightarrow$ quark $L1D45^{(⅓, ⅔)}$
$L_2 = 2\,G_6 = 12\,D_{45} = 207.22526—220.75160 \rightarrow$ quark $L2D45^{(⅓, ⅔)}$
$L_3 = 3\,G_6 = 18\,D_{45} = 310.83790—331.12740 \rightarrow$ quark $L3D45^{(⅓, ⅔)}$
$L_4 = 4\,G_6 = 24\,D_{45} = 414.45053—441.50320 \rightarrow$ quark $L4D45^{(⅓, ⅔)}$
$L_5 = 5\,G_6 = 30\,D_{45} = 518.06316—551.87900 \rightarrow$ quark $L5D45^{(⅓, ⅔)}$

Since Nature does not mix quarks from different intervalic symmetry when assembling baryons and mesons, the number of allowed baryons made by the family of intervalic quarks with {D45} symmetry will be the group SU(7) with intervalic structure symmetry:

7 V 7 V 7 = 343 baryons

INTERVALIC BARYONS WITH {D30} SYMMETRY

$L_{⅓} = ⅓\,G_9 = 3\,D_{30} = 116.56421—124.17278 \rightarrow$ quark $L⅓D30^{(⅓)}$
$L_{⅔} = ⅔\,G_9 = 6\,D_{30} = 233.12843—248.34556 \rightarrow$ quark $L⅔D30^{(⅔)}$
$L_1 = 1\,G_9 = 9\,D_{30} = 349.69264—372.51833 \rightarrow$ quark $L1D30^{(⅓)}$
$L_2 = 2\,G_9 = 18\,D_{30} = 699.38520—745.03666 \rightarrow$ quark $L2D30^{(⅔)}$
$L_3 = 3\,G_9 = 27\,D_{30} = 1{,}049.0779—1{,}117.5550 \rightarrow$ quark $L3D30^{(⅓)}$
$L_4 = 4\,G_9 = 36\,D_{30} = 1{,}398.7705—1{,}490.0733 \rightarrow$ quark $L4D30^{(⅔)}$
$L_5 = 5\,G_9 = 45\,D_{30} = 1{,}748.4631—1{,}862.5916 \rightarrow$ quark $L5D30^{(⅓)}$

The number of allowed baryons made by the family of intervalic quarks with {D30} symmetry is described by the group SU(7):

7 V 7 V 7 = 343 baryons

INTERVALIC BARYONS WITH {D18} SYMMETRY

$L_{⅓} = ⅓\ G_{15} = 5\ D_{18} = 539.64917 - 574.87400 \rightarrow$ quark $L⅓D18^{(⅓)}$
$L_{⅔} = ⅔\ G_{15} = 10\ D_{18} = 1,079.2982 - 1,149.7480 \rightarrow$ quark $L⅔D18^{(⅔)}$
$L_{1} = 1\ G_{15} = 15\ D_{18} = 1,618.9475 - 1,724.6220 \rightarrow$ quark $L1D18^{(⅓)}$
$L_{2} = 2\ G_{15} = 30\ D_{18} = 3,237.8950 - 3,449.2443 \rightarrow$ quark $L2D18^{(⅔)}$
$L_{3} = 3\ G_{15} = 45\ D_{18} = 4,856.8425 - 5,173.8660 \rightarrow$ quark $L3D18^{(⅓)}$
$L_{4} = 4\ G_{15} = 60\ D_{18} = 6,475.7900 - 6,898.4886 \rightarrow$ quark $L4D18^{(⅔)}$
$L_{5} = 5\ G_{15} = 75\ D_{18} = 8,094.7375 - 8,623.1100 \rightarrow$ quark $L5D18^{(⅓)}$

The number of allowed baryons made by the family of intervalic quarks with {D18} symmetry is equally described by the group SU(7):

7 V 7 V 7 = 343 baryons

INTERVALIC STRUCTURE AND FLAVOUR IN THE MODEL OF QUARKS

Henceforth, it is easy to understand why never have been detected baryons composed by any combination among the supposed quark bottom and the intended quarks up and down (*b, u, d*). On the contrary, combinations among the supposed quarks up, down, strange and charme are possible whenever the masses of those quarks vary in order to coincide with their real masses shown in each symmetry family, as all these quarks have several different intervalic structures allowed near their intended masses. For example, at first sight they can be easily confounded —without knowing it— with the following intervalic structures:

- Quark up: $L⅔D45^{(⅓,⅔)}\ L1D45^{(⅓,⅔)}\ L2D45^{(⅓,⅔)}\ L3D45^{(⅓,⅔)}$
  $L4D45^{(⅓,⅔)}\ L⅓D30^{(⅓)}\ L⅔D30^{(⅔)}\ L1D30^{(⅓)}$
- Quark down: $L⅓D45^{(⅓)}\ L⅔D45^{(⅓,⅔)}\ L1D45^{(⅓,⅔)}\ L2D45^{(⅓,⅔)}$

L3D45$^{(⅓,⅔)}$ L4D45$^{(⅓,⅔)}$ L⅓D30$^{(⅓)}$ L⅔D30$^{(⅔)}$ L1D30$^{(⅓)}$
- Quark strange: L4D45$^{(⅓,⅔)}$ L5D45$^{(⅓,⅔)}$ L1D30$^{(⅓)}$ L2D30$^{(⅔)}$ L⅓D18$^{(⅓)}$
- Quark charme: L3D30$^{(⅓)}$ L4D30$^{(⅔)}$ L5D30$^{(⅓)}$ L⅔D18$^{(⅔)}$ L1D18$^{(⅓)}$
- Quark bottom: L2D18$^{(⅔)}$ L3D18$^{(⅓)}$ L4D18$^{(⅔)}$ L5D18$^{(⅓)}$

When summing their masses to compose a monteverdino —baryon or meson— the confusion can be inclusive greater, and affect not only to their masses but also to their electric charges. I think this example may be sufficient to understand the terribly chaos in the classification of the model of quarks in SM.

The total sum of the number of allowed baryons below a mass of ~44.000 (MeV/c^2) are, regarding the SU(7) group of intervalic structure symmetry:

$$(7 \vee 7 \vee 7) + (7 \vee 7 \vee 7) + (7 \vee 7 \vee 7) = 1029$$

Of course, the traditional flavour multiplets in the primitive model of quarks introduced by Gell-Mann, Neeman and Zweig are only some few combinations included in the intervalic model of quarks.

INTERVALIC BARYONS WITH {D6} SYMMETRY

$L_{⅓} = ⅓$ $G_{45} = 15$ $D_6 = 14{,}570.527 — 15{,}521.598$ → quark L⅓D6$^{(⅓)}$
$L_{⅔} = ⅔$ $G_{45} = 30$ $D_6 = 29{,}141.055 — 31{,}043.196$ → quark L⅔D6$^{(⅔)}$
$L_1 = 1$ $G_{45} = 45$ $D_6 = 43{,}711.582 — 46{,}564.794$ → quark L1D6$^{(⅓)}$
$L_2 = 2$ $G_{45} = 90$ $D_6 = 87{,}423.165 — 93{,}129.588$ → quark L2D6$^{(⅔)}$
$L_3 = 3$ $G_{45} = 135$ $D_6 = 131{,}134.74 — 139{,}694.38$ → quark L3D6$^{(⅓)}$
$L_4 = 4$ $G_{45} = 180$ $D_6 = 174{,}846.32 — 186{,}259.17$ → quark L4D6$^{(⅔)}$
$L_5 = 5$ $G_{45} = 225$ $D_6 = 218{,}557.91 — 232{,}823.97$ → quark L5D6$^{(⅓)}$

The number of allowed baryons is the same as in {D30} and {D18} symmetries:

$7 \vee 7 \vee 7 = 343$ baryons

INTERVALIC BARYONS WITH {D5} SYMMETRY

$L_{1/3} = 1/3\ G_{54} = 18\ D_5 = 25,177.871 - 26,821.321 \rightarrow$ quark $L1/3D5^{(1/3)}$
$L_{2/3} = 2/3\ G_{54} = 36\ D_5 = 50,355.742 - 53,642.642 \rightarrow$ quark $L2/3D5^{(1/3,\ 2/3)}$
$L_1 = 1\ G_{54} = 54\ D_5 = 75,533.615 - 80,463.964 \rightarrow$ quark $L1D5^{(1/3,\ 2/3)}$
$L_2 = 2\ G_{54} = 108\ D_5 = 151,067.23 - 160,927.93 \rightarrow$ quark $L2D5^{(1/3,\ 2/3)}$
$L_3 = 3\ G_{54} = 162\ D_5 = 226,600.84 - 241,391.89 \rightarrow$ quark $L3D5^{(1/3,\ 2/3)}$
$L_4 = 4\ G_{54} = 216\ D_5 = 302,134.45 - 321,855.85 \rightarrow$ quark $L4D5^{(1/3,\ 2/3)}$
$L_5 = 5\ G_{54} = 270\ D_5 = 377,668.07 - 402,319.82 \rightarrow$ quark $L5D5^{(1/3,\ 2/3)}$

The SU(7) intervalic structure symmetry is the same in all families:

$7 \vee 7 \vee 7 = 343$ baryons

INTERVALIC BARYONS WITH {D3} AND {D2} SYMMETRY

They are just equal to the {D5} intervalic symmetry. Of course the energy needed for the production of heavier baryons is so huge that it will be very hard to get it in our present devices. Nevertheless it is postulated that those states should exist at the Big Bang, immediately decaying and following the intervalic sequence of symmetries: {D2} → {D3} → {D5} → {D6} → {D18} → {D30} → {D45}.

The decomposition of each symmetry group is a complex and laborious task that deserves an independent work. However the most important think is perhaps the theoretical postulation of the existence of a wide but highly precise and finite set of combinations for the baryonic states. Thus, according to IT the so named "zoo of baryons" is an *expected* experimental data, whilst according to SM it is an *unexpected* and disconcerting event which only can be hardly *described* —but never *explained* in a reliable fundamental way— through some assumptions ad hoc, like the flavours, colours and excited states introduced by hand and always *after* the obtaining of unpredicted experimental data. Practised in this way, Physics of SM is not different of Zoology: a merely *descriptive* "science". Of course, the intervalic symmetries of Nature are so powerful to allow that a partial set of them can always be *described* without knowing its underlying foundations through a lot of misleading assumptions and partial ways which give us by chance an incomplete view of the subatomic particles of Nature, as SM really does.

ISOCHARGE SYMMETRY

It can be noted that all the intervalic structures of {D45}, {D5}, {D3} and {D2} symmetries (with the only exception of the first light quark of each set) have two charges *allowed*, ⅓ and ⅔. However both allowed values pertain to a unique and the same intervalic structure, which stay unchanged. This is a remarkable difference between IT and SM, which considered quarks up and down as *two* different flavours, whilst according to IT they are *isoquarks* which have the same intervalic structure and are therefore they are considered as *one* quark with an electric charge degree of freedom. That degree of freedom of the electric charge in those symmetries can be introduced in the usual group formalism, because the value of the electric charge does not affect to the intervalic structure of the quark, which is not similar but *identical* for both charge values. Therefore, we will have got another quantum number which could be appropriately named *isocharge*. This symmetric unitary group will be SU(2) for those intervalic structures which can got both charges, ⅓ *and* ⅔, and a trivial SU(1) for the remaining ones which have got only one charge, ⅓ *or* ⅔. Incorporating this feature the number of allowed baryons in such *dual* families —{D45}, {D5}, {D3} and {D2}— would be described by the group SU(7) x SU(2) contained in SU(14), which yields 2744 allowed baryons (depicting the fact that the first quark $L⅓D45^{(⅓)}$ has forbidden the value ⅔ charge, and the baryons yielded by the {D45} symmetry would have some blank states).

Following the spin-statistics theorem, if we set the isocharge values to +½ for the ⅔ charge, and to -½ for the ⅓ charge, we will obtain the same group of symmetries as in the case of the SU(2) isospin symmetry, as we will see later.

INTERVALIC ISOSPIN SYMMETRY

At first sight we can consider isospin as a *provisional* assumption for a possible underlying physical quantity, not postulated yet, as it is not acceptable the postulation of an isospin 'abstract space', without any direct or indirect reference to a fundamental geometry nor to a physical event. Moreover, isospin has now become irrelevant to explain the independence of charge in strong interaction, as it is fully explained by means of the intervalic interaction between the intervalic structures, as I have demonstrated in other site.

From the intervalic structures of quarks it appears to be clear that the concept of isospin can be related with a real physical quantity, instead to suppose the existence of a doubtful isospin 'abstract space', as SM does. Really, allowed values of isospin are closely related with electric charge and strangeness of quarks. As strangeness is irrelevant in IT, isospin follows the same way (apart from other reasons). Nevertheless, it is now easy to relate *isospin* with a real physical feature of quarks in IT: the *degree of freedom of the electric charge in the intervalic structures of quarks*, previously defined as *isocharge*. In that case, that degree of freedom would make the following symmetric unitary groups of isospin:

- Quarks of {D45} symmetry:
$L_{1/3} = 1/3\ G_6 = 2\ D_{45}$ → quark $L_{1/3}D45^{(1/3)}$ → SU(1) isospin symmetry
$L_{2/3} = 2/3\ G_6 = 4\ D_{45}$ → quark $L_{2/3}D45^{(1/3, 2/3)}$ → SU(2) isospin symmetry
$L_1 = 1\ G_6 = 6\ D_{45}$ → quark $L1D45^{(1/3, 2/3)}$ → SU(2) isospin symmetry
$L_2 = 2\ G_6 = 12\ D_{45}$ → quark $L2D45^{(1/3, 2/3)}$ → SU(2) isospin symmetry
$L_3 = 3\ G_6 = 18\ D_{45}$ → quark $L3D45^{(1/3, 2/3)}$ → SU(2) isospin symmetry
$L_4 = 4\ G_6 = 24\ D_{45}$ → quark $L4D45^{(1/3, 2/3)}$ → SU(2) isospin symmetry
$L_5 = 5\ G_6 = 30\ D_{45}$ → quark $L5D45^{(1/3, 2/3)}$ → SU(2) isospin symmetry

- Quarks of {D30} symmetry:
Every quark → SU(1) isospin symmetry

- Quarks of {D18} symmetry:
Every quark → SU(1) isospin symmetry

- Quarks of {D6} symmetry:
Every quark → SU(1) isospin symmetry

- Quarks of {D5} symmetry:
SU(2) + SU(1) isospin symmetry as in {D45}

- Quarks of {D3} symmetry:
SU(2) + SU(1) isospin symmetry as in {D45}

- Quarks of {D2} symmetry:
SU(2) + SU(1) isospin symmetry as in {D45}

Setting the isospin values for the singlet of SU(1) to 0, and for the doublets of SU(2) to ½ for electric charge ⅔, and to -½ for electric charge ⅓, we obtain just the same traditional isospin values which had the former quarks. (I sincerely do not see any advantage on maintaining such absurd and misleading use, instead to introduce the *isocharge* symmetry). As isospin of the supposed quark charm is not always 0, as it can also have some of the following intervalic structures: $L4D45^{(1/3, 2/3)}$ or $L5D45^{(1/3, 2/3)}$, and likewise the isospin of the supposed quarks up and down is not always ±½, as they can have some of the following intervalic structures: $L\frac{1}{3}D30^{(1/3)}$, $L\frac{2}{3}D30^{(2/3)}$ or $L1D30^{(1/3)}$, it can be easily understood why it is said that the traditional isospin symmetry of SM is not an "exact" symmetry, without knowing the truthful reason for that inexactness. Really, the isospin symmetry with respect to the supposed flavour is not only inexact but a completely disastrous one.

Whilst the mystic eightfold way of the Gell-Mann model of quarks was intended to appear as a mysterious and almost secret clue of Nature, involving hypercharge and isospin, the fact is simply the following: when the underlying physical concepts holding flavours, colours, hypercharge, isospin and other fantastic assumptions are uncovered and related to a real physical feature —all of them derived from the lisztinian *intervalic structure* of quarks—, the model of quarks becomes a poorly naïve system of combinations involving by chance some of the different degrees of freedom derived from the intervalic structures of quarks: some of these degrees of freedom are no other thing than all those mysterious quantum numbers.

INTERVALIC SPIN SYMMETRY

Since gaudino is postulated to be the first *fermion* of Nature, assembled in the intervalic primordial aggregation at the Big Bang, the most simple assumption is to set the spin of all gaudinos to ½. As quarks are made from the assembly of ⅓, ⅔, 1, 2, 3, 4 or 5 gaudinos in an antisymmetric state under interchange (as the assembly of gaudinos in a symmetric state makes no quarks but zero charge massive bosons), their corresponding groups of spin symmetry will lead to the following basic states (not to mention mixed symmetric and antisymmetric groups of the decomposition which may lead to a lot of excited states of quarks with higher spin; they may increase the already large zoo of baryons almost indefinitely along with the own excited states of the previous constituent gaudinos). In resume, the antisymmetric multiplets of the decomposition made from gaudinos' spin determine the basic states of the quarks' spin:

- Quarks L⅓ of all symmetries made from the assembly of:
gaudinos with SU(1) spin symmetry = 2 → basic spin: ½

- Quarks L⅔ of all symmetries made from the assembly of:
gaudinos with SU(1) spin symmetry = 2 → basic spin: ½

- Quarks L1 of all symmetries made from the assembly of:
gaudinos with SU(1) spin symmetry = 2 → basic spin: ½

- Quarks L2 of all symmetries made from the assembly of:
gaudinos with SU(2) spin symmetry = 2 V 2 → basic spin: 0

- Quarks L3 of all symmetries made from the assembly of:
gaudinos with SU(2) spin symmetry = 2 V 2 V 2 → basic spin: ½

- Quarks L4 of all symmetries made from the assembly of:
gaudinos with SU(2) spin symmetry = 2 V 2 V 2 V 2 → basic spin: 0

- Quarks L5 of all symmetries made from the assembly of:
gaudinos with SU(2) spin symmetry = 2 V 2 V 2 V 2 V 2 → basic spin: ½

Therefore *even* quarks, L2 and L4, are not fermions but bosons. It can be noted that these *even* lisztinos are just the principal constituent of mesons, in good agreement with experimental data (cfr. our paper on intervalic meson).

Once more it is clear that another one of the sacred principles in SM, that one which postulated that fermions were source of interaction particles and bosons were the intermediate particles of the interaction is not always truth; what means that it is simply false. Really, we have found at every step of IT a lot of bosons which contradict that supposed principle: spin 0 intervalinos are both source and intermediate particles of the changeless —strong— intervalic interaction, spin 0 dalinos are source particles; spin ½ gaudinos are both leptons and charged massive bosons; lisztinos can have different spin, and they are quarks which can be fermions or bosons, charged massive bosons which are bosons, and nucleus which can be likewise fermions or bosons; finally, monteverdinos are mesons and baryons, being the first ones bosons, but the second ones can be once more fermions as well as bosons. In a word, the traditional interpretation of the supposed high principle of SM which prayed: fermion = source particle, and boson = intermediate particle, has finally resulted to be a total and complete disaster without palliatives as in general there are much more bosons than the intended ones. In IT we can say that this principle is replaced by a deeper comprehension of the mechanism involved in a interaction, as we will see when describing the intervalic interaction.

According to this the spin symmetry of baryons would be described by a doublet and a singlet, in a similar way as the isocharge symmetry. Experimental physicists have a lot of work to do for detecting and writing down these states. However, it is expected that a lot of these allowed baryons will not be able to be detected through our actual primitive devices working at present available energies. Anyway, it will have to be researched why or whether Nature follows a possible main phenomenology sequence in the making of quarks inside each intervalic symmetry. And I am afraid that the answer to this question will not be able to be explained *only* by usual physical principles, as the principle of *minimal energy*, etc. but introducing what perhaps may be the last foundations of Epistemology concerning the still unknown relations between energy, symmetry and form. This fascinating matter would lead us out of the frontiers of Physics, which is far beyond the subject of this paper.

INTERVALIC STRUCTURE OF THE SUPPOSED OCTET OF BARYONS SU(3) OF THE STANDARD MODEL

The particles composing this supposed octet can be totally explained by their intervalic structures which match the experimental data with remarkable precision. Although it is clear that the decomposition of the SU(7) groups of *intervalic structure* symmetry will give at last a reliable basis for a general classification of baryonic states, which will make irrelevant the provisional and misleading classification according the traditional SU(3) *flavour* symmetry of SM, we are going to view which are the intervalic structures of one of these multiplets of baryons, in order to see their truthful intervalic structure:

p (938.3) = (L3D45$^{(2/3)}$ L3D45$^{(2/3)}$ L3D45$^{(1/3)}$)
n (939.6) = (L3D45$^{(2/3)}$ L3D45$^{(1/3)}$ L3D45$^{(1/3)}$)
Λ^0 (1115.7) = (L2/3D45$^{(2/3)}$ L5D45$^{(1/3)}$ L5D45$^{(1/3)}$)
Σ^+ (1189.4) = (L3D45$^{(2/3)}$ L4D45$^{(2/3)}$ L4D45$^{(1/3)}$)
Σ^0 (1192.6) = (L3D45$^{(2/3)}$ L4D45$^{(1/3)}$ L4D45$^{(1/3)}$)
Σ^- (1197.4) = (L3D45$^{(1/3)}$ L4D45$^{(1/3)}$ L4D45$^{(1/3)}$)
Ξ^0 (1314.8) = (L4D45$^{(2/3)}$ L4D45$^{(1/3)}$ L4D45$^{(1/3)}$)
Ξ^- (1321.3) = (L4D45$^{(1/3)}$ L4D45$^{(1/3)}$ L4D45$^{(1/3)}$)

The mass due to the intervalic energy, $I = c^{\pm 2} h \, Q^{-2}$, fully explains why an intervalic structure with electric charge ⅓ has slightly more mass than the same intervalic structure with charge ⅔. This phenomenon appears in a lot of baryons, such as in $p\,n$, $\Sigma^+ \Sigma^0 \Sigma^-$, $\Xi^0 \Xi^-$, etc. In the case of nucleon, the difference of mass between proton and neutron have been explained by IT with unbelievably precision which not only fit up to the last decimal of the best experimental data, but makes a prediction with an accuracy of two powers beyond such data. The difference of mass between the remaining particles can be now easily explained according to the Intervalic Theory in the same way. This is an important achievement of IT which allow us to understand the masses and other features of subatomic particles.

INTERVALIC STRUCTURE OF THE SUPPOSED DECUPLET OF BARYONS SU(3) OF THE STANDARD MODEL

According to the intervalic structures involved, the supposed decuplet of SU(3) flavour symmetry shows a chaotic classification of baryons where no simple sequence is founded, but the baryons may be remote or partially unrelated regarding their intervalic structures. This can be easily understood since the 6 supposed flavours of SM has been replaced by a wider set of intervalic structures: 7 intervalic structures in each one of the intervalic symmetries —namely {D45}, {D30}, {D18}, {D6}, {D5}, {D3}, {D2}—, totalling 7 x 7 = 49 intervalic structures. The SU(6) x SU(3) supposed *flavour* and *colour* symmetry has been replaced by a SU(7) + SU(7) + SU(7) + SU(7) + SU(7) + SU(7) + SU(7) *intervalic structure* symmetry. By this reason the partial classification of baryons made by the first appears as a unique and ridicule thrown of the dice among the whole set of possibilities realized by the second. Even worst, the vast majority of baryonic combinations —SU(6) x SU(3) = 5832 baryons— predicted by the supposed *flavour* and *colour* symmetry are completely wrong because Nature does not mix intervalic symmetries. It is curious how the group of SU(3) *colour* symmetry is never written down in the traditional model of quarks. Perhaps its incoherence becomes more when it is written down.

Needless to say that according to the SU(7) groups of *intervalic structure* symmetry, the SU(6) group of *flavour* symmetry and the group SU(3) of colour symmetry are pure fantasy, and therefore their derived quantum numbers, as *i.e.* the hypercharge. Thus the traditional classification of baryonic states in SM is entirely disappointing as from the three physical quantities involved in the usual representation of the multiplets —electric charge, hypercharge and isospin— only the first one of them stay as a reliable magnitude inside IT. Besides, flavour, colour, electric charge, spin and isospin are intended to be independent and unrelated physical quantities in SM. On the contrary, inside IT it has to be remarked that all quantum numbers and physical quantities (including

INTERVALIC THEORY:
The Intervalic Structures of Subatomic Particles and the Last Foundations of Physics

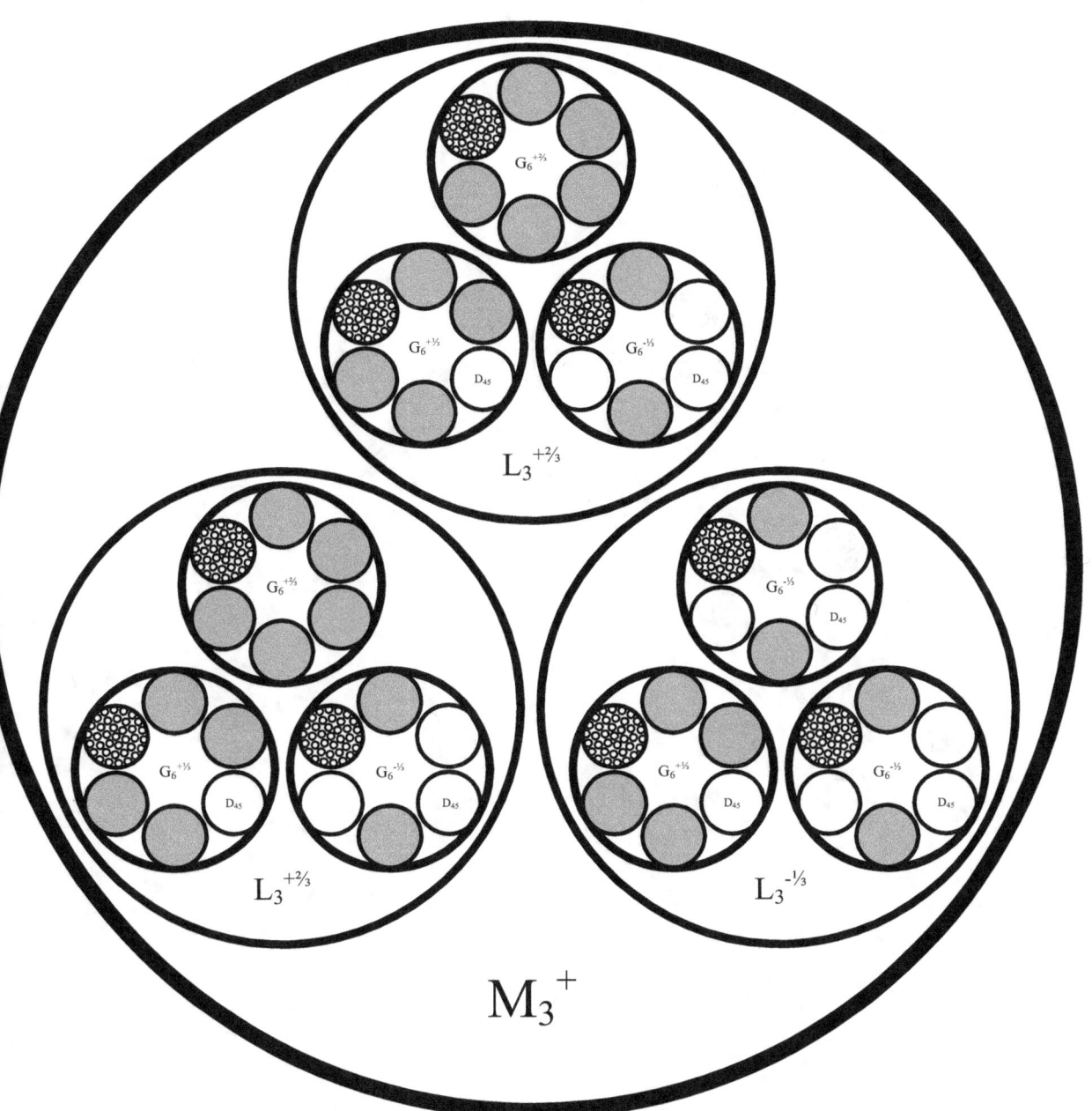

Figured intervalic structure of *proton* (unlike charges shadowed):
$$M_3^+ = 3\ L_3\ (L_3^{+2/3}, L_3^{-1/3}, L_3^{+2/3}) =$$
$$9\ G_6\ (2G_6^{+2/3}, 4G_6^{-1/3}, 3G_6^{+1/3}) = 54\ D_{45}\ (30D_{+45}, 24D_{-45}) = 2430\ \mathbf{I}$$

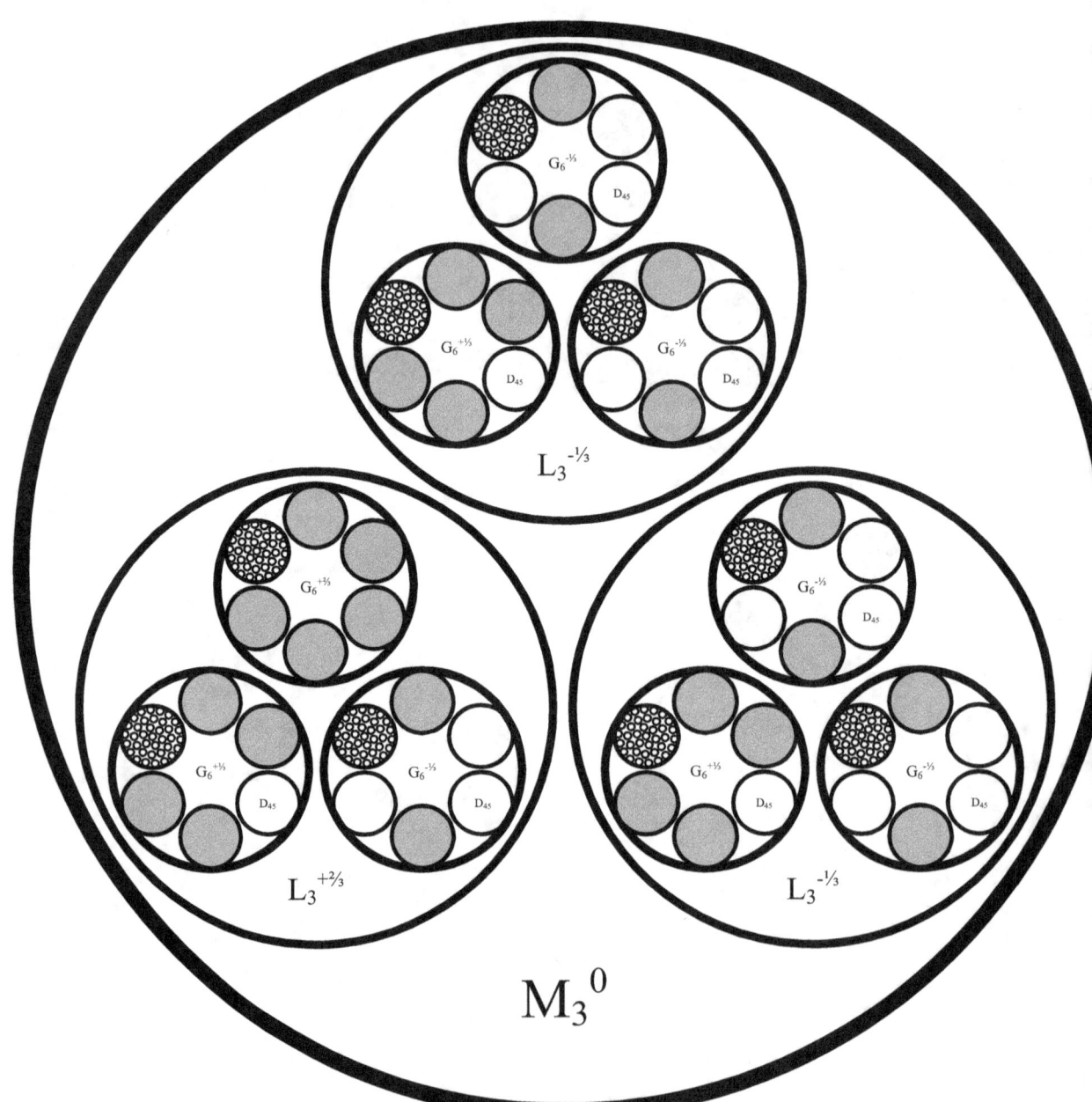

Figured intervalic structure of *neutron* (unlike charges shadowed):
$$M_3^0 = 3\,L_3\,(L_3^{-\frac{1}{3}},\,L_3^{+\frac{2}{3}},\,L_3^{-\frac{1}{3}}) =$$
$9\,G_6\,(1G_6^{+\frac{2}{3}},\,5G_6^{-\frac{1}{3}},\,3G_6^{+\frac{1}{3}}) = 54\,D_{45}\,(27D_{+45},\,27D_{-45}) = 2430\,\mathbf{I}$

mass, intervalic energy, electromagnetic energy, spin energy, etc.) are completely derived from a unique concept: the *intervalic structure*, whose intervalic symmetries determine all the quantum numbers and physical quantities allowed to any subatomic particles. In resume, the classification of particles according to SM is a complete mess which can no longer be maintained.

$$\Delta^{++} (1232) = (L\tfrac{2}{3}D30^{(\tfrac{2}{3})} L\tfrac{2}{3}D30^{(\tfrac{2}{3})} L2D30^{(\tfrac{2}{3})})$$
$$(L1D45^{(\tfrac{1}{3},\tfrac{2}{3})} L5D45^{(\tfrac{1}{3},\tfrac{2}{3})} L5D45^{(\tfrac{1}{3},\tfrac{2}{3})})$$
$$\Delta^{+} (1232) = (L\tfrac{2}{3}D30^{(\tfrac{2}{3})} L\tfrac{2}{3}D30^{(\tfrac{2}{3})} L2D30^{(\tfrac{2}{3})})$$
$$(L1D45^{(\tfrac{1}{3},\tfrac{2}{3})} L5D45^{(\tfrac{1}{3},\tfrac{2}{3})} L5D45^{(\tfrac{1}{3},\tfrac{2}{3})})$$
$$\Delta^{0} (1232) = (L\tfrac{2}{3}D30^{(\tfrac{2}{3})} L\tfrac{2}{3}D30^{(\tfrac{2}{3})} L2D30^{(\tfrac{2}{3})})$$
$$(L1D45^{(\tfrac{1}{3},\tfrac{2}{3})} L5D45^{(\tfrac{1}{3},\tfrac{2}{3})} L5D45^{(\tfrac{1}{3},\tfrac{2}{3})})$$
$$\Delta^{-} (1232) = (L\tfrac{2}{3}D30^{(\tfrac{2}{3})} L\tfrac{2}{3}D30^{(\tfrac{2}{3})} L2D30^{(\tfrac{2}{3})})$$
$$(L1D45^{(\tfrac{1}{3},\tfrac{2}{3})} L5D45^{(\tfrac{1}{3},\tfrac{2}{3})} L5D45^{(\tfrac{1}{3},\tfrac{2}{3})})$$
$$\Sigma^{+} (1385) = (L3D45^{(\tfrac{1}{3},\tfrac{2}{3})} L5D45^{(\tfrac{1}{3},\tfrac{2}{3})} L5D45^{(\tfrac{1}{3},\tfrac{2}{3})})$$
$$(L4D45^{(\tfrac{1}{3},\tfrac{2}{3})} L4D45^{(\tfrac{1}{3},\tfrac{2}{3})} L5D45^{(\tfrac{1}{3},\tfrac{2}{3})})$$
$$\Sigma^{0} (1385) = (L3D45^{(\tfrac{1}{3},\tfrac{2}{3})} L5D45^{(\tfrac{1}{3},\tfrac{2}{3})} L5D45^{(\tfrac{1}{3},\tfrac{2}{3})})$$
$$(L4D45^{(\tfrac{1}{3},\tfrac{2}{3})} L4D45^{(\tfrac{1}{3},\tfrac{2}{3})} L5D45^{(\tfrac{1}{3},\tfrac{2}{3})})$$
$$\Sigma^{-} (1385) = (L3D45^{(\tfrac{1}{3},\tfrac{2}{3})} L5D45^{(\tfrac{1}{3},\tfrac{2}{3})} L5D45^{(\tfrac{1}{3},\tfrac{2}{3})})$$
$$(L4D45^{(\tfrac{1}{3},\tfrac{2}{3})} L4D45^{(\tfrac{1}{3},\tfrac{2}{3})} L5D45^{(\tfrac{1}{3},\tfrac{2}{3})})$$
$$\Xi^{0} (1533) = (L4D45^{(\tfrac{1}{3},\tfrac{2}{3})} L5D45^{(\tfrac{1}{3},\tfrac{2}{3})} L5D45^{(\tfrac{1}{3},\tfrac{2}{3})})$$
$$\Xi^{-} (1533) = (L4D45^{(\tfrac{1}{3},\tfrac{2}{3})} L5D45^{(\tfrac{1}{3},\tfrac{2}{3})} L5D45^{(\tfrac{1}{3},\tfrac{2}{3})})$$
$$(L\tfrac{1}{3}D30^{(\tfrac{1}{3})} L1D30^{(\tfrac{1}{3})} L3D30^{(\tfrac{1}{3})})$$
$$(L\tfrac{1}{3}D30^{(\tfrac{1}{3})} L2D30^{(\tfrac{2}{3})} L2D30^{(\tfrac{2}{3})})$$
$$(L\tfrac{2}{3}D30^{(\tfrac{2}{3})} L\tfrac{2}{3}D30^{(\tfrac{2}{3})} L3D30^{(\tfrac{1}{3})})$$
$$\Omega^{-} (1672) = (L5D45^{(\tfrac{1}{3},\tfrac{2}{3})} L5D45^{(\tfrac{1}{3},\tfrac{2}{3})} L5D45^{(\tfrac{1}{3},\tfrac{2}{3})})$$
$$(L\tfrac{1}{3}D18^{(\tfrac{1}{3})} L\tfrac{1}{3}D18^{(\tfrac{1}{3})} L\tfrac{1}{3}D18^{(\tfrac{1}{3})})$$

INTERVALIC STRUCTURES OF BARYONS

Every known baryon can be fully explained through a determine intervalic structure. Sceptics please note that the probability to reach such a total match with the set of intervalic symmetries is practically zero. Moreover, it has to be pointed out incessantly that *no one parameter* has been introduced or chosen by hand in the Intervalic Theory, as all the intervalic symmetries are logical and unavoidably derived from the intervalic units, which at once are derived exclusively from the two fundamental constants of Nature, c and h. Thus, the whole IT and the whole IT in Particle Physics has been postulated without introducing by hand any parameter or constant. Really, the theory runs inclusive without setting c and ℏ to 1 because the working of the intervalic dimensions does not depend on that setting.

For the SM way of thinking, the obsessive search for further baryons and other subatomic particles was a very strong necessity, as this was the only way SM had to advance blindly across a lot of subatomic particles whose existence, mass and the other physical features did not understand, but only could poorly *describe* ad hoc in a partial and misleading way. Of course, then there was perhaps no other ways to progress in Physics. But once the Intervalic Theory was postulated such obsessive search for baryons leaves obsession and has to be interpreted in a new manner. Now we have a whole *theory* that predicts with detailed precision all the allowed baryonic states. At present there is enough evidence for the existence of the intervalic symmetries {D270}, {D45}, {D30} and {D18} in all sectors, and enough evidence for the existence of {D6} and {D5} symmetries in the lepton-massive boson sector. Therefore once we will have reached full evidence for the existence of the remaining symmetries: {D6} and {D5} symmetries in the quark sector, {D3} and {D2} symmetries in both sectors, there will be no reason to mortify us searching *all* the baryonic states allowed by the intervalic symmetries, but we will be able to address our efforts to other more fruitful tasks, as IT predicts a lot of possibilities for technologic advances which before the postulation of IT could never be imagined.

In resume, we are going to list the baryons which existence has the greatest experimental evidence, showing their intervalic structure. The vast majority of detected baryons pertain to the {D45} and {D30} symmetries, as they are the sets which small energies are nowadays available.

It has to be remembered that the mass energy due to the electric charge of the quarks is not included in the masses of the intervalic structures of quarks given in our tables. Therefore, such mass energy have to be added to those masses, and the total mass will be slightly greater that the sum of the constituent quarks masses according to the quark ratio between intervalic energy and electromagnetic energy (really no one quark —including possible heavier ones due to an excited gaudinar structure— reaches a mass according to the electron ratio, written next to the first, although we have written them down for pure systematic way of work).

It can also be noted the powerful constraints imposed by the intervalic structure to the resultant electric charge of baryons, which allows a reduced number of combinations to yield certain values as $\frac{2}{3} + \frac{2}{3} + \frac{2}{3} = 2$. However, it can be seen that the intervalic structures involved in these cases (for example, the double elementary charge of Δ) are just those *even* ones, but not the *odd* ones. And as we can expect, the masses of such even intervalic structures fit justly with experimental data.

The intervalic symmetries allows to grouping the constituent gaudinos of a baryon into different lisztinos —quarks— without varying the number of constituent gaudinos. In close relation with the preceding paragraph, these different combinations of gaudinos may differ in the electric charge and spin, which allowed magnitudes are anyway strongly constrained by the own intervalic structures. With independence of this, the grouping of gaudinos may be responsible of the existence of the excited states of baryons, along with the own intervalic structures of gaudinos. Some of those possible different groupings have been written down in the following list, which shows the most probable intervalic structure of experimentally well detected baryons.

Finally, we can remember once more that, to determine an allowed baryon, SM has to deal with a minimum of five supposedly independent physical quantities: flavour, charge, spin, isospin and colour; whilst in IT all allowed baryonic states are fully determined by a unique feature: the *intervalic structure*, from which derives the remaining symmetries of spin and isocharge, as well as all the remaining physical quantities of monteverdinos.

$N(939) = (\text{L3D45}^{(\frac{1}{3},\frac{2}{3})} \text{L3D45}^{(\frac{1}{3},\frac{2}{3})} \text{L3D45}^{(\frac{1}{3},\frac{2}{3})})$

$N(1440) = (\text{L3D45}^{(\frac{1}{3},\frac{2}{3})} \text{L5D45}^{(\frac{1}{3},\frac{2}{3})} \text{L5D45}^{(\frac{1}{3},\frac{2}{3})})$
$\qquad (\text{L4D45}^{(\frac{1}{3},\frac{2}{3})} \text{L4D45}^{(\frac{1}{3},\frac{2}{3})} \text{L5D45}^{(\frac{1}{3},\frac{2}{3})})$
$\qquad (\text{L}\frac{1}{3}\text{D30}^{(\frac{1}{3})} \text{L}\frac{2}{3}\text{D30}^{(\frac{2}{3})} \text{L3D30}^{(\frac{1}{3})})$
$\qquad (\text{L1D30}^{(\frac{1}{3})} \text{L1D30}^{(\frac{1}{3})} \text{L2D30}^{(\frac{2}{3})})$

$N(1520) = (\text{L4D45}^{(\frac{1}{3},\frac{2}{3})} \text{L5D45}^{(\frac{1}{3},\frac{2}{3})} \text{L5D45}^{(\frac{1}{3},\frac{2}{3})})$

$N(1535) = (\text{L4D45}^{(\frac{1}{3},\frac{2}{3})} \text{L5D45}^{(\frac{1}{3},\frac{2}{3})} \text{L5D45}^{(\frac{1}{3},\frac{2}{3})})$

$N(1650) = (\text{L}\frac{2}{3}\text{D30}^{(\frac{2}{3})} \text{L2D30}^{(\frac{2}{3})} \text{L2D30}^{(\frac{2}{3})})$

$N(1675) = (\text{L}\frac{2}{3}\text{D30}^{(\frac{2}{3})} \text{L2D30}^{(\frac{2}{3})} \text{L2D30}^{(\frac{2}{3})})$

$N(1680) = (\text{L}\frac{1}{3}\text{D30}^{(\frac{1}{3})} \text{L}\frac{1}{3}\text{D30}^{(\frac{1}{3})} \text{L4D30}^{(\frac{2}{3})})$

$N(1700) = (\text{L}\frac{1}{3}\text{D30}^{(\frac{1}{3})} \text{L}\frac{1}{3}\text{D30}^{(\frac{1}{3})} \text{L4D30}^{(\frac{2}{3})})$

$N(1710) = (\text{L}\frac{2}{3}\text{D30}^{(\frac{2}{3})} \text{L1D30}^{(\frac{1}{3})} \text{L3D30}^{(\frac{1}{3})})$

$N(1720) = (\text{L}\frac{2}{3}\text{D30}^{(\frac{2}{3})} \text{L1D30}^{(\frac{1}{3})} \text{L3D30}^{(\frac{1}{3})})$

$N(1900) = (\text{L}\frac{1}{3}\text{D30}^{(\frac{1}{3})} \text{L1D30}^{(\frac{1}{3})} \text{L4D30}^{(\frac{2}{3})})$
$\qquad (\text{L}\frac{1}{3}\text{D30}^{(\frac{1}{3})} \text{L2D30}^{(\frac{2}{3})} \text{L3D30}^{(\frac{1}{3})})$

$N(1990) = (\text{L}\frac{2}{3}\text{D30}^{(\frac{2}{3})} \text{L1D30}^{(\frac{1}{3})} \text{L4D30}^{(\frac{2}{3})})$
$\qquad (\text{L}\frac{2}{3}\text{D30}^{(\frac{2}{3})} \text{L2D30}^{(\frac{2}{3})} \text{L3D30}^{(\frac{1}{3})})$

$N(2000) = (\text{L}\frac{2}{3}\text{D30}^{(\frac{2}{3})} \text{L1D30}^{(\frac{1}{3})} \text{L4D30}^{(\frac{2}{3})})$
$\qquad (\text{L}\frac{2}{3}\text{D30}^{(\frac{2}{3})} \text{L2D30}^{(\frac{2}{3})} \text{L3D30}^{(\frac{1}{3})})$

$N(2080) = (\text{L}\frac{2}{3}\text{D30}^{(\frac{2}{3})} \text{L1D30}^{(\frac{1}{3})} \text{L4D30}^{(\frac{2}{3})})$
$\qquad (\text{L}\frac{2}{3}\text{D30}^{(\frac{2}{3})} \text{L2D30}^{(\frac{2}{3})} \text{L3D30}^{(\frac{1}{3})})$

$N(2190) = (\text{L}\frac{1}{3}\text{D30}^{(\frac{1}{3})} \text{L}\frac{2}{3}\text{D30}^{(\frac{2}{3})} \text{L5D30}^{(\frac{1}{3})})$

$N(2200) = (\text{L}\frac{1}{3}\text{D30}^{(\frac{1}{3})} \text{L}\frac{2}{3}\text{D30}^{(\frac{2}{3})} \text{L5D30}^{(\frac{1}{3})})$

$N(2220) = (\text{L1D30}^{(\frac{1}{3})} \text{L1D30}^{(\frac{1}{3})} \text{L4D30}^{(\frac{2}{3})})$

$N(2250) = (\text{L1D30}^{(\frac{1}{3})} \text{L2D30}^{(\frac{2}{3})} \text{L3D30}^{(\frac{1}{3})})$

$N(2600) = (\text{L}\frac{1}{3}\text{D30}^{(\frac{1}{3})} \text{L3D30}^{(\frac{1}{3})} \text{L4D30}^{(\frac{2}{3})})$
$\qquad (\text{L}\frac{1}{3}\text{D30}^{(\frac{1}{3})} \text{L2D30}^{(\frac{2}{3})} \text{L5D30}^{(\frac{1}{3})})$

$N(2700) = (\text{L}\frac{2}{3}\text{D30}^{(\frac{2}{3})} \text{L3D30}^{(\frac{1}{3})} \text{L4D30}^{(\frac{2}{3})})$
$\qquad (\text{L}\frac{2}{3}\text{D30}^{(\frac{2}{3})} \text{L2D30}^{(\frac{2}{3})} \text{L5D30}^{(\frac{1}{3})})$

$\Delta(1232) = (\text{L}\frac{2}{3}\text{D30}^{(\frac{2}{3})} \text{L}\frac{2}{3}\text{D30}^{(\frac{2}{3})} \text{L2D30}^{(\frac{2}{3})})$
$\qquad (\text{L1D45}^{(\frac{1}{3},\frac{2}{3})} \text{L5D45}^{(\frac{1}{3},\frac{2}{3})} \text{L5D45}^{(\frac{1}{3},\frac{2}{3})})$

$\Delta(1600) = (\text{L5D45}^{(\frac{1}{3},\frac{2}{3})} \text{L5D45}^{(\frac{1}{3},\frac{2}{3})} \text{L5D45}^{(\frac{1}{3},\frac{2}{3})})$

INTERVALIC THEORY:
The Intervalic Structures of Subatomic Particles and the Last Foundations of Physics

$\Delta\,(1620) = (\; L5D45^{(1/3,\,2/3)}\; L5D45^{(1/3,\,2/3)}\; L5D45^{(1/3,\,2/3)}\;)$

$\Delta\,(1700) = (\; L2/3D30^{(2/3)}\; L2D30^{(2/3)}\; L2D30^{(2/3)}\;)$

$\Delta\,(1900) = (\; L2/3D30^{(2/3)}\; L2/3D30^{(2/3)}\; L4D30^{(2/3)}\;)$

$\Delta\,(1905) = (\; L2/3D30^{(2/3)}\; L2/3D30^{(2/3)}\; L4D30^{(2/3)}\;)$

$\Delta\,(1910) = (\; L2/3D30^{(2/3)}\; L2/3D30^{(2/3)}\; L4D30^{(2/3)}\;)$

$\Delta\,(1920) = (\; L2/3D30^{(2/3)}\; L2/3D30^{(2/3)}\; L4D30^{(2/3)}\;)$

$\Delta\,(1930) = (\; L2/3D30^{(2/3)}\; L2/3D30^{(2/3)}\; L4D30^{(2/3)}\;)$

$\Delta\,(1950) = (\; L2/3D30^{(2/3)}\; L2/3D30^{(2/3)}\; L4D30^{(2/3)}\;)$

$\Delta\,(2000) = (\; L1/3D30^{(1/3)}\; L1/3D30^{(1/3)}\; L5D30^{(1/3)}\;)$

$\Delta\,(2300) = (\; L1/3D30^{(1/3)}\; L3D30^{(1/3)}\; L3D30^{(1/3)}\;)$

$\Delta\,(2400) = (\; L2/3D30^{(2/3)}\; L2D30^{(2/3)}\; L4D30^{(2/3)}\;)$

$\Delta\,(2420) = (\; L2/3D30^{(2/3)}\; L3D30^{(1/3)}\; L3D30^{(1/3)}\;)$

$\Delta\,(2750) = (\; L2/3D30^{(2/3)}\; L3D30^{(1/3)}\; L4D30^{(2/3)}\;)$

$\Delta\,(2950) = (\; L1/3D30^{(1/3)}\; L3D30^{(1/3)}\; L5D30^{(1/3)}\;)$

$\Lambda\,(1116) = (\; L2/3D45^{(1/3,\,2/3)}\; L5D45^{(1/3,\,2/3)}\; L5D45^{(1/3,\,2/3)}\;)$

$\Lambda\,(1405) = (\; L3D45^{(1/3,\,2/3)}\; L5D45^{(1/3,\,2/3)}\; L5D45^{(1/3,\,2/3)}\;)$
$(\; L4D45^{(1/3,\,2/3)}\; L4D45^{(1/3,\,2/3)}\; L5D45^{(1/3,\,2/3)}\;)$

$\Lambda\,(1520) = (\; L4D45^{(1/3,\,2/3)}\; L5D45^{(1/3,\,2/3)}\; L5D45^{(1/3,\,2/3)}\;)$

$\Lambda\,(1600) = (\; L5D45^{(1/3,\,2/3)}\; L5D45^{(1/3,\,2/3)}\; L5D45^{(1/3,\,2/3)}\;)$

$\Lambda\,(1670) = (\; L2/3D30^{(2/3)}\; L1D30^{(1/3)}\; L3D30^{(1/3)}\;)$

$\Lambda\,(1690) = (\; L1/3D30^{(1/3)}\; L1/3D30^{(1/3)}\; L4D30^{(2/3)}\;)$

$\Lambda\,(1800) = (\; L1/3D30^{(1/3)}\; L2/3D30^{(2/3)}\; L4D30^{(2/3)}\;)$

$\Lambda\,(1810) = (\; L1/3D30^{(1/3)}\; L2/3D30^{(2/3)}\; L4D30^{(2/3)}\;)$

$\Lambda\,(1820) = (\; L1D30^{(1/3)}\; L2D30^{(2/3)}\; L2D30^{(2/3)}\;)$

$\Lambda\,(1830) = (\; L1D30^{(1/3)}\; L2D30^{(2/3)}\; L2D30^{(2/3)}\;)$

$\Lambda\,(1890) = (\; L1/3D30^{(1/3)}\; L2D30^{(2/3)}\; L3D30^{(1/3)}\;)$

$\Lambda\,(2100) = (\; L2/3D30^{(2/3)}\; L2D30^{(2/3)}\; L3D30^{(1/3)}\;)$

$\Lambda\,(2110) = (\; L1/3D30^{(1/3)}\; L1/3D30^{(1/3)}\; L5D30^{(1/3)}\;)$

$\Lambda\,(2325) = (\; L1/3D30^{(1/3)}\; L3D30^{(1/3)}\; L3D30^{(1/3)}\;)$

$\Lambda\,(2350) = (\; L2/3D30^{(2/3)}\; L3D30^{(1/3)}\; L3D30^{(1/3)}\;)$

$\Sigma\,(1193) = (\; L1D45^{(1/3,\,2/3)}\; L5D45^{(1/3,\,2/3)}\; L5D45^{(1/3,\,2/3)}\;)$
$(\; L3D45^{(1/3,\,2/3)}\; L3D45^{(1/3,\,2/3)}\; L5D45^{(1/3,\,2/3)}\;)$

$$(\text{L3D45}^{(\frac{1}{3},\frac{2}{3})} \text{L4D45}^{(\frac{1}{3},\frac{2}{3})} \text{L4D45}^{(\frac{1}{3},\frac{2}{3})})$$
$$\Sigma(1385) = (\text{L3D45}^{(\frac{1}{3},\frac{2}{3})} \text{L5D45}^{(\frac{1}{3},\frac{2}{3})} \text{L5D45}^{(\frac{1}{3},\frac{2}{3})})$$
$$(\text{L4D45}^{(\frac{1}{3},\frac{2}{3})} \text{L4D45}^{(\frac{1}{3},\frac{2}{3})} \text{L5D45}^{(\frac{1}{3},\frac{2}{3})})$$
$$\Sigma(1560) = (\text{L5D45}^{(\frac{1}{3},\frac{2}{3})} \text{L5D45}^{(\frac{1}{3},\frac{2}{3})} \text{L5D45}^{(\frac{1}{3},\frac{2}{3})})$$
$$\Sigma(1580) = (\text{L5D45}^{(\frac{1}{3},\frac{2}{3})} \text{L5D45}^{(\frac{1}{3},\frac{2}{3})} \text{L5D45}^{(\frac{1}{3},\frac{2}{3})})$$
$$\Sigma(1620) = (\text{L5D45}^{(\frac{1}{3},\frac{2}{3})} \text{L5D45}^{(\frac{1}{3},\frac{2}{3})} \text{L5D45}^{(\frac{1}{3},\frac{2}{3})})$$
$$\Sigma(1660) = (\text{L}\tfrac{2}{3}\text{D30}^{(\frac{2}{3})} \text{L2D30}^{(\frac{2}{3})} \text{L2D30}^{(\frac{2}{3})})$$
$$\Sigma(1670) = (\text{L}\tfrac{2}{3}\text{D30}^{(\frac{2}{3})} \text{L1D30}^{(\frac{1}{3})} \text{L3D30}^{(\frac{1}{3})})$$
$$\Sigma(1690) = (\text{L}\tfrac{2}{3}\text{D30}^{(\frac{2}{3})} \text{L1D30}^{(\frac{1}{3})} \text{L3D30}^{(\frac{1}{3})})$$
$$\Sigma(1750) = (\text{L}\tfrac{1}{3}\text{D30}^{(\frac{1}{3})} \text{L}\tfrac{2}{3}\text{D30}^{(\frac{2}{3})} \text{L4D30}^{(\frac{2}{3})})$$
$$\Sigma(1775) = (\text{L1D30}^{(\frac{1}{3})} \text{L2D30}^{(\frac{2}{3})} \text{L2D30}^{(\frac{2}{3})})$$
$$\Sigma(1880) = (\text{L}\tfrac{2}{3}\text{D30}^{(\frac{2}{3})} \text{L}\tfrac{2}{3}\text{D30}^{(\frac{2}{3})} \text{L4D30}^{(\frac{2}{3})})$$
$$\Sigma(1915) = (\text{L}\tfrac{1}{3}\text{D30}^{(\frac{1}{3})} \text{L2D30}^{(\frac{2}{3})} \text{L3D30}^{(\frac{1}{3})})$$
$$\Sigma(1940) = (\text{L}\tfrac{1}{3}\text{D30}^{(\frac{1}{3})} \text{L}\tfrac{1}{3}\text{D30}^{(\frac{1}{3})} \text{L5D30}^{(\frac{1}{3})})$$
$$\Sigma(2030) = (\text{L}\tfrac{2}{3}\text{D30}^{(\frac{2}{3})} \text{L1D30}^{(\frac{1}{3})} \text{L4D30}^{(\frac{2}{3})})$$
$$\Sigma(2080) = (\text{L}\tfrac{2}{3}\text{D30}^{(\frac{2}{3})} \text{L2D30}^{(\frac{2}{3})} \text{L3D30}^{(\frac{1}{3})})$$
$$\Sigma(2250) = (\text{L}\tfrac{1}{3}\text{D30}^{(\frac{1}{3})} \text{L1D30}^{(\frac{1}{3})} \text{L5D30}^{(\frac{1}{3})})$$
$$\Sigma(2455) = (\text{L2D30}^{(\frac{2}{3})} \text{L2D30}^{(\frac{2}{3})} \text{L3D30}^{(\frac{1}{3})})$$
$$\Sigma(2620) = (\text{L}\tfrac{1}{3}\text{D30}^{(\frac{1}{3})} \text{L3D30}^{(\frac{1}{3})} \text{L4D30}^{(\frac{2}{3})})$$

$$\Xi(1318) = (\text{L4D45}^{(\frac{1}{3},\frac{2}{3})} \text{L4D45}^{(\frac{1}{3},\frac{2}{3})} \text{L4D45}^{(\frac{1}{3},\frac{2}{3})})$$
$$(\text{L3D45}^{(\frac{1}{3},\frac{2}{3})} \text{L4D45}^{(\frac{1}{3},\frac{2}{3})} \text{L5D45}^{(\frac{1}{3},\frac{2}{3})})$$
$$(\text{L2D45}^{(\frac{1}{3},\frac{2}{3})} \text{L5D45}^{(\frac{1}{3},\frac{2}{3})} \text{L5D45}^{(\frac{1}{3},\frac{2}{3})})$$
$$\Xi(1530) = (\text{L4D45}^{(\frac{1}{3},\frac{2}{3})} \text{L5D45}^{(\frac{1}{3},\frac{2}{3})} \text{L5D45}^{(\frac{1}{3},\frac{2}{3})})$$
$$(\text{L}\tfrac{1}{3}\text{D30}^{(\frac{1}{3})} \text{L1D30}^{(\frac{1}{3})} \text{L3D30}^{(\frac{1}{3})})$$
$$(\text{L}\tfrac{1}{3}\text{D30}^{(\frac{1}{3})} \text{L2D30}^{(\frac{2}{3})} \text{L2D30}^{(\frac{2}{3})})$$
$$(\text{L}\tfrac{2}{3}\text{D30}^{(\frac{2}{3})} \text{L}\tfrac{2}{3}\text{D30}^{(\frac{2}{3})} \text{L3D30}^{(\frac{1}{3})})$$
$$\Xi(1690) = (\text{L}\tfrac{1}{3}\text{D30}^{(\frac{1}{3})} \text{L}\tfrac{1}{3}\text{D30}^{(\frac{1}{3})} \text{L4D30}^{(\frac{2}{3})})$$
$$(\text{L}\tfrac{2}{3}\text{D30}^{(\frac{2}{3})} \text{L1D30}^{(\frac{1}{3})} \text{L3D30}^{(\frac{1}{3})})$$
$$(\text{L}\tfrac{2}{3}\text{D30}^{(\frac{2}{3})} \text{L2D30}^{(\frac{2}{3})} \text{L2D30}^{(\frac{2}{3})})$$
$$\Xi(1820) = (\text{L}\tfrac{1}{3}\text{D30}^{(\frac{1}{3})} \text{L}\tfrac{2}{3}\text{D30}^{(\frac{2}{3})} \text{L4D30}^{(\frac{2}{3})})$$
$$(\text{L1D30}^{(\frac{1}{3})} \text{L1D30}^{(\frac{1}{3})} \text{L3D30}^{(\frac{1}{3})})$$
$$(\text{L1D30}^{(\frac{1}{3})} \text{L2D30}^{(\frac{2}{3})} \text{L2D30}^{(\frac{2}{3})})$$
$$\Xi(1950) = (\text{L}\tfrac{1}{3}\text{D30}^{(\frac{1}{3})} \text{L1D30}^{(\frac{1}{3})} \text{L4D30}^{(\frac{2}{3})})$$

$\quad\quad\quad\quad$ (L⅓D30$^{(⅓)}$ L2D30$^{(⅔)}$ L3D30$^{(⅓)}$)
$\quad\quad\quad\quad$ (L⅔D30$^{(⅔)}$ L⅔D30$^{(⅔)}$ L4D30$^{(⅔)}$)
$\Xi\,(2030) = $ (L⅔D30$^{(⅔)}$ L1D30$^{(⅓)}$ L4D30$^{(⅔)}$)
$\quad\quad\quad\quad$ (L⅔D30$^{(⅔)}$ L2D30$^{(⅔)}$ L3D30$^{(⅓)}$)
$\Xi\,(2250) = $ (L⅓D30$^{(⅓)}$ L3D30$^{(⅓)}$ L3D30$^{(⅓)}$)
$\quad\quad\quad\quad$ (L⅓D30$^{(⅓)}$ L2D30$^{(⅔)}$ L4D30$^{(⅔)}$)
$\quad\quad\quad\quad$ (L⅓D30$^{(⅓)}$ L1D30$^{(⅓)}$ L5D30$^{(⅓)}$)
$\Xi\,(2370) = $ (L⅔D30$^{(⅔)}$ L3D30$^{(⅓)}$ L3D30$^{(⅓)}$)
$\quad\quad\quad\quad$ (L⅔D30$^{(⅔)}$ L2D30$^{(⅔)}$ L4D30$^{(⅔)}$)
$\quad\quad\quad\quad$ (L⅔D30$^{(⅔)}$ L1D30$^{(⅓)}$ L5D30$^{(⅓)}$)

$\Omega\,(1672) = $ (L⅓D18$^{(⅓)}$ L⅓D18$^{(⅓)}$ L⅓D18$^{(⅓)}$)
$\quad\quad\quad\quad$ (L5D45$^{(⅓,\,⅔)}$ L5D45$^{(⅓,\,⅔)}$ L5D45$^{(⅓,\,⅔)}$)
$\Omega\,(2250) = $ (L⅓D18$^{(⅓)}$ L⅓D18$^{(⅓)}$ L⅔D18$^{(⅔)}$)
$\quad\quad\quad\quad$ (L⅓D30$^{(⅓)}$ L3D30$^{(⅓)}$ L3D30$^{(⅓)}$)

$\Lambda_c\,(2285) = $ (L⅓D30$^{(⅓)}$ L1D30$^{(⅓)}$ L5D30$^{(⅓)}$)
$\quad\quad\quad\quad$ (L⅓D30$^{(⅓)}$ L3D30$^{(⅔)}$ L3D30$^{(⅔)}$)
$\Lambda_c\,(2593) = $ (L⅓D30$^{(⅓)}$ L3D30$^{(⅓)}$ L4D30$^{(⅔)}$)
$\Lambda_c\,(2625) = $ (L⅓D30$^{(⅓)}$ L3D30$^{(⅓)}$ L4D30$^{(⅔)}$)

$\Sigma_c\,(2455) = $ (L⅔D30$^{(⅔)}$ L1D30$^{(⅓)}$ L5D30$^{(⅓)}$)
$\quad\quad\quad\quad$ (L⅔D30$^{(⅔)}$ L3D30$^{(⅓)}$ L3D30$^{(⅓)}$)
$\Sigma_c\,(2520) = $ (L1D30$^{(⅓)}$ L1D30$^{(⅓)}$ L5D30$^{(⅓)}$)
$\quad\quad\quad\quad$ (L1D30$^{(⅓)}$ L3D30$^{(⅓)}$ L3D30$^{(⅓)}$)

$\Xi_c\,(2465) = $ (L⅔D30$^{(⅔)}$ L1D30$^{(⅓)}$ L5D30$^{(⅓)}$)
$\quad\quad\quad\quad$ (L⅔D30$^{(⅔)}$ L3D30$^{(⅓)}$ L3D30$^{(⅓)}$)
$\Xi_c\,(2645) = $ (L⅓D30$^{(⅓)}$ L3D30$^{(⅓)}$ L4D30$^{(⅔)}$)

$\Omega_c\,(2704) = $ (L⅓D18$^{(⅓)}$ L⅔D18$^{(⅔)}$ L⅔D18$^{(⅔)}$)
$\quad\quad\quad\quad$ (L⅓D18$^{(⅓)}$ L⅓D18$^{(⅓)}$ L1D18$^{(⅓)}$)
$\quad\quad\quad\quad$ (L⅔D30$^{(⅔)}$ L3D30$^{(⅓)}$ L4D30$^{(⅔)}$)

Chapter 19

INTERVALIC π MESON

INTERVALIC π MESON

$\pi^{\pm}$ MESON INTERVALIC STRUCTURE

As we will see when studying the intervalic changeless —strong— interaction, the intervalic structure of the $\pi^{\pm}$ meson which explains in the most simple way the interaction between quarks inside nucleon is the following one:

$$M_2^{\pm} = (L_1, L_{1/3}) \; (1 \, G_6^{\pm 2/3}, 1/3 \, G_6^{\pm 1/3}) = 8 \, D_{45} \, (7 \, D_{\pm 45}, 1 \, D_{\pm 45}) = 360 \, I$$

Please note that lisztinos L⅓, L⅔ and L1 are identical to gaudinos G⅓, G⅔ and G1 because they are just composed by an unique gaudino.

Nevertheless it can be proposed other alternative structures which fit equally good, such as:

$$M_2^{\pm} = 2 \, L_{2/3} \; 2 \, 2/3 \, G_6 \, (2/3 \, G_6^{\pm 1/3}, 2/3 \, G_6^{\pm 2/3}) = 8 \, D_{45} \, (7 \, D_{\pm 45}, 1 \, D_{\pm 45}) = 360 \, I$$

Please note that their dalinar structures are completely identical in both cases. The intervalic energy of the alternative structure of the $\pi^{\pm}$ meson: $L\tfrac{1}{3}D45^{(\tfrac{1}{3})} L1D45^{(\tfrac{2}{3})} = 8\,D_{45} = 360\,I$, is exactly equal to that of the basic one, $L\tfrac{2}{3}D45^{(\tfrac{1}{3})} L\tfrac{2}{3}D45^{(\tfrac{2}{3})} = 8\,D_{45} = 360\,I$. Both structures can be equally postulated, and the only difference between them lies in the manner in which their 8 constituent dalinos are grouped, respectively: 2+6 or 4+4, a difference which can vary the electromagnetic mass energy of $\pi^{\pm}$ meson. As the reader may check, there are no other isocharge values of the intervalic structures of $\pi^{\pm}$ meson than the ones above written.

The intervalic structure of pion is in deep and meaningful relation with the intervalic structure of nucleon. This beautiful feature will be explained with detail in the chapter on the changeless —strong— intervalic interaction. However at this moment we are going to describe the $\pi^{\pm}$ meson basic features without making reference to that relation.

π^0 MESON INTERVALIC STRUCTURE

At first sight we find at least three possible intervalic structures for the π^0 meson. They are, showing its compositeness and structurefulness:

$$L_2 = 2\,G_5 = 10\,D_{54} = (5\,D_{+54},\,5\,D_{-54}) = 540\,I$$
$$M_2^0 = (L_1, L_{\tfrac{1}{3}})\ (1\,G_6^{\pm\tfrac{2}{3}},\,\tfrac{1}{3}\,G_6^{\pm\tfrac{1}{3}}) = 8\,D_{45}\,(4\,D_{\pm45},\,4\,D_{\pm45}) = 360\,I$$
$$M_2^0 = 2\,L_{\tfrac{2}{3}}\ 2\tfrac{2}{3}\,G_6 = 8\,D_{45}\,(4\,D_{+45},\,4\,D_{-45}) = 360\,I$$

The last one seems to be the most appealing of them, and its isocharge has two possible states:

$$M_2^0 = 2\,L_{\tfrac{2}{3}}\ 2\tfrac{2}{3}\,G_6\,(\tfrac{2}{3}\,G_6^{+\tfrac{2}{3}},\,\tfrac{2}{3}\,G_6^{-\tfrac{2}{3}}) = 8\,D_{45}\,(4\,D_{+45},\,4\,D_{-45}) = 360\,I$$
$$M_2^0 = 2\,L_{\tfrac{2}{3}}\ 2\tfrac{2}{3}\,G_6\,(\tfrac{2}{3}\,G_6^{+\tfrac{1}{3}},\,\tfrac{2}{3}\,G_6^{-\tfrac{1}{3}}) = 8\,D_{45}\,(4\,D_{+45},\,4\,D_{-45}) = 360\,I$$

In other words, both identical constituent isoquarks, $L\tfrac{2}{3}D45^{(\tfrac{1}{3},\,\tfrac{2}{3})}$, can have got at once isocharge value $\tfrac{1}{3}$ or $\tfrac{2}{3}$:

$L\frac{2}{3}D45^{(1/3)}L\frac{2}{3}D45^{(1/3)}$ or $L\frac{2}{3}D45^{(2/3)}L\frac{2}{3}D45^{(2/3)}$

The first point to inquiry is the value of the isocharge of the constituent quarks of π^0 meson among the two allowed ones. Since we experimentally know that the mass of π^0 is smaller than the mass of $\pi^\pm$, and the mass energy of particles with electric charge ⅓ is ever greater than the mass energy of particles with ⅔ charge (because the intervalic energy always predominates over the electromagnetic one at quantum scale, as we already have seen), we can postulate that the intervalic structure of π^0 meson is:

π^0 meson: $L\frac{2}{3}D45^{(2/3)} L\frac{2}{3}D45^{(2/3)}$

If, on the contrary, it was: $L\frac{2}{3}D45^{(1/3)} L\frac{2}{3}D45^{(1/3)}$, we would have m$(\pi^0)$ > m$(\pi^\pm)$, a phenomenon which is not realized by Nature. (Of course, that structure could still be possible if the electromagnetic energy was as smaller as to match with the experimental mass, but it is more logic to think that both mass energies go in a similar way).

π MESON INTERVALIC DECAY

An important and immediate consequence of the intervalic structure of π^0 meson is the deduction of its intervalic decay. Quark $L\frac{2}{3}D45^{(1/3)}$ only can decay into the lightest quark $L\frac{1}{3}D45^{(1/3)}$, but quark $L\frac{2}{3}D45^{(2/3)}$ has not a lighter quark to decay into, and therefore it only can decay via electromagnetic interaction:

$L\frac{2}{3}D45^{(2/3)} L\frac{2}{3}D45^{(2/3)} \rightarrow \gamma\gamma$

Therefore, the intervalic structure of π^0 meson explains why it can only decay electromagnetically, and vice versa. We could question what

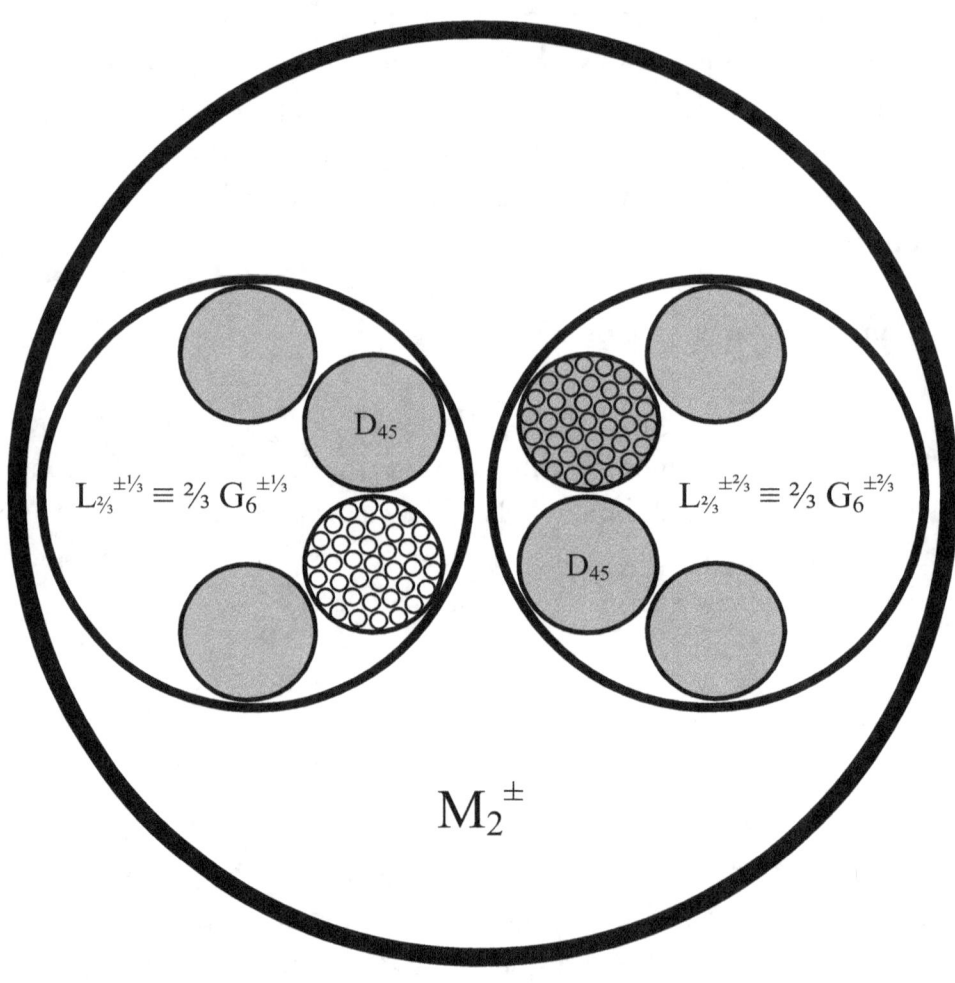

Figured intervalic structure of $\pi^{\pm}$ *meson*:
$M_2^{\pm} = 2\, L_{2/3} \equiv 2\, 2/3\, G_6\, (2/3\, G_6^{\pm 1/3},\, 2/3\, G_6^{\pm 2/3}) = 8\, D_{45}\, (7\, D_{\pm 45},\, 1\, D_{\pm 45}) = 360\, \mathbf{I}$

INTERVALIC THEORY:
The Intervalic Structures of Subatomic Particles and the Last Foundations of Physics

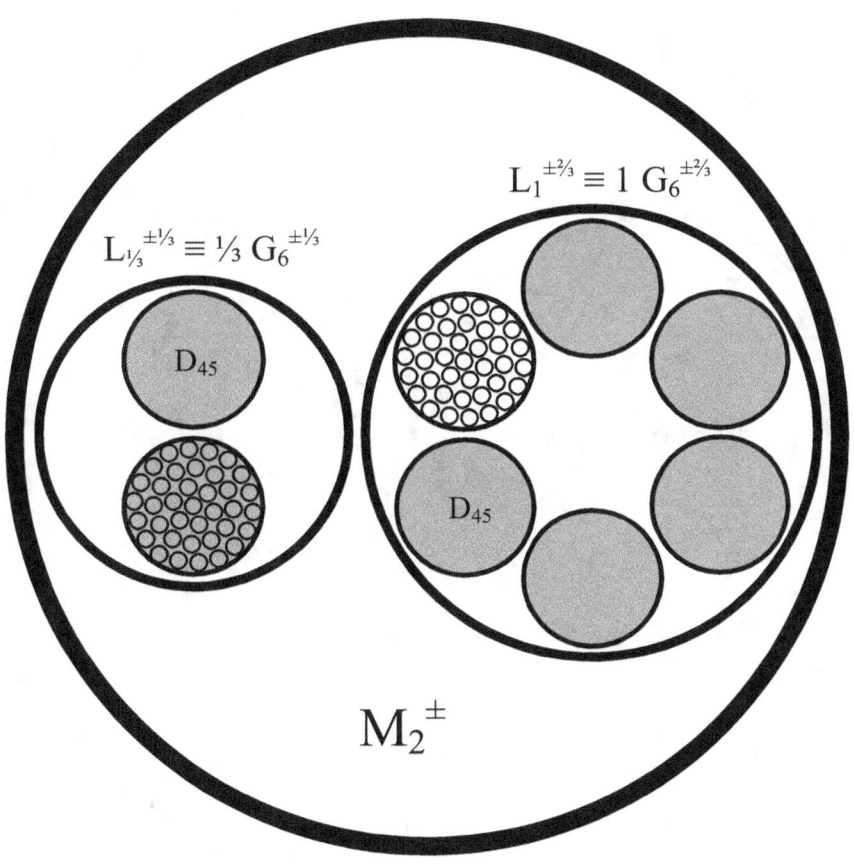

Figured intervalic structure of $\pi^{\pm}$ *meson*:
$M_2^{\pm} = (L_1, L_{⅓}) \equiv (1\ G_6^{\pm ⅔},\ ⅓\ G_6^{\pm ⅓}) = 8\ D_{45}\ (7\ D_{\pm 45},\ 1\ D_{\pm 45}) = 360\ \mathbf{I}$

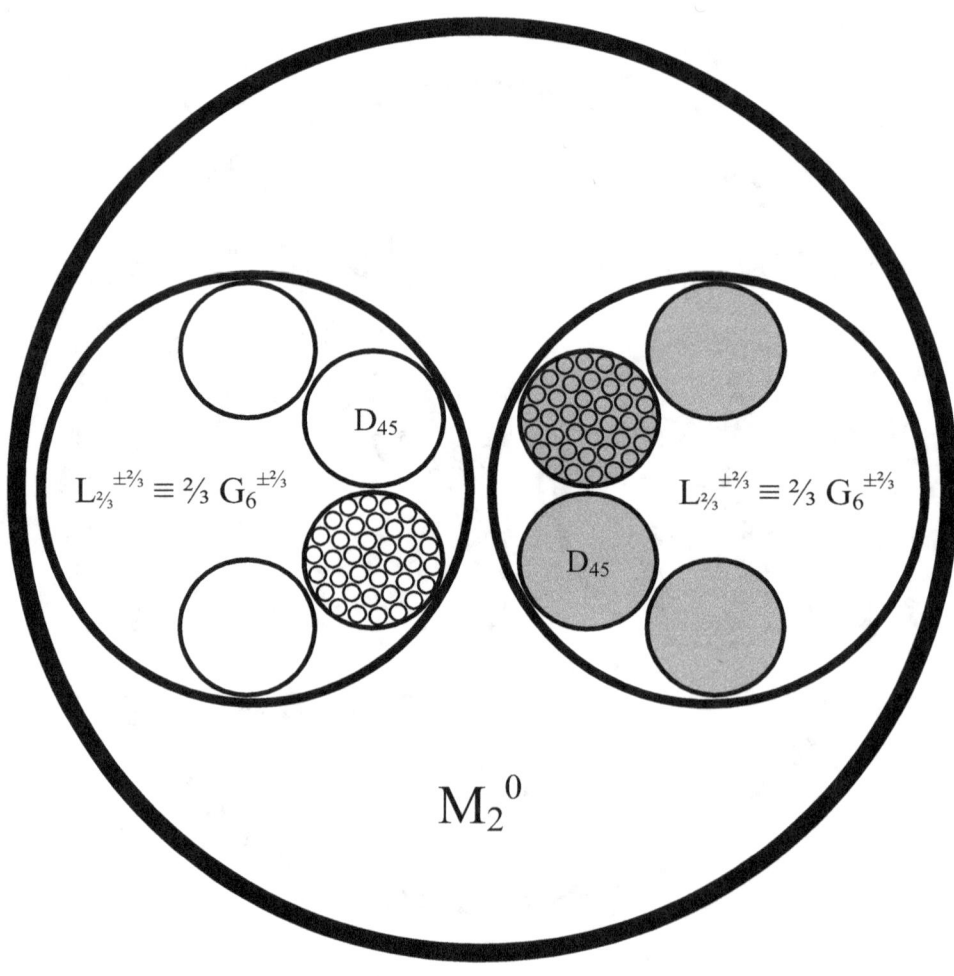

Figured intervalic structure of π^0 *meson*:
$M_2^0 = 2\,L_{2/3} \equiv 2\,{}^{2}\!/_{3}\,G_6\,({}^{2}\!/_{3}\,G_6^{+2/3},\,{}^{2}\!/_{3}\,G_6^{-2/3}) = 8\,D_{45}\,(4\,D_{+45},\,4\,D_{-45}) = 360\,\mathbf{I}$

INTERVALIC THEORY:
The Intervalic Structures of Subatomic Particles and the Last Foundations of Physics

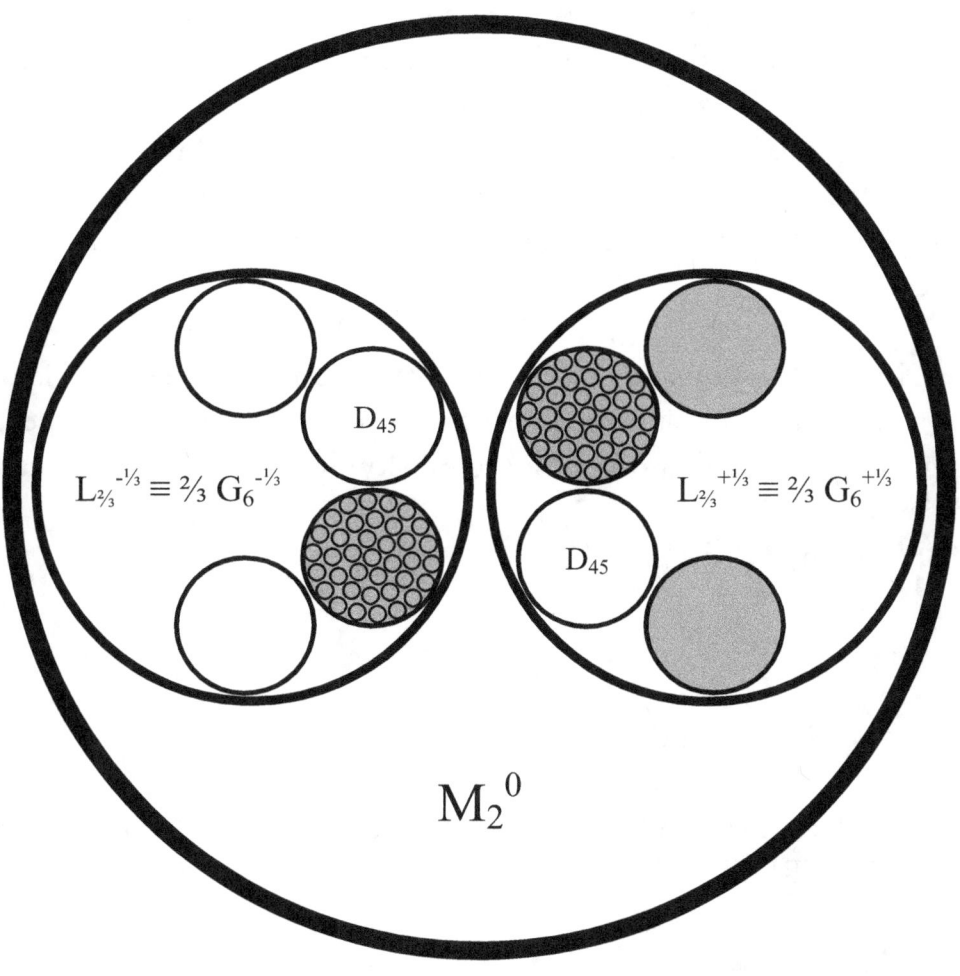

Figured intervalic structure of π^0 meson:
$M_2^0 = 2\ L_{2/3} \equiv 2\ ^2/_3\ G_6\ (^2/_3\ G_6^{+1/3},\ ^2/_3\ G_6^{-1/3}) = 8\ D_{45}\ (4\ D_{+45},\ 4\ D_{-45}) = 360\ \mathbf{I}$

would happen if its intervalic structure had got the isocharge ⅓. In this case the decay of π^0 meson would be:

$$L⅔D45^{(⅓)} L⅔D45^{(⅓)} \rightarrow L⅓D45^{(⅓)} L⅓D45^{(⅓)} \rightarrow \gamma\gamma$$

But the $L⅓D45^{(⅓)} L⅓D45^{(⅓)}$ meson, with mass ~69 (MeV/c²), has not been found in the decay of π^0. Therefore, the intervalic structure of its constituent quarks must have isocharge ⅔.

In a similar way, the intervalic structure of $\pi^\pm$ explains its decay and why it does not decay electromagnetically. We will understand better the decay of $\pi^\pm$ and π^0 after studying the chapter devoted to the Intervalic Decay.

Now we are going to describe the intervalic and electromagnetic energy of the π meson at every level of the intervalic structure. For pedagogic purposes (and also in order to brevity) we have chosen the intervalic structure $M_2^0 = 2\ L_{⅔}\ 2\ ⅔\ G_6 = 8\ D_{45} = 360\ I$, which can be common to both $\pi^\pm$ and π^0 mesons.

π MESON INTERVALIC ENERGY

π MESON INTERVALIC ENERGY AT MONTEVERDIC STRUCTURE LEVEL

We obviously have:

$I(\pi^0)_M = I(M_2^0) = 0$
$I(\pi^\pm)_M = I(M_2^\pm) = c^{\pm 2}\ \hbar\ e^{-2} = c^{\pm 2}\ \hbar\ (270\ \mathbf{q_I})^{-2} = c^{\pm 2}\ \hbar\ [270\ \sqrt{-(c^{-1}\hbar)}]^{-2}$
$= 270^{-2}\ c^{-1} = 4.575639166 \cdot 10^{-14}\ (J) = 0.285588809\ (MeV/c^2)$

π MESON INTERVALIC ENERGY AT LISZTINIAN-GAUDINAR STRUCTURE LEVEL

The intervalic energy of π^0 meson at lisztinian level of the intervalic structure is the sum of its two constituent lisztinos:

$$I(\pi^0)_L = 2\, I(L_{2/3}^{\pm 2/3}) = 2\, c^{\pm 2}\hbar\, (2/3\, e)^{-2} = 2\, c^{\pm 2}\hbar\, (180\, q_I)^{-2} = 2\, c^{-1} 180^{-2} =$$
$$= 2.059037625 \cdot 10^{-13}\, (J) = 1.285149644\, (MeV/c^2)$$

And the intervalic energy of $\pi^\pm$ is likewise:

$$I(\pi^\pm)_L = I(L_{2/3}^{\pm 2/3}) + I(L_{2/3}^{\pm 1/3}) = c^{\pm 2}\hbar\, (180\, q_I)^{-2} + c^{\pm 2}\hbar\, (90\, q_I)^{-2} =$$
$$= c^{-1} 180^{-2} + c^{-1} 90^{-2} = 5.147594062 \cdot 10^{-13}\, (J) = 3.212874109\, (MeV/c^2)$$

π MESON INTERVALIC ENERGY AT GAUDINAR-LISZTINIAN STRUCTURE LEVEL

The gaudinar structure of π meson is:

$$2/3\, G_6^{\pm 2/3} = (\, D_{\pm 45}\, D_{\pm 45}\, D_{\pm 45}\, D_{\pm 45}\,)$$
$$2/3\, G_6^{+1/3} = (\, D_{+45}\, D_{+45}\, D_{+45}\, D_{-45}\,)$$
$$2/3\, G_6^{-1/3} = (\, D_{-45}\, D_{-45}\, D_{-45}\, D_{+45}\,)$$

But since the constituent fractional lisztinos of π meson are at once the fractional gaudinos: $L_{2/3} = 2/3\, G_6 = 4\, D_{45}$, the gaudinar structure have already been subsumed into the lisztinian one:

$$I(\pi^0)_G = 2\, I(2/3\, G_6^{\pm 2/3})$$
$$I(\pi^\pm)_G = I(2/3\, G_6^{\pm 2/3}) + I(2/3\, G_6^{\pm 1/3})$$

(Alternatively, we could also say that π meson is a lisztino instead of a monteverdino, as like as electron is a dalino instead of a gaudino, but in order to a better understanding of the general structure and classification of subatomic particles I think it is better to maintain such *limit cases* of structures which are identical to the preceding level of the intervalic structure. Thus, we say that electron is a dalino and, a fortiori, a limit case of gaudino —which adds nothing to the dalinar structure—, or that the two constituent gaudinos of π meson are, a fortiori, limit cases of lisztinos —which adds nothing to the gaudinar structure—).

π MESON INTERVALIC ENERGY AT DALINAR STRUCTURE LEVEL

As both π^0 and $\pi^\pm$ meson are composed by eight dalinos 45, their intervalic energies at the dalinar level are identical:

$$I(\pi^0)_D \quad I(\pi^\pm)_D = 8\ I(D_{\pm 45}) = 8\ c^{\pm 2} h\ (45\ q_I)^{-2} = 8\ c^{-1} 45^{-2} =$$
$$= 1.31778408 \cdot 10^{-11}\ (J) = 82.24957719\ (MeV/c^2)$$

π MESON TOTAL INTERVALIC ENERGY

The total intervalic energy of π meson will be the sum its constituent levels:

$$I(\pi^0)_{tot} = I(\pi^0)_M + [\ I(\pi^0)_L \quad I(\pi^0)_G\ ] + I(\pi^0)_D =$$
$$= 1.338374456 \cdot 10^{-11}\ (J) = 83.53472683\ (MeV/c^2)$$
$$I(\pi^\pm)_{tot} = I(\pi^\pm)_M + [\ I(\pi^\pm)_L \quad I(\pi^\pm)_G\ ] + I(\pi^\pm)_D =$$
$$= 1.37383566 \cdot 10^{-11}\ (J) = 85.74804011\ (MeV/c^2)$$

π MESON ELECTROMAGNETIC ENERGY

π MESON TOTAL ELECTROMAGNETIC ENERGY

Staring from the previously known magnitude of the intervalic energy, and according to the intervalic principle of energy balance for subatomic particles, we have that the total electromagnetic energy is:

$U(\pi^0)_{tot} = c^{\pm 2}m(\pi^0) - I(\pi^0) = 8.242445041 \cdot 10^{-12}$ (J) =
= 51.44527317 (MeV/c²)
$U(\pi^{\pm})_{tot} = c^{\pm 2}m(\pi^{\pm}) - I(\pi^{\pm}) = 8.623232399 \cdot 10^{-12}$ (J) =
= 53.82195989 (MeV/c²)

The ratios among their constituent mass energies and the total mass are:

$I(\pi^0) / U(\pi^0) = 1.623759029$
$I(\pi^0) / E(\pi^0)_{mass} = 0.618867438$
$U(\pi^0) / E(\pi^0)_{mass} = 0.381132561$

$I(\pi^{\pm}) / U(\pi^{\pm}) = 1.593179444$
$I(\pi^{\pm}) / E(\pi^{\pm})_{mass} = 0.614373003$
$U(\pi^{\pm}) / E(\pi^{\pm})_{mass} = 0.385626996$

π MESON ELECTROMAGNETIC ENERGY AT DALINAR STRUCTURE LEVEL

Supposing that the electromagnetic energy at the dalinar level of π meson follows the same of quark down's ratio: $I(d)/U(d) = 1.650243062$, we would have:

$U(\pi^0)_D = I(\pi^0)_D / [I(d)/U(d)] = 49.84088652 \ (MeV/c^2)$

$U(\pi^\pm)_D = I(\pi^\pm)_D / [I(d)/U(d)] = 49.84088652 \ (MeV/c^2)$

It is sure the these magnitudes are close to the exact value, but in any case, since the electromagnetic energy at this level is equal for both mesons, we can make some deductions only starting from that result.

π MESON ELECTROMAGNETIC ENERGY AT GAUDINAR-LISZTINIAN STRUCTURE LEVEL

As it is experimentally found: $\Delta[m(\pi)] = m(\pi^\pm) - m(\pi^0) = 139.57 - 134.98 = 4.59 \ (MeV/c^2)$. We can also deduce theoretically the difference between the total intervalic energy of both mesons:

$\Delta[I(\pi)] = I(\pi^\pm)_{tot} - I(\pi^0)_{tot} = 2.21331328 \ (MeV/c^2)$

Therefore the remaining difference between both values is due to the electromagnetic energy, $\Delta[m(\pi)] - \Delta[I(\pi)] = \Delta[U(\pi)]$, being:

$\Delta[U(\pi)] = U(\pi^\pm)_{tot} - U(\pi^0)_{tot} = 2.37668672 \ (MeV/c^2)$

That magnitude may be shared among the actual structure level and the deeper ones. However, since the dalinar structure is *identical* for both mesons, regardless of their electric charges, we have the certainty that the difference of mass between both zero and charged π mesons is exclusively due to the different electromagnetic energy at the lisztinian-gaudinar level of the intervalic structure.

The total mass of π meson is simply the sum of its constituent intervalic and electromagnetic energies. From those contributions we can deduce the value of the electromagnetic energy at the gaudinar-lisztinian level:

$$E(\pi^0)_{mass} = I(\pi^0)_{tot} + U(\pi^0)_G + U(\pi^0)_D$$
$$E(\pi^\pm)_{mass} = I(\pi^\pm)_{tot} + U(\pi^\pm)_G + U(\pi^\pm)_D$$

$$U(\pi^0)_G = 1.60438665 \; (MeV/c^2)$$
$$U(\pi^\pm)_G = 3.98107337 \; (MeV/c^2)$$

As can be seen:

$$U(\pi^\pm)_G - U(\pi^0)_G = U(\pi^\pm)_{tot} - U(\pi^0)_{tot} = 2.37668672 \; (MeV/c^2)$$

From the difference of electromagnetic energy between π mesons, previously deduced by intervalic geometrical means, we can deduce the average *distance* between quarks inside π meson, d_q:

$$\Delta[U(\pi)] = E_q(L\tfrac{2}{3}D45^{(\tfrac{1}{3})} L\tfrac{2}{3}D45^{(\tfrac{2}{3})}) - E_q(L\tfrac{2}{3}D45^{(\tfrac{2}{3})} L\tfrac{2}{3}D45^{(\tfrac{2}{3})}) =$$
$$= [(\tfrac{1}{3} \cdot \tfrac{2}{3}) - (\tfrac{2}{3} \cdot \tfrac{2}{3})] (1/4\pi\varepsilon_0) \, e^2 / d_q = -(2/9) (1/4\pi\varepsilon_0) \, e^2 / d_q$$

$$d_q = 1.346378179 \cdot 10^{-15} \; (m)$$

This magnitude is very similar to the distance between nucleonic quarks inside nucleon, $d_q = 1.37610877 \cdot 10^{-15}$ (m), as expected. The minus sign indicates that the radius of $\pi^\pm$ meson is smaller that the radius of π^0. As the electromagnetic energy of meson at lisztinian-gaudinar level is due to the sum of the electromagnetic potential energy of its constituent quarks, we can write the relation:

$$U(\pi^\pm)_G / U(\pi^0)_G = [(\tfrac{1}{3} + \tfrac{2}{3}) / r_q(\pi^\pm)] / [(\tfrac{2}{3} + \tfrac{2}{3}) / r_q(\pi^0)]$$
$$r_q(\pi^0) = 3.308490398 \; r_q(\pi^\pm)$$

And the radius of the two equal constituent quarks of π^0 meson will be:

$$r_q(\pi^0) = \tfrac{1}{2} (\tfrac{2}{3} + \tfrac{2}{3}) (1/4\pi\varepsilon_0) \, e^2 / U(\pi^0)_G = 5.983443839 \cdot 10^{-16} \; (m)$$

As the two constituent quarks of $\pi^\pm$ meson are unlike, their radius may be different and the following magnitude would be their average radius:

$$<r_q(\pi^\pm)> = \tfrac{1}{2}(\tfrac{1}{3} + \tfrac{2}{3})(1/4\pi\varepsilon_0)\, e^2 / U(\pi^\pm)_G =$$
$$= 1.808511774 \cdot 10^{-16} \,(m)$$

π MESON ELECTROMAGNETIC ENERGY AT MONTEVERDIC STRUCTURE LEVEL

$U(\pi^0)_M = 0$

As in the case of proton, we can be tempted to write for the $\pi^\pm$ meson:

$$U(\pi^\pm)_M = U(M_2^\pm) = \tfrac{1}{2}(1/4\pi\varepsilon_0)(270\, q_I)^2 / r_\pi$$

However this electromagnetic energy is just the energy of the electromagnetic *field*, and therefore it is not manifested as *mass* energy.

$U(\pi^\pm)_M = 0$

On the contrary, the intervalic energy at this level is the "equivalent energy" of the total charge which is always manifested as mass energy.

INTERVALIC THEORY:
The Intervalic Structures of Subatomic Particles and the Last Foundations of Physics

INTERVALIC STRUCTURE OF π MESON

$\pi^{\pm}$ meson = $M_2^{\pm}$ = 2 $L_{2/3}$ ≡ 2 ⅔ G_6 (⅔ $G_6^{\pm 1/3}$, ⅔ $G_6^{\pm 2/3}$) = 8 D_{45} (7 $D_{\pm 45}$, 1 $D_{\pm 45}$) = 360 **I** = 720 γ = 1440 S

π^0 meson = M_2^0 = 2 $L_{2/3}$ ≡ 2 ⅔ G_6 (⅔ $G_6^{+2/3}$, ⅔ $G_6^{-2/3}$) = 8 D_{45} (4 D_{+45}, 4 D_{-45}) = 360 **I** = 720 γ = 1440 S

Intervalic structure levels: 1 Intervalic String (S), 2 Photon (γ), 3 Intervalino (**I**), 4 Dalino (D), 5 Gaudino (G), 6 Lisztino (L), 7 Monteverdino (M)

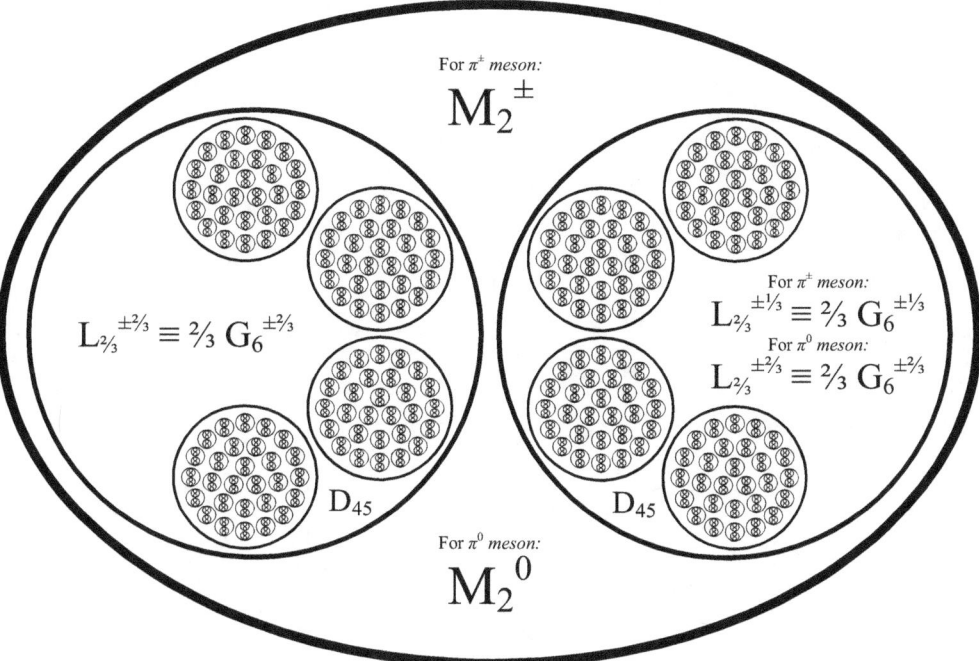

Chapter 20

INTERVALIC MESON

INTERVALIC MONTEVERDINO: MESON

Now we are going to introduce a general view on the intervalic structure of mesons. As in the case of intervalic baryon, to the SU(7) symmetry of *intervalic structure* it have to be added an important symmetry: the *isocharge* symmetry, which can be now interpreted as the underlying physical quantity of the former 'abstract' *isospin* symmetry. This set of symmetries composes the combinations which lead to the principal classification of the intervalic model of mesons. Considering that the three families of dalinar symmetries {D45}, {D30} and {D18} make "light" quarks, there will be just 7 x 3 = 21 intervalic structures of light quarks, the possible combinations of pairs of intervalic quarks to yield light mesons are much more wide than in SM, which worked with the ancient 5 bizarre flavours. Of course, the intervalic structure of quarks can explain in a fundamental manner all the experimentally detected mesons pertaining to the zoo of particles which are waiting for a theoretical satisfactory explanation. In the case of "heavy" mesons the difference between the pretty richness of the four families of dalinar symmetries {D6}, {D5}, {D3} and {D2}, and the unique isolated heavy flavour showed by the delusive SM is still more apparent.

Finally, we will find that it can be proposed an alternative quarkless structure for a lot of mesons which would be monteverdinos composed by 2^n quarks, with n = 1, 2, 3, mainly. This leads to an alternative description of some mesons which is not postulated but that is interesting to study.

INTERVALIC MESONS WITH {D45} SYMMETRY

As we have seen opportunely, every set of quark symmetries is composed by 7 intervalic structures. Those corresponding to {D45} symmetry are the following:

$L_{1/3} = 1/3\ G_6 = 2\ D_{45} = 34.537545 - 36.791934 \rightarrow$ quark $L1/3D45^{(1/3)}$
$L_{2/3} = 2/3\ G_6 = 4\ D_{45} = 69.075090 - 73.583868 \rightarrow$ quark $L2/3D45^{(1/3,\ 2/3)}$
$L_1 = 1\ G_6 = 6\ D_{45} = 103.61263 - 110.37580 \rightarrow$ quark $L1D45^{(1/3,\ 2/3)}$
$L_2 = 2\ G_6 = 12\ D_{45} = 207.22526 - 220.75160 \rightarrow$ quark $L2D45^{(1/3,\ 2/3)}$
$L_3 = 3\ G_6 = 18\ D_{45} = 310.83790 - 331.12740 \rightarrow$ quark $L3D45^{(1/3,\ 2/3)}$
$L_4 = 4\ G_6 = 24\ D_{45} = 414.45053 - 441.50320 \rightarrow$ quark $L4D45^{(1/3,\ 2/3)}$
$L_5 = 5\ G_6 = 30\ D_{45} = 518.06316 - 551.87900 \rightarrow$ quark $L5D45^{(1/3,\ 2/3)}$

Since Nature does not mix quarks from different intervalic symmetry when assembling baryons and mesons, the number of allowed mesons made by the family of intervalic quarks with {D45} symmetry will be the group SU(7) with intervalic structure symmetry:

7 V 7 = 49 mesons

In a similar way as we have done with the description of intervalic baryons, we are going to write down all the quark families involved in the assembly of intervalic mesons.

INTERVALIC MESONS WITH {D30} SYMMETRY

$L_{1/3} = 1/3\ G_9 = 3\ D_{30} = 116.56421 - 124.17278 \to$ quark $L1/3D30^{(1/3)}$
$L_{2/3} = 2/3\ G_9 = 6\ D_{30} = 233.12843 - 248.34556 \to$ quark $L2/3D30^{(2/3)}$
$L_1 = 1\ G_9 = 9\ D_{30} = 349.69264 - 372.51833 \to$ quark $L1D30^{(1/3)}$
$L_2 = 2\ G_9 = 18\ D_{30} = 699.38520 - 745.03666 \to$ quark $L2D30^{(2/3)}$
$L_3 = 3\ G_9 = 27\ D_{30} = 1,049.0779 - 1,117.5550 \to$ quark $L3D30^{(1/3)}$
$L_4 = 4\ G_9 = 36\ D_{30} = 1,398.7705 - 1,490.0733 \to$ quark $L4D30^{(2/3)}$
$L_5 = 5\ G_9 = 45\ D_{30} = 1,748.4631 - 1,862.5916 \to$ quark $L5D30^{(1/3)}$

The number of allowed mesons made by the family of intervalic quarks with {D30} symmetry is described by the group SU(7):

7 V 7 = 49 mesons

INTERVALIC MESONS WITH {D18} SYMMETRY

$L_{1/3} = 1/3\ G_{15} = 5\ D_{18} = 539.64917 - 574.87400 \to$ quark $L1/3D18^{(1/3)}$
$L_{2/3} = 2/3\ G_{15} = 10\ D_{18} = 1,079.2982 - 1,149.7480 \to$ quark $L2/3D18^{(2/3)}$
$L_1 = 1\ G_{15} = 15\ D_{18} = 1,618.9475 - 1,724.6220 \to$ quark $L1D18^{(1/3)}$
$L_2 = 2\ G_{15} = 30\ D_{18} = 3,237.8950 - 3,449.2443 \to$ quark $L2D18^{(2/3)}$
$L_3 = 3\ G_{15} = 45\ D_{18} = 4,856.8425 - 5,173.8660 \to$ quark $L3D18^{(1/3)}$
$L_4 = 4\ G_{15} = 60\ D_{18} = 6,475.7900 - 6,898.4886 \to$ quark $L4D18^{(2/3)}$
$L_5 = 5\ G_{15} = 75\ D_{18} = 8,094.7375 - 8,623.1100 \to$ quark $L5D18^{(1/3)}$

The number of allowed mesons made by the family of intervalic quarks with {D18} symmetry is equally described by the group SU(7):

7 V 7 = 49 mesons

INTERVALIC STRUCTURE AND FLAVOUR IN THE MODEL OF QUARKS

Henceforth, it is easy to understand why never have been detected mesons composed by any combination among the supposed quark bottom and the intended quarks up and down (b, u, d). On the contrary, combinations among the supposed quarks up, down, strange and charme are possible whenever the masses of those quarks vary in order to coincide with their real masses shown in each symmetry family, as all these quarks have several different intervalic structures allowed near their intended masses. For example, at first sight they can be easily confounded —without knowing it— with the following intervalic structures:

- Quark up: $L\frac{2}{3}D45^{(1/3, 2/3)}$ $L1D45^{(1/3, 2/3)}$ $L2D45^{(1/3, 2/3)}$ $L3D45^{(1/3, 2/3)}$ $L4D45^{(1/3, 2/3)}$ $L\frac{1}{3}D30^{(1/3)}$ $L\frac{2}{3}D30^{(2/3)}$ $L1D30^{(1/3)}$
- Quark down: $L\frac{1}{3}D45^{(1/3)}$ $L\frac{2}{3}D45^{(1/3, 2/3)}$ $L1D45^{(1/3, 2/3)}$ $L2D45^{(1/3, 2/3)}$ $L3D45^{(1/3, 2/3)}$ $L4D45^{(1/3, 2/3)}$ $L\frac{1}{3}D30^{(1/3)}$ $L\frac{2}{3}D30^{(2/3)}$ $L1D30^{(1/3)}$
- Quark strange: $L4D45^{(1/3, 2/3)}$ $L5D45^{(1/3, 2/3)}$ $L1D30^{(1/3)}$ $L2D30^{(2/3)}$ $L\frac{1}{3}D18^{(1/3)}$
- Quark charm: $L3D30^{(1/3)}$ $L4D30^{(2/3)}$ $L5D30^{(1/3)}$ $L\frac{2}{3}D18^{(2/3)}$ $L1D18^{(1/3)}$
- Quark bottom: $L2D18^{(2/3)}$ $L3D18^{(1/3)}$ $L4D18^{(2/3)}$ $L5D18^{(1/3)}$

When summing their masses to compose a monteverdino —meson or baryon— the confusion can be inclusive greater, and affect not only their masses but also their electric charges. I think this example may be sufficient to understand the terribly chaos which stand out in the classification of the model of quarks in SM.

The total sum of the number of allowed light mesons are, regarding the SU(7) group of intervalic structure symmetry:

$$(7 \vee 7) + (7 \vee 7) + (7 \vee 7) = 147$$

INTERVALIC THEORY:
The Intervalic Structures of Subatomic Particles and the Last Foundations of Physics

Of course, the traditional flavour multiplets in the primitive model of quarks introduced by Gell-Mann, Neeman and Zweig are only some few combinations included in the intervalic model of quarks.

INTERVALIC MESONS WITH {D6} SYMMETRY

$L_{1/3} = 1/3\ G_{45} = 15\ D_6 = 14{,}570.527 - 15{,}521.598 \rightarrow$ quark $L\frac{1}{3}D6^{(1/3)}$
$L_{2/3} = 2/3\ G_{45} = 30\ D_6 = 29{,}141.055 - 31{,}043.196 \rightarrow$ quark $L\frac{2}{3}D6^{(2/3)}$
$L_1 = 1\ G_{45} = 45\ D_6 = 43{,}711.582 - 46{,}564.794 \rightarrow$ quark $L1D6^{(1/3)}$
$L_2 = 2\ G_{45} = 90\ D_6 = 87{,}423.165 - 93{,}129.588 \rightarrow$ quark $L2D6^{(2/3)}$
$L_3 = 3\ G_{45} = 135\ D_6 = 131{,}134.74 - 139{,}694.38 \rightarrow$ quark $L3D6^{(1/3)}$
$L_4 = 4\ G_{45} = 180\ D_6 = 174{,}846.32 - 186{,}259.17 \rightarrow$ quark $L4D6^{(2/3)}$
$L_5 = 5\ G_{45} = 225\ D_6 = 218{,}557.91 - 232{,}823.97 \rightarrow$ quark $L5D6^{(1/3)}$

The number of allowed mesons is the same as in {D30} and {D18} symmetries:

7 V 7 = 49 mesons

INTERVALIC MESONS WITH {D5} SYMMETRY

$L_{1/3} = 1/3\ G_{54} = 18\ D_5 = 25{,}177.871 - 26{,}821.321 \rightarrow$ quark $L\frac{1}{3}D5^{(1/3)}$
$L_{2/3} = 2/3\ G_{54} = 36\ D_5 = 50{,}355.742 - 53{,}642.642 \rightarrow$ quark $L\frac{2}{3}D5^{(1/3, 2/3)}$
$L_1 = 1\ G_{54} = 54\ D_5 = 75{,}533.615 - 80{,}463.964 \rightarrow$ quark $L1D5^{(1/3, 2/3)}$
$L_2 = 2\ G_{54} = 108\ D_5 = 151{,}067.23 - 160{,}927.93 \rightarrow$ quark $L2D5^{(1/3, 2/3)}$
$L_3 = 3\ G_{54} = 162\ D_5 = 226{,}600.84 - 241{,}391.89 \rightarrow$ quark $L3D5^{(1/3, 2/3)}$
$L_4 = 4\ G_{54} = 216\ D_5 = 302{,}134.45 - 321{,}855.85 \rightarrow$ quark $L4D5^{(1/3, 2/3)}$
$L_5 = 5\ G_{54} = 270\ D_5 = 377{,}668.07 - 402{,}319.82 \rightarrow$ quark $L5D5^{(1/3, 2/3)}$

The SU(7) intervalic structure symmetry is the same in all families:

7 V 7 = 49 mesons

INTERVALIC MESONS WITH {D3} AND {D2} SYMMETRY

They are just equal to the {D5} intervalic symmetry. Of course the energy needed for the production of heavier mesons is so huge that it will be very hard to get them in our present laboratories. Nevertheless it is postulated that those states should exist at the Big Bang, immediately decaying and following the intervalic sequence of symmetries: {D2} → {D3} → {D5} → {D6} → {D18} → {D30} → {D45}, as we will see opportunely.

As in the case of the 3-quarks systems (baryons), in the case of 2-quarks systems (mesons) it has to be added the SU(2) *isocharge* symmetry to the *intervalic structure* symmetry, which makes for each intervalic family, roughly, the product SU(7) x SU(2), contained in the group SU(14), which yields 196 mesons.

As we may already suppose and will see later, mesons can be fully explained by the intervalic structures of their constituent quarks. Nevertheless, as the constituent gaudinos of quarks are postulated to be *fermions* with spin ½ —as dalinos and intervalinos are both *bosons*—, the basic spin of those quarks composed by *even* lisztinos, namely L_2 and L_4, would be 0 (being spin 1 and 2 excited states). This surprising result —the existence of quarks with integer spin, which have a share of 2/7 among the total allowed quarks of every family, as the quarks composed by lisztinos L⅓, L⅔, L1, L3 and L5 are all of them fermions—, lead us to a question: could be those quarks with spin 0 the principal constituents of some mesons?

INTERVALIC STRUCTURE OF THE SUPPOSED NONET OF MESONS SU(3) OF THE STANDARD MODEL

Apart from the alternative bilisztinian structure, L_2, uncannily applicable to a majority of mesons, the particles composing the supposed flavoured SU(3) nonet of mesons have got the typical quarkful intervalic structure, which matches with experimental data with remarkable accuracy:

$\pi^\pm$ (139.57) = (L⅔D45$^{(⅔)}$ L⅔D45$^{(⅓)}$)
(L1D45$^{(⅔)}$ L⅓D45$^{(⅓)}$)
π^0 (134.98) = (L⅔D45$^{(⅔)}$ L⅔D45$^{(⅔)}$)
(L5D90$^{(⅓)}$ L5D90$^{(⅓)}$)s
(L2D54$^{(0)}$)
$K^\pm$ (493.68) = (L⅔D45$^{(⅔)}$ L4D45$^{(⅓)}$)
(L⅔D45$^{(⅓)}$ L4D45$^{(⅔)}$)
K^0, $\underline{K}^0$ (497.67) = (L⅔D45$^{(⅓)}$ L4D45$^{(⅓)}$)
η^0 (547.3) = (L⅓D45$^{(⅓)}$ L5D45$^{(⅓)}$)
$\eta^{'}$ (958) = (L4D45$^{(⅓, ⅔)}$ L5D45$^{(⅓, ⅔)}$)

QUARKFUL INTERVALIC STRUCTURES OF EXPERIMENTALLY DETECTED MESONS

All experimentally detected mesons have a typical quarkful structure, apart from their alternative bilisztinian one (L_2). This multiplicity of possible intervalic structures can explain partially the existence of excited states in the zoo of particles detected experimentally. In any case, this stunning coincidence only reconfirm strongly the reliability and fruitfulness of the intervalic structures, although it may be sometimes disconcerting until a new classification of mesons is made. As they fit impressively with those of Nature, we believe that the symmetries of Nature have been really made in an intervalic mode, or in other words, that the intervalic symmetries are the genuine symmetries of Nature.

$K^*_2 (1430) = (L1D30^{(1/3)} L3D30^{(1/3)})$
$\qquad\qquad (L2D30^{(2/3)} L2D30^{(2/3)})$
$K_2 (1770) = (L2D30^{(2/3)} L3D30^{(1/3)})$
$K^*_3 (1780) = (L1D30^{(1/3)} L4D30^{(2/3)})$

$\rho (770) = (L3D45^{(1/3, 2/3)} L4D45^{(1/3, 2/3)})$
$\omega (783) = (L3D45^{(1/3, 2/3)} L4D45^{(1/3, 2/3)})$
$\eta' (958) = (L4D45^{(1/3, 2/3)} L5D45^{(1/3, 2/3)})$
$f_0 (975) = (L4D45^{(1/3, 2/3)} L5D45^{(1/3, 2/3)})$
$a_0 (980) = (L4D45^{(1/3, 2/3)} L5D45^{(1/3, 2/3)})$
$\Phi (1020) = (L5D45^{(1/3, 2/3)} L5D45^{(1/3, 2/3)})$
$h_1 (1190) = (L1/3D30^{(1/3)} L3D30^{(1/3)})$
$b_1 (1235) = (L1/3D30^{(1/3)} L3D30^{(1/3)})$
$f_2 (1270) = (2L3D45^{(1/3, 2/3)} 2L3D45^{(1/3, 2/3)})$
$a_1 (1270) = (2L3D45^{(1/3, 2/3)} 2L3D45^{(1/3, 2/3)})$
$f_1 (1285) = (2L3D45^{(1/3, 2/3)} 2L3D45^{(1/3, 2/3)})$
$f_0 (1300) = (L2/3D30^{(2/3)} L3D30^{(1/3)})$
$\pi (1300) = (2L3D45^{(1/3, 2/3)} 2L3D45^{(1/3, 2/3)})$
$a_2 (1320) = (L2/3D30^{(2/3)} L3D30^{(1/3)})$
$f_1 (1420) = (L2D30^{(2/3)} L2D30^{(2/3)})$
$\eta (1440) = (L1D30^{(1/3)} L3D30^{(1/3)})$
$f'_2 (1525) = (L1/3D30^{(1/3)} L4D30^{(2/3)})$
$f_0 (1590) = (L1/3D30^{(1/3)} L4D30^{(2/3)})$
$\rho (1600) = (L1/3D30^{(1/3)} L4D30^{(2/3)})$
$\omega_3 (1670) = (L2/3D30^{(2/3)} L4D30^{(2/3)})$
$\pi_2 (1680) = (L2/3D30^{(2/3)} L4D30^{(2/3)})$
$\Phi (1680) = (L2/3D30^{(2/3)} L4D30^{(2/3)})$
$\rho_3 (1690) = (L2/3D30^{(2/3)} L4D30^{(2/3)})$
$f_2 (1720) = (L2/3D30^{(2/3)} L4D30^{(2/3)})$
$\Phi_J (1850) = (L1/3D30^{(1/3)} L5D30^{(2/3)})$
$f_4 (2030) = (L2/3D30^{(2/3)} L5D30^{(1/3)})$

η_c (2980) = (L3D30$^{(⅓)}$ L5D30$^{(⅓)}$)
            (L4D30$^{(⅔)}$ L4D30$^{(⅔)}$)
X_0 (3415) = (L1D18$^{(⅓)}$ L1D18$^{(⅓)}$)
X_1 (3510) = (L5D30$^{(⅓)}$ L5D30$^{(⅓)}$)
X_2 (3555) = (L5D30$^{(⅓)}$ L5D30$^{(⅓)}$)
ψ (4415) = (2L3D30$^{(⅓)}$ 2L3D30$^{(⅓)}$)
X_{b0} (9860) = (L2D18$^{(⅔)}$ L4D18$^{(⅔)}$)
X_{b1} (9895) = (L3D18$^{(⅓)}$ L3D18$^{(⅓)}$)
X_{b2} (9915) = (L1D18$^{(⅓)}$ L5D18$^{(⅓)}$)

INTERVALIC STRUCTURE OF QUARKONIUMS

It is intended that the six supposed bizarre quarks of SM make six basic mesons composed by pairs quark-antiquark, which could be named *quarkoniums, q$\underline{q}$*. They can be identified with the following intervalic structures:

$u\underline{u}$ (upomium) → ρ (770) = (L3D45$^{(⅔)}$ L4D45$^{(⅔)}$)
$d\underline{d}$ (downomium) → ω (783) = (L3D45$^{(⅓)}$ L4D45$^{(⅓)}$)
$s\underline{s}$ (strangenium) → Φ (1,020) = (L5D45$^{(⅓)}$ L5D45$^{(⅓)}$)
$c\underline{c}$ (charmomium) → J/ψ (3,097) = (L4D30$^{(⅔)}$ L4D30$^{(⅔)}$)
$b\underline{b}$ (bottomium) → Y (9,460) = (L3D18$^{(⅓)}$ L3D18$^{(⅓)}$)
$t\underline{t}$ (topomium) → T (?) = (L4D6$^{(⅔)}$ L4D6$^{(⅔)}$)

It can be easily seen that quarkoniums follow just the *principal intervalic sequence* of quarks:

- quark L3D45$^{(⅓, ⅔)}$ → former quark up, down
- quark L4D30$^{(⅔)}$ → former quark charm
- quark L3D18$^{(⅓)}$ → former quark bottom

- quark L4D6$^{(2/3)}$ → former quark top
- quark L3D5$^{(1/3, 2/3)}$ → predicted new heavy quark

The only lack is the first quarkonium of {D45} symmetry:

L3D45$^{(1/3)}$ L3D45$^{(1/3)}$

which mass would roughly be 626 (MeV/c^2). This meson would be just composed by two of the three constituent quarks of nucleons, but in its place we find the "fatter" states ρ, ω, Φ, a phenomenon which deserves further research to be fully explained at a fundamental level. By some reason the "normal" {D45} quarkonium is not realized at low temperatures, but only are favoured a few of their fatter states. Probably this will be in close relation with the extreme stability of the intervalic structure of nucleon, which may avoid the mesonic state of the nucleonic quarks.

In any case and as can be easily seen, any classification of mesons or any other particles according to the vicissitudes of SM is an entirely partial and chaotic task, since it only has found a few particles by chance and without a minimal knowledge of the underlying intervalic symmetries which governs the structures of Nature in the subatomic world. Nevertheless, and until we have got a whole new intervalic classification of mesons, we can see some moderately interesting features in that intervalic structures of disperse mesons, as for example the decay of strangenium, charmonium and bottomium in two pairs of mesons, respectively, mesons of the families K, D and B. It can be checked that the intervalic symmetries of either strangenium, charmonium and bottomium and their corresponding two pairs of mesons of the decay are, respectively, {D45}, {D30} and {D18}:

$K^\pm$ (493.68) = (L⅔D45$^{(2/3)}$ L4D45$^{(1/3)}$)
                   (L⅔D45$^{(1/3)}$ L4D45$^{(2/3)}$)
K^0, $\underline{K}^0$ (497.67) = (L⅔D45$^{(1/3)}$ L4D45$^{(1/3)}$)

D^0, $\underline{D}^0$ (1864.5) = (L⅓D30$^{(1/3)}$ L5D30$^{(1/3)}$)

$$D^\pm (1869.3) = \begin{pmatrix} (\ 2L\tfrac{2}{3}D30^{(2/3)}\ 2L2D30^{(2/3)}\ ) \\ (\ L1D30^{(1/3)}\ L4D30^{(2/3)}\ ) \\ (\ L2D30^{(2/3)}\ L3D30^{(1/3)}\ ) \end{pmatrix}$$

$$D_s^\pm (1968.5) = (\ L\tfrac{2}{3}D30^{(2/3)}\ L5D30^{(1/3)}\ )$$

$$B^\pm (5279.0) = (\ L1D18^{(1/3)}\ L2D18^{(2/3)}\ )$$
$$B^0,\ \underline{B}^0\ (5279.4) = (\ L\tfrac{1}{3}D18^{(1/3)}\ L3D18^{(1/3)}\ )$$
$$B_s^0\ (5369.6) = (\ 2L\tfrac{2}{3}D18^{(2/3)}\ 2L1D18^{(1/3)}\ )$$

This information may be of interest when we will study the intervalic decay of subatomic particles, an important feature which follows rigorously the order of the chain of quarks and leptons-massive bosons according to their masses and intervalic structures, whilst on the contrary, SM has no any *fundamental* explanation for the decay of particles and can not give any theoretical *prediction* about the features of the zoo of particles made in any decay, but only to name some of them as "X" —a symbol of great help to understand the structure of mesons—.

THE ALTERNATIVE QUARKLESS INTERVALIC STRUCTURE OF MESONS

We will suppose that meson is the subatomic particle that intermediates the changeless —strong— intervalic interaction from the lisztinian level of the intervalic structure an onwards. After the postulation of the intervalic structure of quarks, it is clear that the intervalic structure of mesons could easily be deduced through SM, since we only have to combine the intervalic structures of quarks according to traditional rules. Nevertheless we should check if some mesons could have a quarkless structure. Since mesons are intended to be composed by a pair quark-antiquark, we could explore those intervalic structures composed by two lisztinos in a systematic way. After all, nobody can assure us that there were no mesons out of the quarkic structure. Of course, since every lisztino 1, L_1, makes the intervalic structure of a quark, any lisztino 2 can be also interpreted as having a quarkful as well as an alternative quarkless structure.

Following this way, the most simple kind of lisztinos is the combination of a two unlike charged gaudinos with elementary charge for composing a lisztino 2 (L_2) with zero charge, integrated by 2 x 270 = 540 constituent intervalinos. If it is supposed, in a rough approximation, that the ratio intervalic energy / electromagnetic energy of dalinos is similar to that of the electron, ~5/4, the masses of the whole set of the 16 possible lisztinos 2 are as follows (in MeV/c^2):

$L_2 = 2\,G_1 = 2\,D_{270} = 1.0219982$
$L_2 = 2\,G_2 = 4\,D_{135} = 8.1759856$
$L_2 = 2\,G_3 = 6\,D_{90} = 27.593950$
$L_2 = 2\,G_5 = 10\,D_{54} = 127.74977$
$L_2 = 2\,G_6 = 12\,D_{45} = 220.75160$
$L_2 = 2\,G_9 = 18\,D_{30} = 745.03666$
$L_2 = 2\,G_{10} = 20\,D_{27} = 1{,}021.9982$
$L_2 = 2\,G_{15} = 30\,D_{18} = 3{,}449.2440$
$L_2 = 2\,G_{18} = 36\,D_{15} = 5{,}960.2932$
$L_2 = 2\,G_{27} = 54\,D_{10} = 20{,}115.990$
$L_2 = 2\,G_{30} = 60\,D_9 = 27{,}593.950$
$L_2 = 2\,G_{45} = 90\,D_6 = 93{,}129.588$
$L_2 = 2\,G_{54} = 108\,D_5 = 160{,}927.93$
$L_2 = 2\,G_{90} = 180\,D_3 = 745{,}036.66$
$L_2 = 2\,G_{135} = 270\,D_2 = 2{,}514{,}498.8$
$L_2 = 2\,G_{270} = 540\,D_1 = 11{,}242{,}489$

Since the majority of intervalic mesons has a lisztinian 2 structure, it could be postulated that the spin energy scarcely intervenes in the energy mass balance of the particle. Therefore the structural energy mass balance for the intervalic meson would be:

$I - I^{-1} \approx 0$

That is to say, in the case of meson the electromagnetic mass energy is almost equal to the intervalic mass energy. According to this new balance we would have the following masses of lisztinos 2. These magnitudes of the mass are the maximum allowed values for those particles (in MeV/c^2):

$L_2 = 2\,G_1 = 2\,D_{270} \leq 1.1423552$
$L_2 = 2\,G_2 = 4\,D_{135} \leq 9.1388424$
$L_2 = 2\,G_3 = 6\,D_{90} \leq 30.843592$
$L_2 = 2\,G_5 = 10\,D_{54} \leq 142.79441$
$L_2 = 2\,G_6 = 12\,D_{45} \leq 246.74873$
$L_2 = 2\,G_9 = 18\,D_{30} \leq 832.77698$
$L_2 = 2\,G_{10} = 20\,D_{27} \leq 1{,}142.3552$
$L_2 = 2\,G_{15} = 30\,D_{18} \leq 3{,}855.4490$
$L_2 = 2\,G_{18} = 36\,D_{15} \leq 6{,}662.2158$
$L_2 = 2\,G_{27} = 54\,D_{10} \leq 22{,}484.978$
$L_2 = 2\,G_{30} = 60\,D_9 \leq 30{,}843.592$
$L_2 = 2\,G_{45} = 90\,D_6 \leq 104{,}097.12$
$L_2 = 2\,G_{54} = 108\,D_5 \leq 179{,}879.83$
$L_2 = 2\,G_{90} = 180\,D_3 \leq 832{,}776.98$
$L_2 = 2\,G_{135} = 270\,D_2 \leq 2{,}810{,}622.3$
$L_2 = 2\,G_{270} = 540\,D_1 \leq 11{,}242{,}489$

In this set we find a lot of already detected mesons, as we show below. And the remaining ones are pairs of the lisztino 2, that is to say, the majority of mesons appears to have an *even* lisztinian structure. Therefore, as in the intervalic structure of quarks, it can be seen a *principal intervalic sequence* in the intervalic structure of mesons.

PRINCIPAL INTERVALIC SEQUENCE OF MESONS

In the alternative quarkless structure of mesons, the *principal intervalic sequence* for mesons is focused on the intervalic structures $\{2^n L_2\}$, being $n = 0, 1, 2, 3,...$ as can be seen below:

$M_1 = L_2 = 2\, G_5 = 10\, D_{54} \leq 142.79441 \to \pi^0$ meson
$M_2 = 2\, L_2 = 4\, G_5 = 20\, D_{54} \leq 285.58882 \to$
$M_4 = 4\, L_2 = 8\, G_5 = 40\, D_{54} \leq 571.17764 \to \eta^0$ meson
$M_8 = 8\, L_2 = 16\, G_5 = 80\, D_{54} \leq 1{,}142.35528 \to \Phi$ meson

$M_1 = L_2 = 2\, G_6 = 12\, D_{45} \leq 246.74873 \to$
$M_2 = 2\, L_2 = 4\, G_6 = 24\, D_{45} \leq 493.49746 \to$ K mesons
$M_4 = 4\, L_2 = 8\, G_6 = 48\, D_{45} \leq 986.99492 \to \eta', f_0, a_0$ mesons
$M_8 = 8\, L_2 = 16\, G_6 = 96\, D_{45} \leq 1{,}973.9898 \to$ D mesons

$M_1 = L_2 = 2\, G_9 = 18\, D_{30} \leq 832.77698 \to \rho, \omega$ mesons
$M_2 = 2\, L_2 = 4\, G_9 = 36\, D_{30} \leq 1{,}665.5539 \to \eta, f_0$ mesons
$M_4 = 4\, L_2 = 8\, G_9 = 72\, D_{30} \leq 3{,}331.1079 \to J/\psi$ meson
$M_8 = 8\, L_2 = 16\, G_9 = 144\, D_{30} \leq 6{,}662.2158 \to$

$M_1 = L_2 = 2\, G_{10} = 20\, D_{27} \leq 1{,}142.3552 \to \Phi$ meson
$M_2 = 2\, L_2 = 4\, G_{10} = 40\, D_{27} \leq 2{,}284.7104 \to$ several mesons
$M_4 = 4\, L_2 = 8\, G_{10} = 80\, D_{27} \leq 4{,}569.4208 \to \psi$ mesons
$M_8 = 8\, L_2 = 16\, G_{10} = 160\, D_{27} \leq 9{,}138.8416 \to$

$M_1 = L_2 = 2\, G_{15} = 30\, D_{18} \leq 3{,}855.449 \to \chi$ mesons
$M_2 = 2\, L_2 = 4\, G_{15} = 60\, D_{18} \leq 7{,}710.898 \to$
$M_4 = 4\, L_2 = 8\, G_{15} = 120\, D_{18} \leq 15{,}421.796 \to$
$M_8 = 8\, L_2 = 16\, G_{15} = 240\, D_{18} \leq 30{,}843.592 \to$

Obviously, mesons placed in the symmetries $\{D54\}$ and $\{D27\}$ can't be composed by quarks because such symmetries are not allowed in the intervalic structure of quarks.

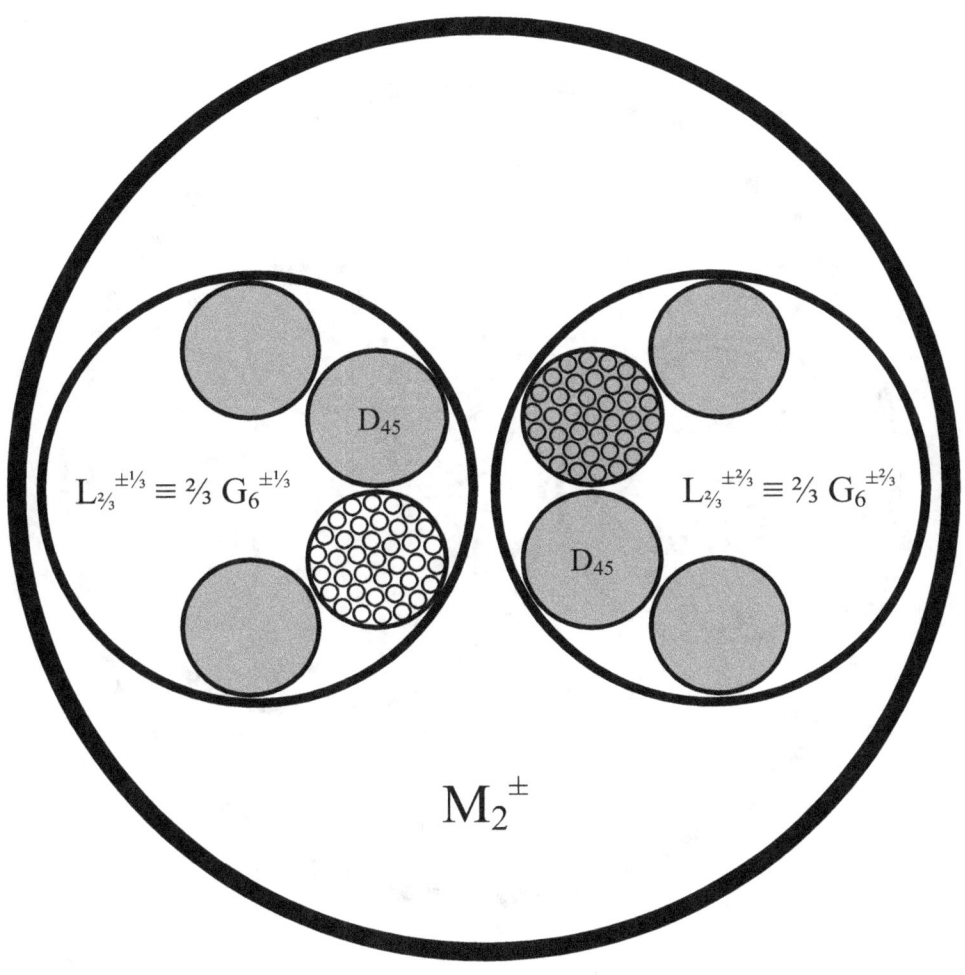

Figured intervalic structure of $\pi^\pm$ *meson*:
$M_2^\pm = 2\,L_{2/3} \equiv 2\,{}^2\!/_3\,G_6\,({}^2\!/_3\,G_6^{\pm 1/3},\,{}^2\!/_3\,G_6^{\pm 2/3}) = 8\,D_{45}\,(7\,D_{\pm 45},\,1\,D_{\pm 45}) = 360\,\mathbf{I}$

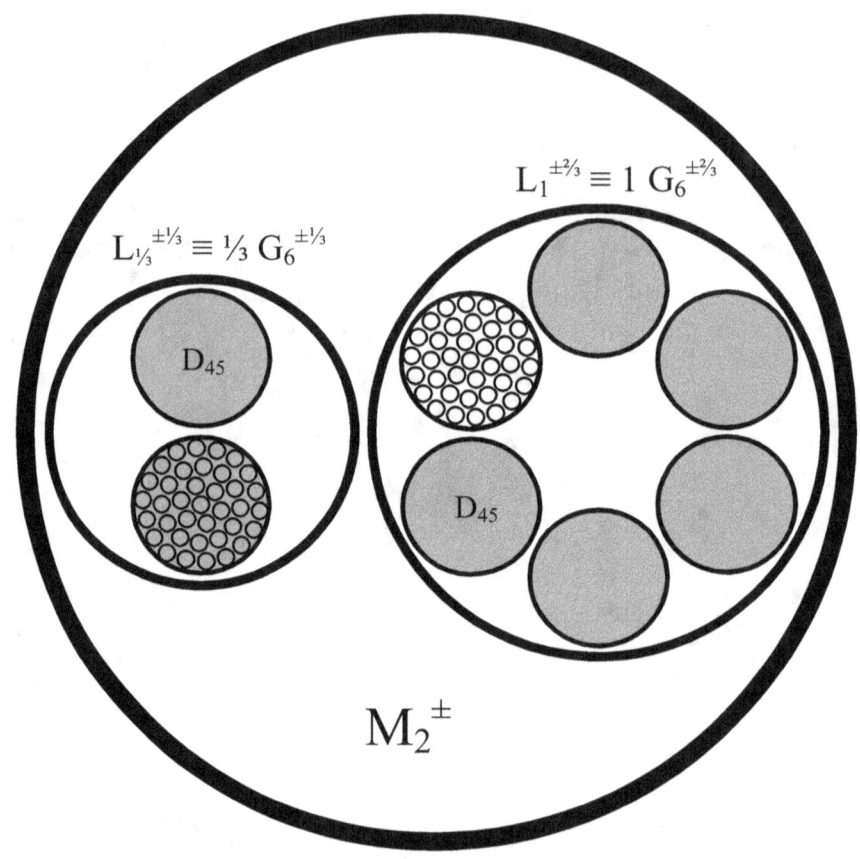

Figured intervalic structure of $\pi^{\pm}$ *meson*:
$$M_2^{\pm} = (L_1, L_{1/3}) \equiv (1\ G_6^{\pm 2/3},\ 1/3\ G_6^{\pm 1/3}) = 8\ D_{45}\ (7\ D_{\pm 45},\ 1\ D_{\pm 45}) = 360\ \mathbf{I}$$

INTERVALIC THEORY:
The Intervalic Structures of Subatomic Particles and the Last Foundations of Physics

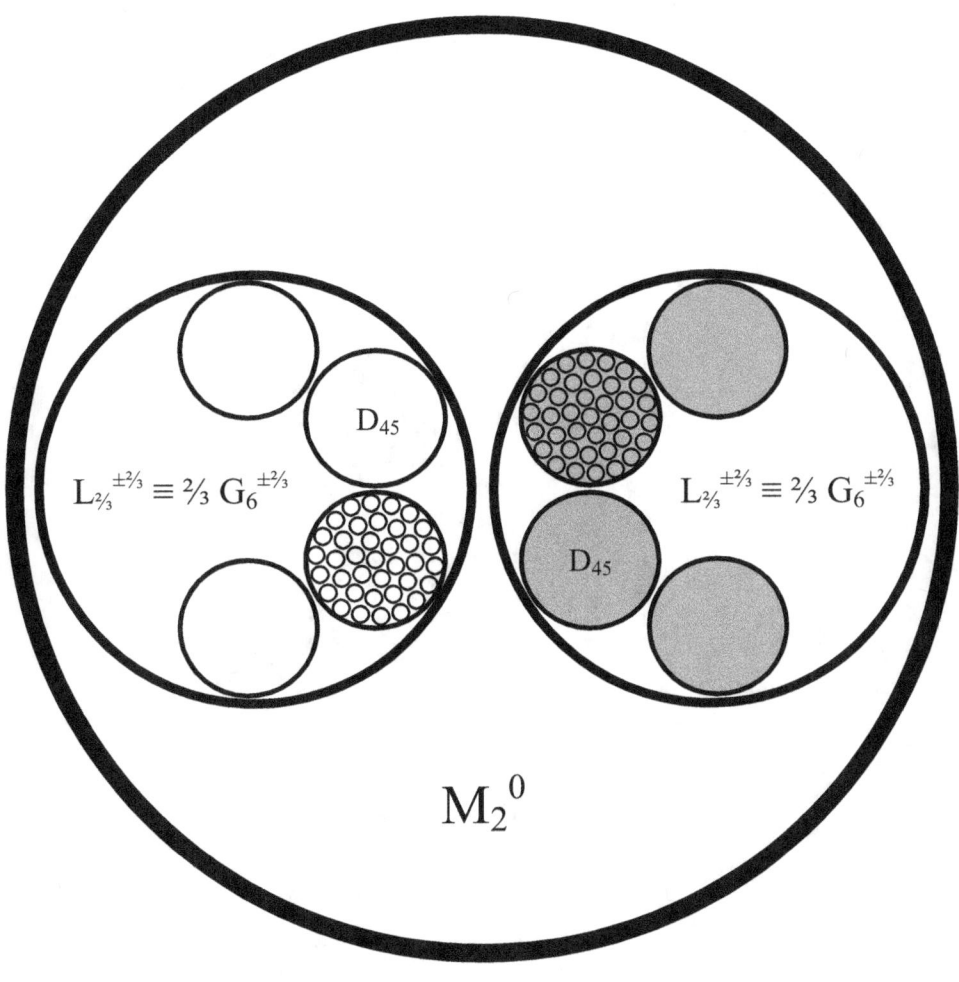

Figured intervalic structure of π^0 *meson*:
$M_2^0 = 2\,L_{2/3} \equiv 2\,{}^{2}\!/_{3}\,G_6\,({}^{2}\!/_{3}\,G_6^{+2/3},\,{}^{2}\!/_{3}\,G_6^{-2/3}) = 8\,D_{45}\,(4\,D_{+45},\,4\,D_{-45}) = 360\,\mathbf{I}$

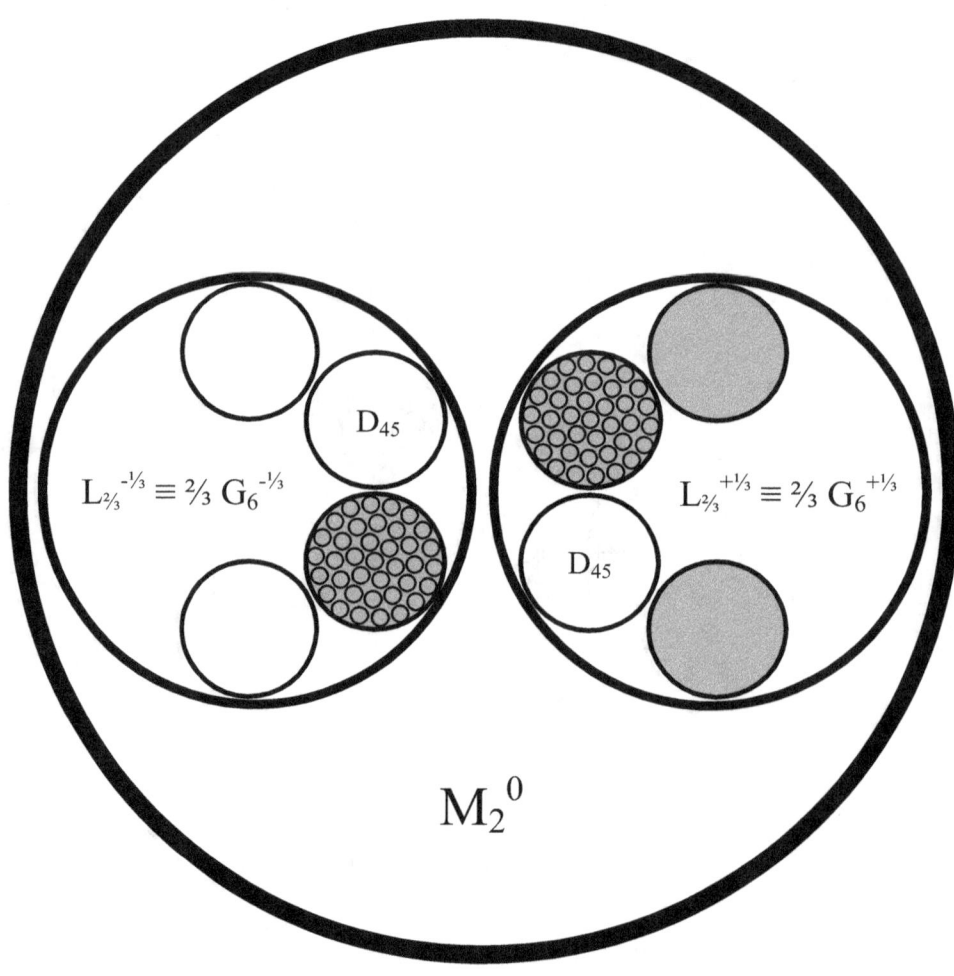

Figured intervalic structure of π^0 *meson*:
$M_2^0 = 2\,L_{2/3} \equiv 2\,\tfrac{2}{3}\,G_6\,(\tfrac{2}{3}\,G_6^{+1/3},\, \tfrac{2}{3}\,G_6^{-1/3}) = 8\,D_{45}\,(4\,D_{+45},\, 4\,D_{-45}) = 360\,\mathbf{I}$

www.ingramcontent.com/pod-product-compliance
Lightning Source LLC
Chambersburg PA
CBHW082011230526
45468CB00022B/1834